T0329599

TESTING FOR
EMC COMPLIANCE

TESTING FOR EMC COMPLIANCE
Approaches and Techniques

MARK I. MONTROSE
EDWARD M. NAKAUCHI

IEEE Electromagnetic Compatibility Society, *Sponsor*

IEEE PRESS

A JOHN WILEY & SONS, INC., PUBLICATION

Library of Congress Cataloging-in-Publication Data:

Montrose, Mark I.
    Testing for EMC compliance : approaches and techniques / Mark I. Montrose, Edward
M. Nakauchi.
        p. cm.
    Includes bibliographical references and index.
    ISBN 0-471-43308-X (cloth)
        1. Electromagnetic compatibility.   2. Electromagnetic interference.   Nakauchi, Edward
M.   II. Title.

TK7867.?M66 2004
621.383'24—dc22   2003063488

To my family
Margaret,
Maralena,
and
Matthew

—Mark—

To my family
Linda,
Michael,
Caryn,
and
Pamela,

—Ed—

CONTENTS

6 Conducted Testing

PREFACE

Testing for EMC Compliance: Approaches and Techniques is another book in a series by author Mark I. Montrose and first-time co-author Edward M. Nakauchi. The reason for this book lies in the fact that the topic of electromagnetic compatibility (EMC) and regulatory compliance, currently in the public domain, misses a particular combined aspect of EMC—how to perform both emission and immunity tests efficiently and what does one do "*if*" the product is already designed, manufactured, and then fails EMC tests.

Testing for EMC Compliance: Approaches and Techniques is written to provide value to the working engineer regardless of education or experience. Although the focus in this book is products classified as information technology, telecommunication, industrial, scientific, and medical, other categories of products will find use from the material presented. In addition, this book provides significant value to *non-EMC engineers*. Those who work in EMC may already know this material but require a comprehensive review on instrumentation, probes, and test techniques. There is always something new to learn.

When looking at available publications dealing with applied EMC engineering, we noticed that authors generally focus on product design at the conceptual or engineering stage, rarely postmanufacturing (after the fact), or discuss the field of electromagnetics appropriate only for university students using high levels of mathematics. Certain books cover a topic briefly while other authors examine a subject on a specific area of interest in great detail . Books that deal with EMC testing generally describe only the test environment, setup, and procedures directly from the standards themselves without interpretation or analysis. For this reason, we have gathered information from many resources, including personal experience, and put these together into one comprehensive reference on diagnosing and troubleshooting for EMC.

It has been our observation that many engineers are working in the field of EMC by virtue of being selected at random without regard to experience. Many times, someone may have to perform in-house testing with rental equipment and no knowledge on how to use the system or how to configure a product. For those companies that send equipment blindly to a test laboratory hoping for the best and get their unit returned with failing results, it becomes the responsibility of someone back at that factory to tackle this problem. Guidance provided herein is for those engineers who become delegated to fix a problem without extensive knowledge on what to do and how to do it. For example, there are many types of probes; which one is optimal for an intended functional use can be daunting for an inexperienced engineer.

For those who have worked all night at a test site to get a product to pass conformity tests using trial-and-error techniques along with many rolls of copper tape and ferrite clamps yet never reduced the emission levels by more than 1 or 2 dB, guidance is provided that allows one to quickly discover the problem area and apply corrective action. Troubleshooting is both a skill and an art that must be mastered over the course of time. There are many companies that fail a test and then decide to rent test equipment in an effort to resolve the problem in-house, saving the cost of spending time at test facilities, which may far exceed the cost of rental equipment.

The intended audiences for this guide are engineers and technicians who design, develop, or test electronic systems. These engineers may focus on analog, digital, or system-level products. Regardless of specialty, all designers must develop a product suitable for production. Frequently, more emphasis is placed on functionality than on system integration. System integration is usually assigned to product engineers, mechanical engineers, or others within the organization, with the EMC engineer generally the last person to get involved in the product design. Design engineers must now consider other aspects of product design, including the layout and production of printed circuit boards, enclosures, and cable assemblies. Considerations include cognizance of the manner in which electromagnetic fields transfer from circuit boards to the chassis and/or case structure, including interconnects, both internal and external. Those engineers and technicians that work at test laboratories may find the information herein valuable in supplementing their daily work assignments.

A fundamental understanding regarding field propagation is required before grabbing a sniffer probe to search for the illusive signal or problem area. Although one may quickly determine where the source of radiated/conducted energy is coming from using probes, a lack of understanding on how to apply corrective measures can cause many hours of frustration to remain.

With this in mind, the following topics are presented. Chapter 1 provides insight into the field of electromagnetic compatibility, why requirements exist, basic definitions, an overview on product testing, and testing methodologies that will be expanded upon in later chapters. Chapter 2 presents different types of electromagnetic and static fields and how these fields propagate. Common and differential modes are examined along with the coupling mechanism between assemblies. The contents of this chapter set the tone for the rest of the book. Chapter 3 details different types of common measurement equipment used for both testing and troubleshooting for EMC. Tests must be performed somewhere. Chapter 4 provides information on various facilities in common use. Chapter 5 provides guidance and information on various types of transducers used to both measure emissions and inject RF energy for immunity testing. Knowing which transducer to use for a particular application is important for both testing and troubleshooting. All forms of conducted tests are examined, both emissions and immunity. Chapter 6 presents basic test procedures and methodologies. Chapter 7 is identical to Chapter 6 except it covers radiated testing and immunity. Testing a product is one thing; how to perform it correctly is another. Chapter 8 provides insights into testing and how one should approach this arena of engineering along with problems that may develop. Chapter 9 is the heart of the book that ties all previous chapters together. Testing and troubleshooting are detailed herein using conventional and nonconventional techniques. Information is targeted toward in-house test personnel and inexperienced engineers that need guidance during the design cycle or after a product failure when time and money are not available. Commercial EMC facilities should already be knowledgeable with this chapter's contents, although new techniques may be introduced to EMC engineers and technicians. Appendix A shows how to build simple homemade probes and transducers. Appendix B is an important part of the book where actual procedures are presented for those who need to perform tests for all aspects of EMC, based on international standards. Contents are generally what a commercial test laboratory uses on a daily basis to perform a variety of tests.

It is the intent of this book to present *applied engineering concepts and principles along with hands-on techniques* to get the job done. Information is presented in a format that is easy to understand and implement. Those interested in Maxwell's equations or the more highly technical aspects of circuit theory, field propagation, and numerical modeling/simulation are guided to the bibliography at the end of the book. The reference section at the end of each chapter and the bibliography provide a listing that examines other aspects of EMC. These references minimize discussion of technical material and concepts beyond the scope "and" target audience of this book.

We do not discuss management and legal issues, such as how much testing should one do to ensure compliance with the EMC Directive, or how to perform EMC tests in great detail. Specific information on testing is available from test standards themselves and from the references provided. Test standards are not detailed since each product requires different certification approvals and test requirements constantly change.

The topics discussed herein include

1. radiated and conducted emissions,
2. radiated and conducted immunity,
3. electrostatic discharge,
4. fast transient/burst,
5. surge,
6. low-frequency power line magnetic fields, and
7. AC mains dips, dropouts, harmonics, and flicker.

Our focus as EMC consultants is to assist and advise in the design and testing of high-technology products at minimal cost. Implementing suppression techniques saves money, enhances performance, increases reliability, and achieves first-time compliance with emissions and immunity requirements. The electronics industry has given us the opportunity to participate in state-of-the-art designs as we move into the future. Use of nanotechnology in product design is rapidly approaching. The importance of efficient product development will escalate in this competitive marketplace. Regulatory compliance will always be mandatory. Companies that design and build high-technology system in 5 months and then take an additional 6 months to pass regulatory requirements will lose their competitive edge. On the other hand, those who integrate EMC into their design process will optimize their productivity and their financial return.

MARK I. MONTROSE
EDWARD M. NAKAUCHI

Santa Clara, California
Westminister, California
January 2004

ACKNOWLEDGMENTS

This book is dedicated to my wonderful and understanding family. They have put up with all my faults and foibles without complaints. For my wife, Linda, and my children, Michael, Caryn, and Pamela, who were always inquiring about "What is it that I do?" I hope you all now know what it is I do. Besides my family, I am fortunate to have the friendship of many who offered wonderful encouragement, in particular, Michael Oliver, a young, talented colleague; Sharilyn Bratton, a great source of enthusiasm and spirit; the "boys" at G&M Compliance, Thomas, Paul, Carlos, Rob, Greg, and Marciela; and all the other "voices" from the past and present (in particular, Charlie Bayhi, Nancy Kadotani, John Downs, and Neal Williams). I want to give a big thank you to Eddie Pavlu, who was there as a friend during a very difficult time for me. Also, a great big thank you to W. Michael King, a long-time friend and a great mentor of many years. Maybe too many years, right Michael! Finally, many thanks to the reviewers for their time and dedication for spending personal hours and taking the time to make many helpful comments to this book.

E. M. N.

I want to recognize Bill Kimmel and Daryl Gerke for their technical review of the draft manuscript which ensured accuracy and provided comments on structure and format. These experts helped guarantee that the book matches the target audience with appropriate content.

To Elya Joffe, who heavily scrutinized the material with a fine-tooth comb and provided "massive" feedback to ensure technical accuracy while monitoring my writing style, I give my deepest thanks. He kept me focused on achieving the target goal of the book along with a professional manner of presentation. For this I am grateful.

A special acknowledgment is given to W. Michael King. Michael is not only my mentor but also a friend. Without Michael, I would not have been able to achieve the knowledge that has allowed me to become not only a consultant but also an engineer with the ability to view things differently than others in this rapidly changing field and to communicate complex concepts never before researched or published.

My most special acknowledgment is to my family, Margaret, my wife, and Maralena and Matthew, my two teenagers. As with my previous three books and consulting work that takes me all over the world on a frequent basis, their understanding of my passion to write is appreciated. The long hours and months on a keyboard became a common site in my office, and for this I am grateful for their understanding and support.

M. I. M.

CHAPTER 1

INTRODUCTION

1.1 THE NEED TO COMPLY

Electrical and electronic products often (unexpectedly) produce radio-frequency (RF) energy. Every digital device has the potential of causing unintentional interference to other electrical devices. Electrical products are used in every aspect of our lives, such as providing communication, all forms of entertainment, luxurious lifestyles (transportation, appliances, utilities, recreational), and life support, to name a few. Of all items listed, communication systems and life support rank highest in the areas of concern when interference from unintentional sources of RF energy may be observed.

Control of electromagnetic compatibility (EMC) is an increasing necessity. Correct application of design methods ensures reliable operation, minimizes liability risk, reduces project timescales, and helps meet regulatory requirements. The best time to consider all aspects of EMC is during the preliminary design cycle, long before the first circuit is incorporated on a schematic, the first instruction written for a software program, or the outline of a mechanical chassis drawn. Management must also buy into the EMC arena if an early product shipment date is desired.

In North America, interference to communication systems in the 1930s led the U.S. Congress to enact the Communications Act of 1934. The Federal Communications Commission (FCC) was created to oversee enforcement and administration of this act.

Electromagnetic interference (EMI) was also a problem during World War II. The terminology then used was radio-frequency interference (RFI). Later, spectrum signatures of communication transmitters and receivers were developed along with radar systems. The notion of characterizing the spectrum signature of both types of systems, communication and radar, did not evolve until the 1950s. Because of the

Testing for EMC Compliance. By Mark I. Montrose and Edward M. Nakauchi
ISBN 0-471-43308-X © 2004 Institute of Electrical and Electronics Engineers

size and expense of equipment the military owned, the majority of high-technology electronic systems started to have erratic operation.

Following the Korean War, most EMC work was not classified unless it dealt with the specifics of a particular tactical or strategic system such as the ballistic missile, bombers, and similar military and espionage equipment. Conferences on EMI began to be held in the mid-1950s where unclassified information was presented. During this time frame, the Army Signal Corps of Engineers and the U.S. Air Force created strong ongoing programs dealing with EMI, RFI, and related areas of EMC.

In the 1960s, NASA (National Aeronautical and Space Administration) began stepped-up EMI control programs for its launch vehicles and space system projects. Governmental agencies and private corporations became involved with combating EMI emission and susceptibility in equipment such as security systems, church organs, hi-fidelity amplifiers, and the like. All of these devices were analog-based systems. The impetus for this work arose from the U.S. Air Force concerns due to problems caused by the Distant Early Warning (DEW) line radars.

As digital logic devices were increasingly developed for consumer systems, EMI became a wider concern. Research was started to characterize EMI in consumer electronics that included TV sets, common amplitude- and frequency-modulated (AM/FM) radios, medical devices, audio and video recorders, and similar products. Comparatively few of these products were digital, but were becoming so. Analog systems are more susceptible to problems than digital equipment.

In the late 1970s, problems associated with EMC became an issue for additional products. These products include home entertainment systems (TVs, VCRs, camcorders), personal computers, communication equipment, household appliances with digital features, intelligent transportation systems, sophisticated commercial avionics, control systems, audio and video displays, and numerous other applications. During this period, the public became aware of EMC and problems associated with it.

After the public became involved with EMI associated with digital equipment used within residential areas, the FCC in the mid- to late-1970s began to promulgate an emissions standard for personal computers and similar equipment. In Europe, concerns regarding EMC developed during World War II, especially in Germany in the forum of VDE (Verband Deutscher Electrotekniker). One reason why Europe started to consider EMC years before the FCC reflects the different attitudes toward the role of government regulation in the marketplace. The EMC directive 89/336/EEC simplified and alleviated differences among various standards of the NATO countries.

Since personal computers comprised such a huge market, commercial entities became involved in the field of EMC. Now almost all equipment is digital, whether or not it needs to be.

The focus of electronic equipment has now shifted from analog to digital. Another factor that pushed digital devices into regulation status was that in the early days of digital the prevailing wisdom was that digital devices were "not susceptible" to EMI. Because of this perception, the commercial community was surprised to learn that digital devices were actually susceptible to disruption.

The Food and Drug Administration (FDA), however, recognized the threat posed by EMI because of reported problems with patient care and diagnostic electronics. The issue of compliance became a concern when the European Union (EU) through its EMC directive 89/336/EEC imposed emissions and immunity requirements. Another forcing function of EMC compliance is the increasing role played by electronics in power conversion, communications, and control systems where electromechanical systems once were primarily used. In observation, the general population has had to deal with EMC for only 20 plus years, whereas the military, NASA, and RF engineers have been dealing with this issue from day one.

1.2 DEFINITIONS

The following terms and concepts are used throughout this book. A detailed glossary is provided at the end of this book.

Electromagnetic Compatibility. The capability of electrical and electronic systems, equipment, and devices to operate in their intended electromagnetic environment within a defined margin of safety and at design levels or performance without suffering or causing unacceptable degradation as a result of electromagnetic interference [American National Standards Institute (ANSI) C64.14-1992)].

Electromagnetic Interference. The process by which disruptive electromagnetic (EM) energy is transmitted from one electronic device to another via radiated or conducted paths (or both). In common usage, the term refers particularly to RF signals; however, EMI is observed throughout the EM spectrum.

Radio Frequency. A frequency range containing coherent EM radiation of energy useful for communication purposes—roughly the range from 9 kHz to 300 GHz. This energy may be emitted as a by-product of an electronic device's operation. Radio frequency is emitted through two basic mechanisms:

Radiated Emissions. The component of RF energy that is emitted through a medium as an EM field. Although RF energy is usually emitted through free space, other modes of field transmission may be present.

Conducted Emissions. The component of RF energy that is emitted through a medium as a propagating wave generally through a wire or interconnect cables. Line-conducted interference (LCI) refers to RF energy in a power cord or alternating-current (AC) mains input cable. Conducted signals propagate as conducted waves.

Susceptibility. A relative measure of a device or a system's propensity to be disrupted or damaged by EMI exposure to an incident field. It is the lack of immunity.

Immunity. A relative measure of a device or system's ability to withstand EMI exposure while maintaining a predefined performance level.

Electrostatic Discharge (ESD). A transfer of electric charge between bodies of different electrostatic potential in proximity or through direct contact. This definition is observed as a high-voltage pulse that may cause damage or loss of functionality to susceptible devices.

Radiated Immunity. A product's relative ability to withstand EM energy that arrives via free-space propagation.

Conducted Immunity. A product's relative ability to withstand EM energy that penetrates through external cables, power cords, and input–output (I/O) interconnects.

Containment. A process whereby RF energy is prevented from exiting an enclosure, generally by shielding a product within a metal enclosure (Faraday cage or Gaussian structure) or by using a plastic housing with RF conductive coating. Reciprocally, we can also speak of containment in the inverse, as exclusion—preventing RF energy from entering the enclosure.

Suppression. The process of reducing or eliminating RF energy that exists without relying on a secondary method, such as a metal housing or chassis. Suppression may include shielding and filtering as well.

Voltage Probe. A transducer that measures the voltage level in a transmission line. This probe consists of a series resistor, a direct-current (DC) blocking capacitor, and an inductor to provide a low-impedance input to a receiver. It is used for direct connection to a transmission line and is unaffected by the current level present.

Current Probe. A transducer that measures the current level in a transmission line. This probe consists of a magnetic core material that detects the magnitude of magnetic flux present and presents this field measurement to a receiver.

Sniffer Probe. Any small transducer used to isolate or locate radiating RF energy. Through EM field coupling, calibration of the measurement is not a concern since the process is comparative.

FET Probe. A high-impedance transducer used to measure signal characteristics in a transmission line without adding a capacitive load or affecting performance of the propagating wave.

Spectrum Analyzer. An instrument primarily used to display the power distribution of an incoming signal as a function of frequency. Useful in analyzing the characteristics of electrical waveforms by repetitively sweeping through a frequency range of interest and displaying all components of the signal being investigated.

Oscilloscope. An instrument primarily used for making visible the instantaneous value of one or more rapidly varying electrical quantities as a function of time.

Correlation Analyzer. Similar to a spectrum analyzer, but has two inputs that are frequency and time synchronized to each other. This allows use of digital signal processors for analysis of input signals.

Line Impedance Stabilization Network (LISN). A network inserted in the supply mains load of an apparatus to be tested that provides, in a given frequency range, a specified load impedance for the measurement of disturbance voltages and which may isolate the apparatus from the supply mains in that frequency range. Also identified as an "artificial mains network."

Antenna. A device used for transmitting or receiving EM signals or power. Designed to maximize coupling to an EM field.

Biconical. An antenna consisting of two conical conductors that have a common axis and vertex and are excited or connected to a receiver at the vertex point.

Log Periodic. A class of antennas having a structural geometry such that its impedance and radiation characteristic repeat periodically as the logarithm of frequency.

Bilog. A single antenna that combines the features and EM characteristics of both biconical and log periodic antennas into one assembly.

Loop. An antenna in the shape of a coil that is sensitive to magnetic fields and shielded against electric fields. A magnetic field component perpendicular to the plane of the loop induces a voltage across the coil that is proportional to frequency according to Faraday's law.

Horn. A radiating or receiving aperture having the shape of a horn. Generally used in the frequency range above 1 GHz.

1.3 NATURE OF INTERFERENCE

Electromagnetic compatibility is grouped into two categories: internal and external. The internal category is the result of signal degradation along a transmission path, including parasitic coupling between circuits (i.e., crosstalk) in addition to field coupling between internal subassemblies (such as a power supply to a disk drive).

External interactions are divided into emissions and immunity. Emissions derive, for example, from harmonics of clocks or other periodic signals. Remedies concentrate on containing the periodic signal to as small an area as possible, blocking parasitic coupling paths to the outside world.

Susceptibility to external influences such as ESD or RFI is related initially to propagated fields that couple into I/O lines which then transfer to the inside of the unit and secondarily to case shielding. The principal receptors are transmission lines, critical devices, and sensitive adjacent traces, particularly those terminated with edge-triggered components.

There are five major considerations when performing EMC analysis on a product or design [1]:

1. *Frequency.* Where in the frequency spectrum is the problem observed?
2. *Amplitude.* How strong is the source energy level and how great is its potential to cause harmful interference?

3. *Time.* Is the problem continuous (periodic signals) or does it exist only during certain cycles of operation (e.g., disk drive write operation or network burst transmission)?

4. *Impedance.* What is the impedance of both the source and receptor units and the impedance of the transfer mechanism (related to separation distance, which affects wave impedance) between the two?

5. *Dimensions.* What are the physical dimensions of the emitting device (or device groups) that cause emissions to be observed? The RF currents will produce EM fields that will exit an enclosure through chassis leaks that equal significant fractions of a wavelength or significant fractions of a "rise time distance." For instance, routed trace lengths on a printed circuit board (PCB) have a direct relationship as transmission paths for RF currents. A similar example is for external cables affixed to a system that are physically the same dimension as the wavelength of a propagating field.

Whenever an EMI problem is approached, it is helpful to review this list based on product application. Understanding these five items will clarify much of the mystery of how EMI exists. Applying these five considerations teaches us that design techniques make sense in certain contexts but not in others. For example, single-point grounding is excellent when applied to low-frequency (such as audio) applications, but it is completely inappropriate for RF signals, which is where most EMI problems exist. Engineers may blindly apply single-point grounding for all product designs without realizing that additional and more complex problems are created using this grounding methodology.

When designing a PCB for use within a product, we are concerned with RF current flow. Current is preferable to voltage for a simple reason: Current always travels around a closed-loop circuit following one or more paths. It is to our advantage to direct or steer this current in the manner that is desired for proper system operation. To control the path in which the current flows, we must provide a low-impedance, RF return path back to the source of the energy. Interference current should be diverted away from the load or victim circuit. For applications that require a high-impedance path from the source to the load, all possible paths through which the return current may travel should be considered [2, 3].

1.4   OVERVIEW ON PRODUCT TESTING

To feel comfortable with the concept of product testing and troubleshooting, one must understand how systems work and whether measured data are valid and accurate along with instrumentation problems that may be present.

1.4.1   Test Environment

When trying to identify an EMC event, where the product is physically located may play a significant role in either causing a problem or preventing one from happening. For products not located in an EMC-controlled location, the environment may pose a challenge. Diagnostic techniques may be hard to implement. For this situation, it becomes difficult to ascertain if the compatibility problem is between dissimilar systems or internal to one specific unit, with the cause possibly blamed on an unrelated source. The first part of testing a product or troubleshooting is to figure out if undesired fields are caused by radiated or conducted mechanisms.

The most difficult part in conducting tests in an industrial environment is that other products located in close proximity may be identical or similar in build. These systems could be from different manufacturers. An example is an office complex with many personal computers and networking equipment. If the primary network hub installed in a wiring closet is having functional or EMI problems, it may be impractical to remove the hub and take it back to the factory. The problem might not be with the hub but with a large number of computers working at similar frequencies causing a complex set of radiated emission events. The computers are the cause, but the hub could be blamed. This type of situation is difficult, at best, to diagnose, but may be possible by using a correlation analyzer.

Another example lies with personal computers. There are many manufacturers of desktop and laptop computers, all with Class B radiated emissions approvals. One vendor may be highly compliant and another noncompliant. The same may be true with different models within the same series marketed by a single company. If trying to integrate different systems together into a finished assembly, one subassembly assumed compliant may cause significant problems related to functionality and regulatory compliance. As soon as different assemblies are placed within the same environment, an EMI event may develop.

In Europe, the European Parliament has enacted legislation that legally mandates electrical equipment to comply with both emission and immunity levels of protection. When compliant with test standards, the device is marked with a logo, CE (Conformity European). The fact that CE-compliant assemblies are marked does not mean that they will function compatibly with other subassemblies making the entire system compliant. In addition, a CE-marked product may have been tested in a best-case configuration but installed in a worst-case chassis or environment.

The following pitfalls provide further insight into the issue of previously compliant products.

1. Components may not have been previously tested or certified for EMC, even though they may be provided with the CE mark. Conformity may have been issued for the Low Voltage or another directive.

2. Compliance results and the related documentation may be suspect, particularly in regard to performance tolerances.

3. The component or assembly may not have been tested correctly to the appropriate standard. Upon request of the actual test report, one may find significant problems in the test setup or the test data are from a different configuration and not those that the report describes.

4. Even if the product was tested correctly, the test setup may not have been worst case or as specifically described in the installation manual. Concern

must be made for the environment the product is used in, in addition to having all cables connected to the system operating in a worst-case mode. Data must be physically present on interconnect cables that are properly terminated, not just dangling from the equipment under test (EUT). When installing a system under this pitfall, significant EMI problems are possible.

5. Most units are provided to the test house without proper cables and installation instructions.

6. Components may have been declared compliant based on a certain environment which may be inappropriate for the end-use product. An example is using an assembly tested for a heavy-industrial application but installed in a residential environment.

7. After certification, changes to the product may happen without retesting to ensure continued conformity, assuming that the change has no effect on EMC. No quality control system may exist to verify that all products are copy exact. An example is die-shrink for silicon components.

8. Purchasers may substitute counterfeit components for legitimate ones by mistake or without knowledge of a change in provider if buying through distribution. Both good and poor components may be mixed together in the stockroom.

9. Buying products with a Declaration of Conformity (DOC) does not mean that due diligence was applied in the certification process. Buying a product in good faith is not a legal defense under most European directives, especially when the end-use application is not tested.

10. Mistakes may have happened that allow a product to be certified as compliant when in fact the unit fails. The manufacturer remains liable for the performance of the product in production quantities. It therefore becomes important to use quality test facilities that are accredited or assessed every year by third-party experts.

11. With the points noted above, either emissions or immunity may be compromised by the weakest component. With various subassemblies connected together, a summation of emissions can develop to the point where the system is now not compliant or interconnect cables may be susceptible to externally induced events.

1.4.2   Self-Compatibility

A system may not be compatible with itself. Different subassemblies may cause functional disruption to other portions of the unit. This situation is commonly referred to as self-jamming. If a PCB is functionally disrupted, is the problem due to errors in software, firmware, or hardware? Several engineers may be called upon to investigate the design, each claiming that his or her portion of the design is perfect and that someone else is responsible for fixing the system. At this stage, nothing can be done to solve the problem as everyone may be in denial. Although the problem lies (technically) in hardware, undesired RF energy can cause firmware or software to be disrupted. Disruption is usually caused by "glitches." Glitches are temporary spikes or anomalies within a digital component or associated transmission lines (traces) due to EM field coupling (i.e., crosstalk).

Self-compatibility also refers to radiated fields that propagate between functional sections within a product or between digital components and interconnects or cable harnesses. Determination must be made in advance if logic circuitry or subsections are candidates for both emissions of and susceptibility to internal radiated RF energy. Depending on the placement of components, relative to susceptible circuits or I/O connectors, potential coupling of internal radiated RF energy must be anticipated before finalizing the design of a PCB or routing cables in the vicinity of high-bandwidth switching components. Should this condition develop, sniffer probes become valuable tools in isolating the area where the undesired energy is either developed or propagated.

1.4.3 Validation of Measured Data

Electromagnetic compatibility tests may be performed in nonideal conditions. Susceptibility testing within anechoic chambers is assumed to be an optimal test environment. Preliminary radiated emission testing can be performed that may assist in predicting if a product will pass or fail at an open-area test site. Before relying totally on data taken in environments that are not ideal, become a devil's advocate and critique the data before tearing down the test setup. If one analyzes results the next day in the office, it may be too late and additional expensive retests may be required. The following are guidelines on determining if the data taken are valid:

1. Was the signal measured the correct signal or was it an ambient? (Be suspicious if it is not related to any harmonic of the system's oscillators.)
2. Was the EUT operating in standby mode or fully functional, with all possible options being exercised at the same time?
3. Does the measured RF energy, after applying correction factors, appear to be consistent or are several signals significantly skewed?
4. Was the ambient environment greater than 6 dB below the specification limit for both electric and magnetic field measurements?
5. Are all instrumentation and dynamic ranges optimal for desired operation and are all instruments properly calibrated?
6. Are correction factors for antennas and instrumentation accurate (measurement uncertainty)?
7. Does the engineer know how to properly test a product and are all procedures required for conformance testing being followed correctly?
8. Were all cables properly maximized during the test?
9. Was the system and/or power supply tested under nominal, minimum, or maximum load?

10. Were dynamic/reactive loads used instead of resistive loads when loading down a power distribution system? This item is a primary cause of failure in equipment after testing has been completed and the system is shipped with live assemblies instead of dummy loads.

Automated instrumentation generally prevents some of the concerns noted above provided it is programmed correctly. The best way to ascertain validity of data is having an engineer or technician question the results from automated testing. Problems do occur that may not be noted. An example is a burst of transmitted data happening in a certain time frame when automated equipment has already passed that portion of the frequency spectrum. Manually observing a large spectral bandwidth of frequency over an extended period to determine if random spikes are present is prudent. The same test conditions can be applied to troubleshooting and debugging.

A gross mistake in analyzing data or not taking all data by manual or automated means can result in noncompliance unknown to the manufacturer. A customer, or Original Equipment Manufacturer (OEM), who will be verifying one's test results will easily note this conformity concern. The worst thing for a company selling large quantities of a product is to be informed by its customer that the system is defective and to mandate a recall or retrofit. A greater concern should be manufacturers located in Europe that buy their competitor's products and test them for conformity. Should the product be noncompliant, regardless of how extensive the original testing may have been, notification of failing data will not be given to the manufacturer but to government authorities. The purpose of reporting nonconforming products to authorities forces the authority to investigate if illegal products are being placed into service. This is done in an attempt to have the competing company fined or penalized. If a mandatory recall must happen, ordered by authorities, that company not only will lose the goodwill of its client base but also may lose out on market share and future permanent business opportunities. This now becomes a dual penalty, forcing numerous companies out of business within Europe and taking away their customer base. The primary item one must remember when certifying products for Europe is to perform "due diligence" during product testing and when auditing tests during production. Basically, Europe self-regulates itself.

1.4.4 Problems during Emissions Testing

As with any aspects of EMC, the concept of "Murphy's Law" is useful to follow during engineering design and analysis. Problems can and do happen during testing and debugging. To minimize problems, one must be aware of the following, taking appropriate action as required.

Equipment Setup and Environment. Most products require use of support (auxiliary) systems to ensure functionality. Auxiliary equipment may be located directly adjacent to the EUT or be remote. If remote, routing cables between systems play an important part in the setup. Cables may be routed under the floor or overhead. The following problems commonly happen during equipment setup:

1. *Ambient Assessment.* Parasitic pickup and antenna configuration (horizontal or vertical polarization, dipole, biconical, log periodic, horn, etc.) may cause false readings. Metallic structures near the EUT on an open-field test site can seriously disrupt signal propagation. The coaxial cable between antenna and receiver may develop a leak or break in the shield, skewing results. To minimize cable discontinuities, routing the coax directly on the ground plane helps. It is prudent to change the coax on a test range yearly, especially if the cable is exposed to harsh environments on a full-time basis.

2. *Mismatch and VSWR Errors.* For intentional radiated emission testing, the voltage standing-wave ratio (VSWR) plays a primary part in assessing the magnitude of RF energy being measured. The VSWR relates to the percentage of transmitted power that is reflected back to the source. Another definition of VSWR is the ratio of the maximum to minimum voltage of the standing wave on the transmission line connecting the generator to the antenna, assuming that the generator is matched to the transmission line and the line is lossless.

    It is desired to have a 1 : 1 VSWR for the antenna system, although this is not achieved in practice. The 1 : 1 condition means that there are no losses within the transmission line between source and load, thus ensuring accurate data are recorded. When an antenna approaches or exceeds a quarter wavelength, significant measurement error may be introduced.

3. *Background Noise within Instrumentation Setup.* This problem is usually observed at test sites. A support unit such as an ancillary printer might be the cause of a problem, although the printer carries a certificate indicating compliance. The only way to ascertain if the signal observed is from the EUT or auxiliary unit is to power down the auxiliary. Sometimes it is necessary to turn off all devices one at a time to isolate the unit causing the failure. This is where having a correlation analyzer would be advantageous. With a correlation analyzer, it is possible to discriminate whether a particular frequency is coming from the EUT or nearby support equipment without powering down all units. This advantage exists since a correlation analyzer can determine "coherence" and can be accomplished in real time. Correlation analyzers are discussed elsewhere in this book.

Note: Just because a device has FCC/Department of Commerce (DOC) or CE approval marks applied does *not* mean the unit is compliant! Certain vendors may get one version of a product to pass EMC tests with expensive rework and then ignore the rework within manufacturing in an effort to save cost. When any component is cost reduced or a change occurs, full EMC tests may be required to validate the change. Some companies may not want to retest a product after they have a Certificate of Conformity. The following procedures help in determining if background ambient noise is present.

(a) Remove the coax from the spectrum analyzer. Replace with a direct terminator. This allows one to determine if the receiver is at fault.

(b) Terminate the coax at the antenna end with a 50-Ω load. This evaluates whether the coax is at fault. If a problem is noted, reroute the coax, shorten the length of the cable, or secure it against the ground reference and see if the undesired signal goes away.

(c) Install ferrite cores, or clamps, on the coax. Sometimes, it takes up to 30 ferrite cores to remove the common-mode energy on the shield due to external ambients or a damaged cable shield.

(d) Place ferrite cores on the AC mains cables.

(e) To investigate if the spectrum analyzer has front-end overload, which can be experienced in high-ambient-noise environments, change the amplitude of the analyzer in increments of 10 dB. If the change is nonlinear, then an attenuator is required for the front end of the analyzer, with proper correction factors applied to the measured signal.

(f) If a signal still exists with support units powered off, remove interconnect cables one at a time and/or place a ferrite clamp on the cable. Clamps may be required at both ends. Sometimes, one ferrite clamp may allow more RF energy to propagate on a cable assembly, whereas two ferrite cores, one on each end, may provide significant benefit.

(g) If noise is still present, check the AC mains cable to all devices. Substitute a shielded cable for the regular cable. *Note:* The shielded cable is only good for diagnostic analysis. Always use an unshielded AC mains cord for the final test, especially if the cord is detachable. In Europe, shielded power cords are generally not used, except in special molded assemblies with a special AC mains receptacle.

1.5   TIME-DOMAIN VERSUS FREQUENCY-DOMAIN ANALYSIS

It is common for digital design engineers to think in terms of a time frame or in the time domain. Electromagnetic interference is generally viewed as a frequency spectrum or in the frequency domain. Radio-frequency energy is typically a periodic wave front that propagates through various media. Different wavelengths of a sine wave are recorded as EMI for those products that are not designed to be intentional radiators. It is difficult to understand an EMI problem in the time domain alone. All digital transitions (when viewed in the time domain) produce a spectral distribution of RF energy (frequency domain). Conversely, a series of fast slew rate sine waves appear as a digital transition pulse. In other words, a time-domain waveform may be defined as a set of sine waves and may be combined graphically or mathematically, but not physically, into a time waveform.

Baron Jean Baptiste Joseph Fourier (1768–1830), a French mathematician and physicist, formulated a method for analyzing periodic functions. Fourier proved that any periodic waveform could be decomposed into an infinite series of sine waves each at an integral multiple or harmonic of a fundamental frequency. The composition of the harmonics is determined during a mathematical operation known as a Fourier Transform. Fourier series can easily be calculated for simple waveforms and displayed with modern instrumentation.

Most engineers think in only one domain, time or frequency. Time-domain engineers are those that consider signal integrity and functionally the only item of importance within a product design. Frequency-domain engineers are concerned with getting a product to pass legally mandated limits, such as FCC, CE, or CISPR.* In reality, the two domains are closely related. The only difference is how one views the signature profile of a transmitted EM field with either an oscilloscope or spectrum analyzer as the measurement tool.

Within our physical universe, the only natural waveform is an analog sine wave. This concept is illustrated in Figure 1.1. We can technically replace the word *digital* in our vocabulary with *infinitely fast AC slew rate signal*.

In regard to any digital transition, there is no such thing as a truly digital component. All output drivers and input receivers of components are operational amplifiers (op-amps) driven to saturation. Digital components are analog devices using an infinitely fast AC slew rate signal. An op-amp is a direct-coupled high-gain amplifier to which feedback is added to control its overall response characteristic. These devices are used to perform a wide variety of linear functions (and also some nonlinear operations). This is a versatile, predictable, and economic system building block. A silicon circuit in reality is a printed circuit board that has been shrunk down to microscopic size. There are no differences between a silicon processor and the physical structure on which the device is used (i.e., a PCB); only the element of scale differs.

A typical ideal op-amp has the following characteristics:

1. Input impedance $R_i \approx \infty$ (very high).
2. Output impedance, $R_o \approx 0$ (very low).
3. Voltage gain $A_v = -\infty$.
4. Bandwidth $= \infty$.
5. Perfect balance: $V_s = 0$ with $V_1 = V_2$ (V_1 and V_2 are the two input pins with voltage levels present as shown in Fig. 1.2).

---

*CISPR (Comité International Spécial des Perturbations Radioélectriques or International Special Committee on Radio Interference) is a European organization that deals primarily with limits and measurements of radio interference characteristics of potentially disturbing sources or emissions.

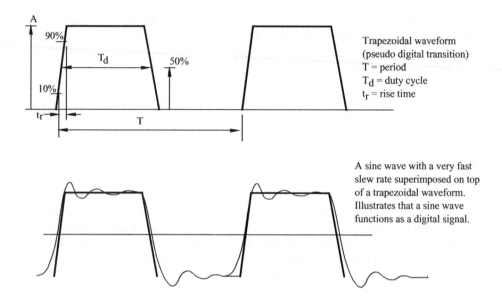

Trapezoidal waveform
(pseudo digital transition)
T = period
T_d = duty cycle
t_r = rise time

A sine wave with a very fast
slew rate superimposed on top
of a trapezoidal waveform.
Illustrates that a sine wave
functions as a digital signal.

Figure 1.1   Sine wave appearing as a digital transition.

A typical inverting op-amp with voltage-shunt feedback is shown in Figure 1.2. Depending on the logic family selected, the configuration will probably be different. Figure 1.2 illustrates a fundamental concept on how a digital signal is sent from a driver to a receiver in a differential mode down a transmission line.

1.6   EMC TESTING METHODOLOGIES

Before discussing how to perform EMC testing, and troubleshooting if necessary, there are considerations that one must be aware of. The most important aspect of EMC engineering lies in understanding fundamental EM theory and being able to apply this theory to product design. The next chapter provides a basic understanding of field theory and why the arena of EMC is one of the more difficult fields of engineering in which one may work.

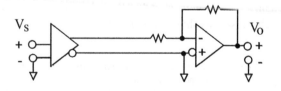

Figure 1.2   Typical inverting operational amplifier configuration.

Products must be conceived with EMC in mind. System analysis and testing must be performed at various stages of development. With regard to EMC, each test has unique technical requirements along with cost and time to perform. To ensure compliance, the following methodologies are recommended.

1.6.1   Development Testing and Diagnostics

Performing tests well ahead of production will save a great deal of time and money throughout all stages of a product's development cycle. When a product has finally been integrated, it can be tested using standard EMC test methods. Standard test methods are not very useful in the early stages of development and evaluation when, for example, microprocessor or digital signal processing (DSP) chips are being specified or chosen.

Standard EMC laboratory test methods provide minimal value late in the stage of a project design when remedial work is required to solve an emissions problem. Standard test methods do not identify where the emissions are coming from, only that they exist. Therefore, it is necessary to use different techniques for development and diagnostic testing over that required for conformity compliance.

1.6.2   Compliance and Precompliance Testing

Many countries require compliance with international regulations before products may be imported. Other countries mandate only specific test laboratories within their jurisdiction. The EMC Directive in Europe requires manufacturers to issue a Declaration of Conformity listing the test standards they have "applied" when using the standards route to conformity. The term "applied" is not well defined. Customs officers in the EU have poorly defined legal rights to insist on seeing any EMC test report or certificate as a requirement for any goods supplied to member states. For most products, use of the CE mark is adequate on the packaging or product; however, EMC Directive enforcement officers may request to see evidence that "due diligence" has been achieved in the conformity of a given product at any time by examining the Declaration of Conformity, Technical Construction File, and/or test report.

While full-compliance EMC testing is not necessarily a burden for manufacturers of systems manufactured in large volumes, it can be extremely expensive for manufacturers of low-cost, custom-engineered, or small-batch products.

There are benefits to performing precompliance testing to discover whether there are EMC concerns before a mass-produced item is submitted for certification. Precompliance testing has the advantage that tests can be stopped at any time, the EUT modified, and the test redone. Full-compliance testing is more expensive per day and typically permits no disruption in the test sequence or involvement by the EUT's designers. On occasion, if precompliance testing is sufficient to pass due diligence requirements, it can be all that is needed for legal sale into Europe. This is good news for manufacturers of low-cost custom or small-batch equipment since the essence of due diligence has been met; however, should a complaint be filed, civil and/or criminal penalties are possible if a formal test report from a competent test laboratory does not exist. It is always best to never certify a product based on a precompliance test.

REFERENCES

1. Gerke, D., and W. Kimmel. 1994. The Designer's Guide to Electromagnetic Compatibility. *EDN,* January 20.
2. Montrose, M. I. 2000. *Printed Circuit Board Design Techniques for EMC Compliance—A Handbook for Designers,* 2nd ed. New York: IEEE/Wiley.
3. Montrose, M. I. 1999. *EMC and the Printed Circuit Board—Design, Theory and Layout Made Simple.* New York: IEEE/Wiley.

CHAPTER 2

ELECTRIC, MAGNETIC, AND STATIC FIELDS

The material in this chapter provides an overview of the types of fields that exist and the manner in which they propagate. Electomagnetic interference is observed through free-space radiation, conducted within interconnects or by propagating EM fields. Included in this category are electrostatic fields or ESD. Coupling is a significant aspect of signal propagation. When performing testing and troubleshooting, the material in this chapter will highlight the types of coupling we deal with and how they affect overall system operation.

A primary concern with EMC lies in understanding that there are two modes of current flow—differential and common-mode. Definition of these two modes is provided herein as well as how they relate to each other. For almost every type of communication, differential mode is desired; however, common mode is generated due to various reasons. Electromagnetic compatibility problems are mainly common mode. When performing testing and troubleshooting, one must be able to distinguish between these two modes and use appropriate tools and measurement techniques to categorize each one and to implement corrective action.

2.1 RELATIONSHIP BETWEEN ELECTRIC AND MAGNETIC FIELDS

Before one performs testing for EMC, especially troubleshooting, an understanding of field theory in simple terms is required. There are two basic types of fields, electric and magnetic. The word *electromagnetic* consists of two root words: *electric* and *magnetic.* Therefore, we are dealing with two types of fields simultaneously. Each field has unique characteristics that one must consider. The differences are considerable yet easy to understand.

Testing for EMC Compliance. By Mark I. Montrose and Edward M. Nakauchi
ISBN 0-471-43308-X © 2004 Institute of Electrical and Electronics Engineers

According to EM theory (Maxwell's equations), a time-variant current within a transmission line develops a time-variant magnetic field, which gives rise to an electric field. These two fields are related to each other mathematically. A static-charge distribution creates electric fields similar in function to that of a capacitor [1]. To understand these fields, we must examine the geometry of a current source and how it affects a radiated signal. In addition, we must be aware that signal strength falls off according to the distance from the source. When we are close to the source, it is called the near-field. This is typically determined to be $\lambda/6$. Any distance greater is called the far-field.

Time-varying currents exist in two configurations:

1. Magnetic sources (flows in closed-loop configurations, represented by a loop antenna)
2. Electric sources (represented by a dipole antenna)

The relationship between near-field (magnetic and electric components) and far-field RF energy is illustrated in Figure 2.1. All waves are a combination of electric and magnetic field components. We call this combination of both electric and mag-

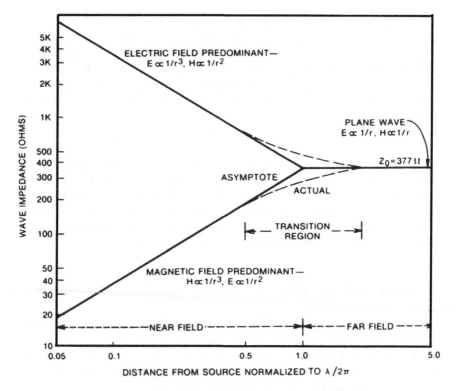

Figure 2.1   Wave impedance versus distance from E and H sources.

netic fields a plane wave or Poynting vector. The Poynting vector is a convenient method for expressing the direction and the power of the EM wave with units of watts per square meter (W/m^2). In the far field, both electric and magnetic field components are at right angles to each other and perpendicular to the direction of propagation. There is no such thing as an electric wave or a magnetic wave by itself.

The reason we see a plane wave is that to a small antenna several wavelengths from the source the wavefront looks nearly planar. This appearance is due to the physical profile that would be observed at the antenna (like ripples in a pond some distance from the source charge). Fields propagate radially from the field point source at the velocity of light, ($c \approx 1/\sqrt{\mu_0 \varepsilon_0} = 3 \times 10^8$ m/s, where $\mu_0 = 4\pi \times 10^{-7}$ H/m and $\varepsilon_0 = 8.85 \times 10^{-12}$ F/m). The electric field component is measured in volts/meter while the magnetic component is in amps/meter. The ratio of the electric field (E) to the magnetic field (H) is identified as the impedance of the EM wave and has units of ohms (Ω). The point to emphasize is that for the Poynting vector the wave impedance Z_0 (characteristic impedance of free space) is constant and does not rely on the characteristics of the source. For a plane wave in free space,

$$Z_0 = \frac{E}{H} = \sqrt{\frac{\mu_0}{\varepsilon_0}} \cong \sqrt{\frac{4\pi \times 10^{-7} \text{ H/m}}{(1/36\pi)(10^{-9}) \text{ F/m}}} = 120\pi \text{ or approximately } 377 \ \Omega \quad (2.1)$$

Power density in the wave front is measured in watts/meter?

We First Examine Magnetic Sources. Consider a circuit containing a clock source (oscillator) and a load (Figure 2.2). Current is flowing in this circuit around a closed loop (trace and RF current return path). We can easily calculate the generated radiated field. Fields produced by this loop are a function of four variables:

1. *Current Amplitude in Loop.* The field is proportional to the current that exists in the transmission line and is sometimes referred to as a low-impedance field.

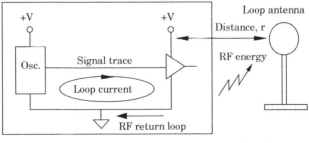

Source of emissions   Reception of emissions

Figure 2.2   RF transmission of a magnetic field.

2. *Orientation of Source Loop Antenna Relative to Device Under Test.* For a signal to be measured or observed, the orientation of the source loop current should match that of the measuring device. For example, if a loop antenna is horizontally polarized, it must be in an identical orientation.

3. *Size of Loop.* If the loop is electrically small (much less than the wavelength of the generated signal at the frequency of interest), the field strength will be proportional to the area of the loop. The larger the loop, the lower the frequency observed at the terminals of the antenna.

4. *Distance.* The rate at which the field strength drops off from the source depends on the distance between source and antenna. In addition, this distance also determines whether the field created is magnetic or electric dominant. When the distance is electrically "close" to the loop source, the magnetic field falls off as the cube of the distance ($1/r^3$). When the distance is electrically *"far,"* we observe an EM plane wave. This plane wave falls off inversely with increasing distance ($1/r$). The point where the magnetic and electric field vectors intersect is at approximately one-sixth of a wavelength (which is also identified as $\lambda/2\pi$). The wavelength is the speed of light divided by frequency. This formula can be simplified to $\lambda = 300/f$, where λ is in meters and f is in megahertz. This one-sixth wavelength applies to a point source.

We Now Examine Electric Sources. Electric sources, in contrast to the closed-loop magnetic source, are modeled by a time-varying electric dipole and is sometimes referred to as a high-impedance field. This means that two separate time-varying point charges of opposite polarity exist in close proximity. The ends of the dipole permit a change in electric charge to be developed. This change in electric charge is accomplished by current flowing throughout the dipole's length. Using the circuit described above, we represent the electric source by an oscillator's output driving an unterminated antenna. When examined in the context of low-frequency circuit theory, we discover that this circuit is not valid. We did not take into account the finite propagation velocity of the signal in the circuit (based on the dielectric constant of the nonmagnetic material) in addition to the RF currents that are created herein. This is because propagation velocity is *finite,* not *infinite!* The assumptions made are that the wire, at all points, contains the same voltage potential and the circuit is at equilibrium.

The fields created by an electric source are a function of four variables:

1. *Current Amplitude in Dipole.* The fields created are proportional to the amount of current flowing in the dipole.

2. *Orientation of Dipole Relative to Measuring Device.* This is equivalent to the magnetic source variable described above.

3. *Size of Dipole.* Fields created are proportional to the length of the current element. This is true if the length of the trace is a small fraction of a wavelength. The larger the dipole, the lower the frequency that is observed at the termi-

nals of the antenna. For a specific physical dimension, the antenna will be resonant at a particular frequency.

4. *Distance.* Electric and magnetic fields are related to each other. Both field strengths fall off with increasing distance. In the far field, the behavior is similar to that of the loop source. When we move in close to the point source, both magnetic and electric fields have a greater dependence on distance from the source.

Propagation of RF energy, both electric and magnetic fields, can be represented as equivalent component models that help describe their propagation mode with an antenna structure as the visual display. A time-varying *electric* field between two conductors can be represented as a capacitor configuration (dipole antenna). A time-varying *magnetic* field between these same two conductors is represented by mutual inductance (loop antenna). Figures 2.3*a,b* illustrate these two coupling configurations.

For these noise-coupling models to be valid, the physical dimensions of the circuit must be small compared to the wavelength of the signals involved. When the model is not truly valid, we use lumped-component representation to explain EMC for the following reasons:

1. Maxwell's equations are difficult to solve for most real-world situations due to complicated boundary conditions. If we have no level of confidence in the validity of the approximation due to lumped modeling, then the model is invalid.

2. Numerical modeling usually does not show all RF energy paths. Those paths are generally perfect in nature and may not represent actual circuit operation. Radio-frequency energy is developed from many aspects of signal-routing topology and configuration. Common-mode currents exist due to an imbalance in the differential-mode transmission of a propagating wave, which is a primary reason for the development of EMI. Other aspects of RF generation are due to parasitic elements within a transmission line, especially unknown capacitive and inductive elements that cannot be anticipated as they exist within the system in a virtual manner—not visually seen on a schematic or identified within a component model. Even if a modeling answer is possible, system-dependent parameters are not clearly identified or shown along with the explanation in item 1 above.

For these two reasons, it is nearly impossible to simulate EMI, especially if common-mode current is the culprit. Too many times, we base results from simulation tools as being accurate at the PCB level when the real transmission paths are interconnects external to the source generator. Without knowledge of unexpected parasitics, which are difficult if not impossible to include within a simplified mode, the results of simulation only provide an insight into how RF energy may or may not exist.

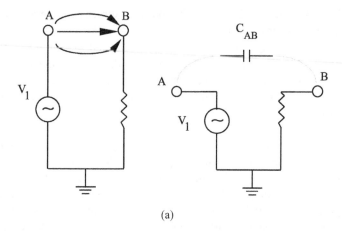

(a)

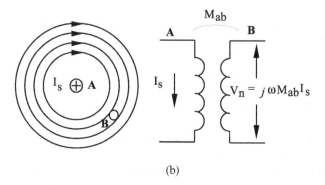

(b)

Figure 2.3a   Noise-coupling method: (*a*) electric field (dipole antenna); (*b*) magnetic field (loop antenna).

Why is this discussion on basic field theory presented in a book on testing and troubleshooting? The answer is simple. We need to know how fields are created and propagated. With this concept, we can determine which fields are present when testing products for EMC compliance. Should a failure happen, one must recognize that there are two basic types of fields such that the proper tool (probe and instrumentation) is selected to determine which field is causing the problem.

2.2   METHODS OF NOISE COUPLING

Why is this section on noise coupling provided in this book? The answer is simple. When testing a product for EMC, a problem with compliance may develop. Where and how this problem developed is described elsewhere [1–5]. Prevention of undesired RF energy must be performed by removing the coupling path.

Situations occur where EMC becomes a primary concern for electronic products. All situations conditionally involve a source and a victim. If either element is missing, then an EMC event will not exist. If both elements are present within the same electrical assembly, we have an "intrasystem" EMC environment. If these elements are located in different assemblies, we have an "intersystem." Regardless of whether the environment is an intra- or intersystem, consideration on functional aspects of product design must be analyzed for proper operation within its intended environment.

A product must be designed for two levels of performance: minimize RF energy leaving an enclosure (emissions) and minimize RF energy entering the unit, thus causing harmful operation (susceptibility or immunity). Both emissions and immunity travel by radiated and/or conductive paths.

Knowledge of the manner in which EM fields propagate must be understood. Reduction in propagated energy is mandatory to reduce interference and ensure signal integrity and/or functionality. Both emissions and immunity must be considered simultaneously; however, during the analysis portion we focus generally on only one aspect of the problem, either emissions or immunity.

When designing products, the emissions profile and susceptibility level of passive and active components should be known ahead of time to determine if a problem with EMC may occur. Adherence to regulatory standards does not mean automatic approval. Regulatory standards were written to minimize interference for a particular environment; not cause harm to communication services. Regulatory standards do not address electric and magnetic field coupling internal to a system or between units connected together in a rack assembly. The rack as a whole is tested and approved, not the individual components required to make one large functional unit.

Almost every electrical system contains elements capable of antenna-like behavior. These elements include external cable assemblies, PCB traces, and internal wiring and mechanical structures. Each element can unintentionally transfer RF energy through electric, magnetic, or electromagnetic means. In practical environments, intrasystem and external coupling can be minimized by providing shielding and incorporating suppression components. Metal partitions can minimize radiated coupling by absorbing the energy or redirecting the propagating field back into itself. Cable-to-cable coupling occurs through either capacitive or inductive means. Dielectric materials may also reduce field propagation, although use of dielectrics is not as effective as proper design techniques.

Every propagation path may contain multiple transfer mechanisms [2]. These are detailed in Figure 2.4 and include the following:

1. Direct radiation from source to receptor (path 1)
2. Direct RF energy radiated from the source transferred to the AC mains cables or signal/control cables of the receptor (path 2)

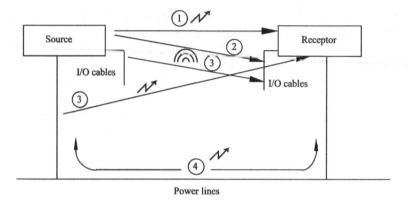

Figure 2.4 Coupling path mechanisms.

3. RF energy radiated by AC mains, signal, or control cables from source to receptor (path 3)
4. RF energy conducted by common electric power supply lines or by common signal/control cables (path 4)

In addition to the four primary coupling paths, there are four transfer mechanisms that exist for each path:

1. Conductive
2. Electromagnetic
3. Magnetic field dominant (subset of the EM identified separately in Figure 2.4)
4. Electric field dominant (subset of the EM identified separately in Figure 2.4)

If the noise-coupling mechanism can be ascertained, a logical solution can be determined to reduce coupling between source and receptor. Use of instrumentation (analyzers, scopes, antennas, sniffer probes, simulation software, etc.) can help isolate problem areas and permits implementation of design techniques and procedures to achieve EMC.

The process of radiation between a source and receptor involves transference of an EM field through a propagation path that does not involve a metallic interconnect. Various modes of radiated coupling include transfer of EM energy from adjacent equipment through free space and natural coupling between similar EM environments such as power lines. A receptor may also receive EM noise or interference from exposure to connectors, signals, or other transmission lines within a product or circuit.

2.2.1 Common-Impedance Coupling

Common-impedance coupling develops when both source and victim share a transmission path through a common impedance. The most frequent example of common-impedance coupling is found in a shared reference connection between modules. This shared reference is usually a return wire within a cable assembly or in the earth ground reference of the power distribution system.

Figure 2.5 illustrates two means of common-impedance coupling in a simplified manner. In the figure, a common connection is used between assemblies for power distribution. There is also an interconnect cable with a 0-V reference between the two. System 1 may induce RF energy, V_{PWR}, into the earth ground connection usually by stray capacitance C_S or by direct means (hard-wired point from the input receptacle).

All interconnects have a finite impedance, even power distribution. Noise voltage V_{IO} is developed across the transmission line due to inductance. This noise voltage can be significant between the two systems. Generally, the return path is the primary source for development of common-mode current.

For almost all system designs, where common references are provided, the coupling mechanism can become extremely complex. This is especially true when multiple assemblies are interconnected by both I/O and power distribution. If we do not observe undesired RF energy, then the noise voltage between assemblies is small in magnitude or proper EMC suppression techniques have been incorporated in the interconnect. Although the earth conductor impedance is nonzero, energy present may not be significant at the operating frequency of the system to cause harm.

The key to minimizing common-impedance coupling is to prevent development of noise voltage across a transmission line that is shared between two or more assemblies and to ensure that interconnects between systems have the least amount of inductance possible. This is a primary reason why a flat braid has superior performance over a round wire. Whenever an EMC problem is noted between two systems remote from each other, the interconnect cable is usually the problem and should be the first item to investigate during the troubleshooting process.

2.2.2 Electromagnetic Field Coupling

Electromagnetic field coupling is a combination of both magnetic and electric fields affecting a circuit simultaneously. Depending on the distance between source and

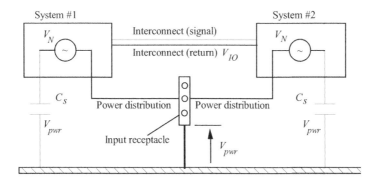

Figure 2.5 Common impedance coupling between systems.

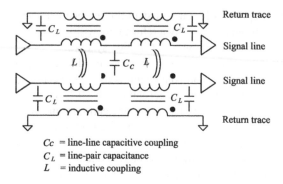

Cc = line-line capacitive coupling
C_L = line-pair capacitance
L = inductive coupling

Figure 2.6   Coupling model for multiple transmission lines.

receptor, the electric field (E) and magnetic field (H) may be operatively dominant, depending on whether we are in the near field or far field. This is the most common transfer mechanism observed by measurement with an antenna.

An occasionally overlooked noise-coupling process is through a metallic conductor: wire, transmission line (PCB trace), or conductive means. A metallic conductor may pick up RF energy from a culprit source and transfer undesired noise to a victim circuit. The easiest way to prevent this transfer is to remove or prevent RF current from a source circuit from being generated or propagated, or prevent the victim circuit from receiving this undesired RF energy.

What happens to a signal propagating down a transmission line from a source to a destination? Figure 2.6 illustrates a model of one propagation path. The signal line connects directly between the source and destination. With this configuration, we have both inductive (L) and capacitive (C) coupling between adjacent circuits.

In Figure 2.6, the output capacitance of the load siphons off a certain percentage of drive current. The inductance of the line helps attenuate the signal, which also couples to adjacent traces. Capacitance between signal traces also diverts RF energy in addition to corrupting signal fidelity. Finally, the capacitance of the load will shunt energy away from the input. Load capacitance will couple EM energy to ground or 0-V reference.

If a transmission line is physically long compared to the rise/fall time of the signal (edge rate transition), distributed effects are observed.[1] The RF energy that has been propagating down the transmission line with characteristic impedance Z_0 will arrive at the load some time later. If the load impedance Z_L is the same as the source impedance Z_0, all EM energy will be absorbed within the load and with desired fidelity. If the load impedance is high, a portion of the transmitted signal will reflect back to the source. Reflections within transmission lines are observed as a ringing

---

[1]An electrically long conductor is defined as a transmission line (l) containing a signal with an edge rate transition (t_r) that is faster than the time it takes for a signal to travel from source to load (t_{pd}) and return from the load to source, causing functionality concerns that include ringing and reflections, mathematically described as $t_r \leq 2t_{pd}L_{max}$.

condition with overshoot (both positive and negative directions in regard to a desired set point) for a digital logic transition.

2.2.3 Conductive Coupling

The process of conduction between a source and receptor involves transference of an EM field through a metallic interconnect. Interference energy can be carried between power supply lines and signal transmission cables. For example, interference between systems plugged into the same electrical outlet may share undesired RF energy, causing harmful interference or disruption of functionality.

Conductive transfer can occur through common-impedance coupling. This happens when both the noise source and susceptible circuits are connected by mutual impedance. A minimum of two connections is required. Two connections are required because noise current must flow from a source to a load and then return to the source. Figure 2.7 illustrates two circuits and a power source. Current from each circuit flows through both the shared impedance of the power subsystem and interconnect wiring, all caused by shared metallic transmission lines. For this figure, the shared connection is the return line.

2.2.4 Radiated Coupling—Magnetic Field

Magnetic coupling occurs when a portion of magnetic flux created by one current loop passes through the flux pattern of a second loop formed by another current path. Magnetic flux coupling exists due to mutual inductance between the two loops. The RF noise voltage inducted in the second loop is $V_2 = M_{12} \, dI_1/dt$, where M_{12} is the mutual coupling factor and dI_1/dt is the time rate of change of current in the trace within current loop I_1. Magnetic flux coupling is illustrated in Figure 2.8.

A coupled voltage appears in series with the desired signal in the victim circuit regardless of circuit impedance. Mutual inductance between two transmission lines is determined by the separation of the conductors and the length that they are in parallel plus the presence of any magnetic screening that surrounds the circuit. A spe-

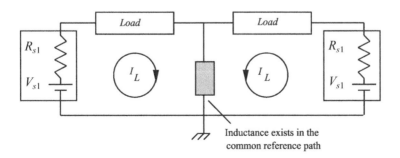

Figure 2.7   Conductive coupling.

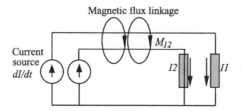

Figure 2.8 Magnetic field coupling.

cial condition of magnetic coupling occurs when multiple transmission lines are bundled into one assembly, such as a cable harness. In this situation, a very long parallel route of the wire may be present, exacerbating the transference of undesired RF energy. For this reason, the recommendation of routing a 0-V reference trace or wire between signal lines should be implemented. In addition, separation of different voltage source bundles is mandatory, such as AC, DC, signal, and control. It is useful to partition functional areas. This should be implemented within the same cable to avoid magnetic flux linkage.

If an RF return path is routed adjacent to a transmission line, the magnetic flux that surrounds the source wire will observe magnetic flux in the return path traveling in the opposite direction. When two transmission lines are in close proximity, with magnetic flux propagating in opposite directions, flux cancellation occurs. The process of flux canceling is the primary concern one should have to minimize development of radiated RF energy and to achieve EMC (suppression of undesired RF energy) [5].

Mutual inductance between cable pairs is inversely proportional to the log of the separation distance. For two individual transmission lines associated with different circuits, mutual inductance is again inversely proportional to the log of the square of their separation distance. Between two transmission lines, mutual capacitance also exists. Figure 2.9 illustrates mutual inductance and capacitance between transmission lines [6].

2.2.5 Radiated Coupling—Electric Field

Electric field coupling dominates in high-impedance circuits and is a counterpart to magnetic field coupling. Magnetic field coupling is created by inductive means whereas electric field coupling is capacitive in nature. When a voltage potential difference exists between two conductors, an electric field is developed. This field will induce a voltage in an adjacent transmission line. The field that is developed is described by Eq. (2.2). Unlike magnetic field coupling, which introduces a differential voltage source in the victim circuit, the load impedance will significantly affect the current in the victim circuits. On the other hand, *E*-field coupling appears as a current source with current division dependent upon the source and load impedance of

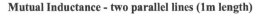

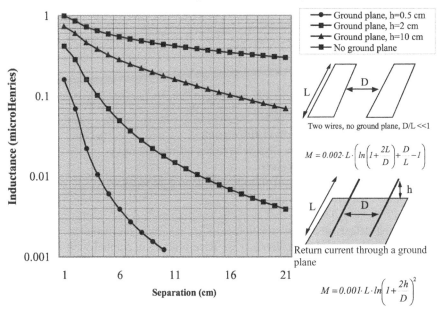

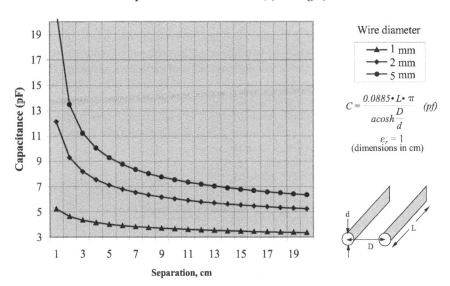

Figure 2.9   Mutual inductance and capacitance between transmission lines. [6]

the victim circuit. The effect of the load impedance in the victim circuit on the current in the load is now a secondary effect. The voltage developed in the victim circuit will be dependent on the device's impedance. High impedance circuits are more susceptible to capacitive coupling. Therefore, both source and load must be referenced together:

$$V_{in} = C_m Z_{in} \frac{dV_s}{dt} \tag{2.2}$$

where V_{in} = induced voltage on victim circuit
$\quad$ C_m = mutual capacitance between two transmission lines
$\quad$ Z_{in} = impedance of receiver
$\quad$ V_s = interfering voltage (of negligible source impedance)
$\quad$ dt = time period of event

Coupling effects are small relative to other coupling mechanisms. For many circuits, there is mutual capacitance if we have high Z_S in parallel with Z_L (Figure 2.10). Capacitive coupling happens when a portion of the electric flux created by one circuit terminates on the conductors of another circuit. Electric flux coupling between two circuits is represented by mutual capacitance. The noise current injected into the susceptible circuit is approximately $I = C(dV/dt)$, where C is the capacitance between sections and dV/dt is the time rate of change of voltage in the trace.

Mutual capacitance between two circuit nodes is dependent on the voltage level present, not the current within the transmission line. Capacitance is affected by separation distance, especially the area of overlap between the two wires. In addition, the dielectric material between the two lines and the presence of other electric field shielding near the circuits can affect the magnitude of capacitive coupling. Because area of overlap or adjacency exists, capacitive coupling tends to be larger between bigger objects than between smaller ones. This is often not a concern, since larger structures tend not to carry high levels of dV/dt. For situations where high levels of dV/dt are present, such as in switch-mode power supply converters, capacitive coupling can become a serious threat for EMC.

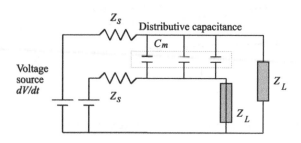

Figure 2.10   Electric field coupling.

2.2.6   Radiated and Conducted Coupling Combined

For most situations, a combination of radiated and conducted interference exists. A radiated field can couple into a cable assembly and cause disruption. In effect, the radiated field appears as a conducted event. Conversely, radiated common-mode energy from an unshielded cable transferring high-speed, high-energy data can cause a field potential to propagate to sensitive electronic circuitry, causing disruption. Examples of combined radiated and conducted interference are as follows:

1. Radiated coupling from transmission lines (power, signal, and control) into any cable assembly or chassis enclosure associated with other electrical equipment
2. Conductive coupling of both electric and magnetic fields from transmission lines between assemblies (component radiation, etc.)
3. Undesired EM fields developed within a system propagating to other electrical equipment. Interference can enter a receptor by either radiated or conducted means

Coupling in cable harnesses, transmission lines, and traces on a PCB occurs through both capacitive and inductive modes. Capacitive coupling (transfer) is a result of high impedance to ground and is more predominant at higher frequencies. Inductive coupling between two currents (current within a closed-loop system) is predominant in low-series-impedance circuits and at lower frequencies. Apart from the reactive transfer of RF energy, or interference, a resistive transfer may also happen through the voltage drop in a common reference path between two devices, circuits, or systems. A voltage drop across the common reference impedance between functional areas, caused by current flow in one circuit, will appear as an interference signal source. Interference current generated is propagated in a conducted manner along the transmission line and is observed at the load terminals of adjacent circuits

Radiated EM energy is developed when cables or signal transmission lines are poorly shielded between functional sections. Radiation may also be observed from exposed wires carrying digital signals. Printed circuit boards are the main source of radiated RF energy. For a transmission line between a source and load, terminated in a fixed arbitrary impedance, three types of energy transference may be observed:

1. The transmission line containing line losses
2. A radiating EM wave representing losses in the surrounding space
3. An axial propagating field between source and load

The first item determines the magnitude of voltage drop in the transmission line. This voltage drop allows common-mode currents to be developed. The second item is radiated coupling into free space, which is the primary source of interference to other systems. This is more significant at higher frequencies when the separation distance between transmission lines approximates a wavelength. Radiation coupling dominates in digital circuits, where very fast edge rate signal transitions are present. The final item describes the existence of a field within a transmission line that may present itself in either a capacitive or an inductive manner to adjacent signals or chassis enclosures.

2.3   COMMON-MODE CURRENTS VERSUS DIFFERENTIAL-MODE CURRENTS

In all circuits both common-mode (CM) and differential-mode (DM) currents are present. Both types of current determine the amount of RF energy propagated between circuits or radiated into free space. There is a significant difference between the two. Given a pair of transmission lines and a return path, one or the other mode will exist, usually both. Differential-mode signals carry data or a signal of interest (information). Common-mode is an undesired side effect from differential-mode transmission and is most troublesome for EMC. For purposes of this discussion, reference to differential (twisted) pair transmission lines is not presented.

When using simulation software to predict emissions, DM analysis is usually the form of analysis used. It is impossible to predict radiated emissions based solely on DM (transmission line) currents. Common-mode currents are the primary source of EMI. If only calculating DM currents, one can severely underpredict anticipated radiated emissions since numerous factors and parasitic parameters are involved in the creation of CM currents from DM voltage sources. These parameters usually cannot be anticipated and are present within a PCB structure dynamically in the formation of power surges in the power and return planes during edge-switching times.

2.3.1   Differential-Mode Currents

Differential-mode current is the component of RF energy present in both the signal and return paths that is equal and opposite to each other. If a 180° phase shift is established precisely, RF differential-mode currents will be canceled. Common-mode effects may, however, be developed because of ground bounce and power plane fluctuation caused by components drawing current from a power distribution network.

Differential-mode signals will (1) convey desired information and (2) cause minimal interference as the fields generated oppose each other and cancel out if properly set up.

Using DM signaling, a device sends out current that is received by a load. An equal value of return current must be present. These two currents, traveling in opposite directions, represent standard DM operation. Fields not coupled to each other will be the source of common-mode EMI. In the battle to control undesired emissions and crosstalk, through the common mode, the key is to control excess energy fields through proper source control and careful handling of the energy-coupling mechanisms.

2.3.2 Common-Mode Currents

Common-mode current is the component of RF energy that is present in both signal and return paths, often in common phase to each other. The measured RF field due to CM currents will be the sum of the currents that exist in both the signal and return path. This summation could be substantial. Common-mode currents are generated by any imbalance in the circuit. Radiated emissions are the result of such imbalance.

Poor flux cancellation may be caused by an imbalance between two transmitted signal paths or excessive impedance in the return path. If DM signals are not exactly opposite and in phase, magnetic flux will not cancel out. The portion of RF current that is not canceled is identified as "common-mode" current.

Common-mode signals (1) are the major sources of cable and interconnect EMI and (2) contain no useful information.

Common-mode currents begin as the result of currents mixing in a shared conductive path such as power and return planes within a PCB, cable assemblies between systems, or the chassis enclosure. Typically this happens because RF currents flow through both intentional and unintentional paths. The key to preventing CM energy is to understand and manage RF return currents.

In any return path RF current will attempt to couple with RF current in the source path (magnetic flux traveling in opposite directions to each other). Flux that is coupled to each other (i.e., 180° out of phase) will cancel, permitting the total magnitude of flux to approach zero. However, if an RF current return path is not provided with symmetry to the source through a path of least impedance, residual common-mode RF currents will be developed across the source of impedance. There will always be some CM currents in any transmission line or system as a finite distant spacing must be present between the signal trace and return path (flux cancellation approaches 100%). The portion of DM return current that is "not" canceled becomes residual CM current.

To make a DM–CM comparison for both magnetic and electric field sources, consider a pair of parallel wires carrying a DM signal. Within this wire, RF currents flow in opposite directions. As a result, all RF fields are contained between the wire pair. This parallel set of wires will act as a balanced transmission line that delivers a clean differential (signal-ended) signal to a load.

Using this same wire pair, look at what happens when CM voltage is placed on this wire. No useful information is transmitted to the load since the wires carry the same voltage potential. This wire pair now functions as a driven dipole antenna with respect to ground. This antenna radiates unwanted CM fields with extreme efficiency. Common-mode currents are generally observed in I/O cables. This is why I/O cables radiate RF energy as well. An illustration of how a PCB and an interconnect cable allow CM and DM current to exist is shown in Figure 2.11.

2.3.3 Example on Difference between Differential- and Common-Mode Currents

To describe how CM and DM currents exists within a transmission line, Figure 2.12 is provided. A very "simplified" analogy (discussion) follows to explain a very

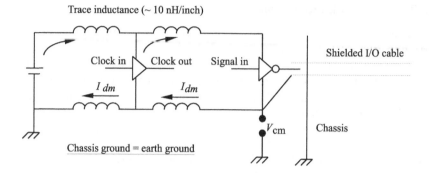

Figure 2.11 System equivalent circuit of DM and CM currents.

complex topic. Although not mathematically or precisely correct, it helps one to visualize the difference between CM and DM currents and how common-mode is the result of an imbalance in differential-mode propagation.

The flow of current from source E to load Z in DC terms is represented as I_1. The alternating current is propagated as an EM field in the same transmission path. Return current (both AC and DC) travels back to the source in I_2. The direct current returns in the primary return path, and alternating current I_2' returns in a different path (identified as the dotted line in Figure 2.12). This secondary RF return path is sometimes called an image plane, ground plane, or 0-V reference when referenced to a PCB.

Differential-Mode Configuration. For a simple analogy, assume 1 A of current is propagated from the source to the load (using I_1 to represent the flow of current). For a DM transmission, 1 A of current must return to the source, represented by I_2. If the amount of outgoing current is the same as the return current without any loss in the network, we have a perfectly balanced transmission line system. The EM field that exists in the outgoing path will couple inductively to the RF "return" path (AC transmission while DC will always travel in the lowest resistance path, I_2). Magnetic flux between these two transmission lines will cancel each other out, as they should be equal and opposite in direction. Thus, there should be no radiated EMI that could cause harmful interference to other circuits or systems if everything is in balance. This analysis assumes that the spacing between opposite conductors is extremely small.

Common-Mode Configuration. Regarding the CM current configuration of Figure 2.12, a different situation is present. Assume 50% of the transmitted current in the transmission line is consumed within the load. This means that 50% of the original current that has not been consumed must still return to its source. Under this situation, we violate Kirchhoff's law that states that the sum of the all currents within a transmission line must equal zero. However, we have a 50% loss. Where is the missing 50%?

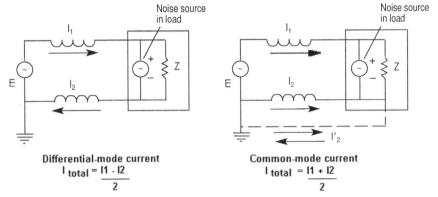

Common-mode current configuration.
I_2' is a virtual return path through free space or other parasitic path.
Not all "desired" return current will flow in I_2 due to inductance or losses in the transmission line.
The remainder of the "desired" return current will therefore flow in I_2'.
A summation of currents must flow in both paths to satisfy Amperer's Law.
A negative current flow will exist in I_2, traveling in the opposite direction to satisfy Ampere's Law.
The "undesired" (or negative) current flow in I_2 is that portion that contributes to common-mode currents.

Figure 2.12 Common and differential-mode current configurations.

Fifty percent of the current returns to the source within the transmission line (I_2) while the remaining 50% propagates through an alternate return path (I_2', the dotted line in Figure 2.12). This alternate return path can be either free space or a metallic interconnect. The dotted line in the CM configuration is a virtual return path. Regardless of the mode of propagation, 100% of the current must return to the source. For this example, more than one return path exists to satisfy Kirchhoff's law.

What we have in essence is that at any particular point in time 50% of the energy travels in one direction (right to left) and 50% propagates in the opposite direction (left to right). The current that travels from right to left is virtual return current with opposite polarity. The summation of both currents (positive and negative at a fixed point in time and space) equals 100% if measured with an ohmmeter.

Taking this simplified explanation, we have

$$I_{\text{total(dm)}} = \tfrac{1}{2}(I_1 - I_2) = \tfrac{1}{2}(1\ \text{A} - 1\ \text{A}) = 0\ \text{A}$$

$$I_{\text{total(cm)}} = \tfrac{1}{2}(I_1 + I_2) = \tfrac{1}{2}(1\ \text{A} + 0.5\ \text{A}) = 0.75\ \text{A}$$

For DM transmission, the electric field component is the difference between I_1 and I_2. If $I_1 = I_2$ exactly, there will be no RF field radiation from the time-variant current that emanates from the circuit (assuming the distance from the point of observation is much larger than the separation between the two current-carrying conductors). A minimal amount of RF energy is developed if the distance separation between I_1 and I_2 is electrically small. As observed, any amount of RF loss within a system or transmission line will cause CM energy to be developed.

To summarize, the total magnitude of imbalance in a DM transmission line system becomes the total magnitude of the CM current! This CM current is what gives us the majority of problems with EMI.

2.3.4   Radiation due to Differential-Mode Currents

Differential-mode radiation is caused by the flow of RF current loops within a system's structure. For a small-loop receiving antenna operating in a field above a ground plane (free space is not a typical environment when considering systems effects), RF energy is described approximately as [4]

$$E = 263 \times 10^{-16}(f^2 A I_s)\left(\frac{1}{r}\right) \qquad \text{V/m} \qquad (2.3)$$

where A = loop area (m²)
   f = frequency (Hz)
   I_s = source current (A)
   r = distance (m) from radiating element to receiving antenna

For most systems, the RF energy is created from currents flowing between assemblies and within the power and 0-V reference structure. Radiated emissions can be modeled as a small-loop antenna carrying RF currents (Figure 2.13). When a signal travels from source to load, a return current must be present in the return system. A small loop is one whose dimensions are smaller than a twentieth of a wavelength ($\lambda/20$) at a particular frequency of interest illuminated by RF current flowing within its structure. This is the linear length of the transmission line. For most PCBs containing transmission lines, trace lengths have dimensions that correspond to frequencies up to the gigahertz range.

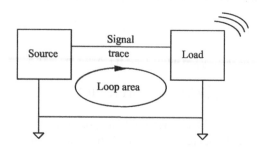

Figure 2.13   Loop area between components.

The maximum loop area that will not exceed a specific specification level is described by Eq. (2.4). Areas is defined by $A = (r^2$, where r is the radius of the circle that is defined by the circumference [4]:

$$A = \frac{380rE}{f^2 I_s} \tag{2.4}$$

Conversely, the maximum field strength created from a closed-loop boundary area is described by

$$E = \frac{Af^2 I_s}{380r} \tag{2.5}$$

where E = radiation limit (μV/m)
  r = distance between loop and measuring antenna (m)
  f = frequency (MHz)
  I_s = current (mA)
  A = loop area (cm^2)

Within free space, radiated energy falls off inversely proportional distance-wise between a source and receiving antenna. The loop perimeter formed by a specific current component within the PCB must be known. The total area of a loop is between the trace and current return path, described by Eqs. (2.4) and (2.5). These equations must be solved for every loop (different loop size area) and for each frequency of interest. This makes for many calculations, suitable for solving only by computer.

Using Eqs. (2.4) and (2.5), we can determine if a particular routing topology needs to have special attention as it relates to the development of radiated emissions. Special attention may involve rerouting traces stripline (internal routing layers of the PCB), changing routing topology, locating source and load components closer to each other, or providing external shielding of the assembly (containment measures).

EXAMPLE

Assume that a convoluted shape exists between two components located on a PCB similar in configuration to Figure 2.13. This loop acts as an antenna with the signal trace at voltage potential and the return trace at ground. The variables for Eq. (2.3) are A = 4 cm^2, I_s = 5 mA, and f = 100 MHz. The field strength is 52.8 μV/m at 10 m distance. Radiated emission limits for EN 55022[2] [4], Class B, is 30 μV/m (quasi-peak). This loop area for a typical application on a PCB is 22.6 μV/m above the limit!

---

[2]European Normalized standard EN 55022, *Limits and methods of measurement of radio disturbance characteristics of information technology equipment.* This is an international test specification for regulatory compliance purposes.

Note: To accurately calculate area, one must determine if the perimeter is a square or circle. For a square, if S is the perimeter, then S_{SQ} is the perimeter of a square current loop while S_{CIR} is the perimeter of a circular current loop. Likewise, if A is area, then A_{SQ} is the area of a square current loop and A_{CIR} is the area of a circular current loop.

For a square:

$$A_{SQ} = \left(\frac{S_{SQ}}{4} \right)^2$$

and for a circle

$$A_{CIR} = \pi \left(\frac{S_{SQ}^2}{2\pi} \right)^2 = \frac{1}{4\pi} S_{CIR}^2$$

Even if $S_{SQ} = S_{CIR}$, this does not guarantee that $A_{SQ} = A_{CIR}$.

For $A_{SQ} = A_{CIR}$

$$A = A_{CIR} = A_{SQ} = \frac{1}{4\pi} S_{CIR}^2 = \left(\frac{S_{SQ}}{4} \right)^2$$

then

$$S_{CIR} = S_{SQ} \sqrt{\frac{\pi}{4}}$$

Thus, for equal areas, the circumference of a circle is smaller than the circumference of an equivalent square (with equal area) by a factor of $\sqrt{\pi/4}$, which is less than 1. In fact, a circle is a geometric shape, which minimizes the circumference for a given area. It follows that for $A_{SQ} = A_{CIR}$ we get $S_{SQ} > S_{CIR}$.

2.3.5 Common-Mode Radiation

Common-mode radiation results from unintentional voltage drops caused by a circuit element rising above the 0-V reference. Cables connected to the affected reference system will act as a dipole antenna when stimulated with a voltage source. The far-field electric term of the field propagated is described by [4]

$$E \approx \frac{1.26(fI_{CM}L)}{r} \qquad \text{(V/m)} \tag{2.6}$$

where L = antenna length (m)
I_{CM} = common-mode current (A)
f = frequency (MHz)
r = distance (m)

With a constant current source and fixed antenna length, the electric field potential at a prescribed distance is directly proportional to the frequency of operation. Unlike DM radiation that is easy to minimize using proper design techniques, CM radiation is a more difficult problem to solve. The only variable available to the designer to solve the problem is reducing the common path impedance for the return current. This is achieved using a sensible grounding scheme. If it is not possible to change the grounding methodology, which may involve major modifications, devices such as CM chokes (baluns—balanced/unbalanced matching transformer) should be implemented.

2.3.6 Conversion between Differential- and Common-Mode Energy

Common-mode currents may be unrelated to an intended signal source (e.g., they may be from other devices or signals developed through coupling). Conversion between DM and CM is performed when two signal conductors both with different impedances are present. These impedances are dominated at radio frequencies by stray capacitance and inductance within a chassis enclosure or between transmission lines (interconnect cables).

To illustrate this effect, Figure 2.14 shows the desired signal of interest, represented by DM current I_{dm} across R_L. Common-mode current I_{cm} will not flow through R_L directly. Common-mode current will flow through impedances Z_a and Z_b and return through the chassis. Impedances Z_a and Z_b are not physical components. This is stray parasitic capacitance or parasitic transfer impedance present within the system. Parasitic capacitance is present because of a transmission line located against an RF return path. Parasitic capacitance includes the distance separation between the power and 0-V reference plane, decoupling capacitors, input capacitance of circuits, interconnect cables, or other numerous factors that are present within a product design. If $Z_a = Z_b$, no voltage drop will be developed across R_L by

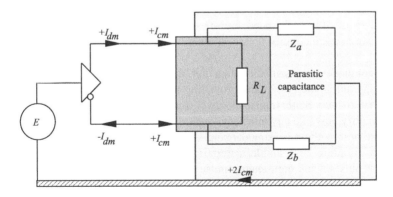

Figure 2.14 Differential to common-mode conversion.

I_{cm}. If any inequality results in the network ($Z_a \neq Z_b$), a voltage difference will be present proportional to the difference in impedance:

$$V_{cm} = I_{cm}Z_a - I_{cm}Z_b = I_{cm}(Z_a - Z_b) \tag{2.7}$$

Because of the need for balanced voltage and return references, circuits with high-frequency signals that tend to corrupt other signal traces or radiate RF energy (video, high-speed data, etc.) or traces susceptible to external influences must be balanced in such a way that stray and parasitic capacitances of each conductor are identical.

2.4 STATIC FIELDS

Electrical products and related circuitry must incorporate protection against strong static fields. These fields are commonly classified as ESD, transient surge, electromagnetic pulse (EMP), or lightning, although other forms of high-energy potentials can exist. Most static events enter through I/O interconnects and enclosure openings. In addition, direct handling of components on a PCB or chassis assembly can cause permanent damage to electrical devices. Upsets from static events arise from the current produced by the discharge developing a voltage gradient along a discharge path, rather than coupling from the static field itself. The design goal of an engineer is to prevent component or system failure resulting from externally induced high-voltage-level impulses. High-voltage impulse levels may affect system operation through both radiated and conducted means.

A static event starts with a very slow buildup of energy (seconds or minutes) that is stored in the capacitance of a structure (e.g., a human body, furniture, or unconnected cable). This charge is followed by a very rapid breakdown (typically nanoseconds). With this pulse in the nanosecond range, the discharged energy can produce EMI in the frequency range of hundreds of Megahertz to beyond 1 GHz. A static event from a human can exhibit rise times ranging from approximately 200 ps to greater than 10 ns with peak impulse currents from a few amperes to greater than 30 A [3]. Because of its high-speed, high-frequency spectral distribution, electrostatic energy can damage circuits, bounce grounds, and even cause upsets through EM coupling.

Once a component is assembled into a unit, devices are generally protected from static damage unless the design and grounding methodology was not implemented in an optimal manner. Static transients can corrupt operation of a microprocessor or clocked circuitry, just like any transient coupled signal injected into the power supply or signal port, without causing damage.

Humans, furniture, and simple materials such as paper or plastic can produce electrostatic pulses. This high-energy pulse may travel through multiple coupling paths, including circuits and grounds, or may even be radiated as a transient EM field. A static event creates multiple failure modes, including damage, upset, lockup, and latent failures.

In addition to current levels the electrostatic rise time is important. For example, ESD is a very fast transient. Two parameters are of great concern: peak level of current and rate of change (*dI/dt*). In the EMI world, rise times (t_r) are associated with frequency spectral distribution based on the Fourier transform, which relates time-domain signals (edge rates) to frequency-domain components:

$$f = \frac{1}{\pi t_r} \approx \frac{0.32}{t_r} \tag{2.8}$$

2.4.1   Electrostatic Discharge Waveforms

A detailed discussion of waveforms and equivalent ESD circuits of humans, furniture, and other materials is complex and depends on many variables. Some of these variables include speed and angle of approach, environmental conditions, and distributed circuit reactance. At lower voltage levels, a "precursor" spike due to the local area of the source (test finger) is produced that has a very fast rise time on the order of a few hundred picoseconds. Although this spike contains a small amount of energy, damaging effects happen, especially with fast digital equipment. Excellent references are provided at the end of this chapter for those interested in both mathematical and technical analysis of this subject.

An ESD event is typically broken down into two primary types of discharge: human (direct) and furniture (air) [3, 7]. Human discharge is characterized by a fast rise of current, approximately 1 ns, up to a peak of 10 A followed by a damped decay back to zero. With this waveform, significant RF energy is present up to 300 MHz. Furniture discharge is a slower rise of current to a peak of 100 A followed by damped oscillations. The RF energy for this waveform is observed up to 30 MHz (Figure 2.15). The primary area of concern lies in the magnitude of energy present, not in the waveform. Furniture discharge is a more severe event, as the area under the curve is significantly greater than the human discharge model.

2.4.2   Triboelectric Series

Electrostatic charge is a natural phenomenon that affects materials of different potential. This difference in potential occurs because of accumulated electric charges. Static electricity is generated when two materials rub against each other, each with a different dielectric constant, and are then separated. When an excessive number of electrons accumulate, they discharge to another material with a lower concentration of electrons to balance out the charge. The effect of the discharge resulting in EM disruption can vary from noise and disturbance in audio or measurement equipment to complete destruction of sensitive components, including electric shock to a person who has been energized.

A buildup of energy must result before discharge occurs. When two materials are rubbed together, with at least one being a dielectric, an accumulation of positive charge is developed on one material, while the other material receives a negative charge. The farther apart on the triboelectric scale, the more readily the charge will

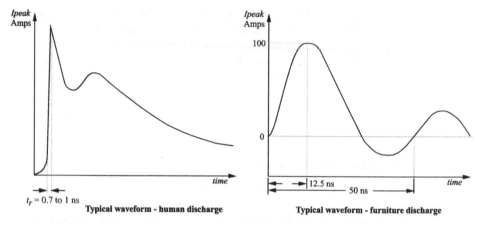

$t_r = 0.7$ to 1 ns

Typical waveform - human discharge

Typical waveform - furniture discharge

Figure 2.15 Electrostatic waveforms.

accumulate. The triboelectric scale (Table 2.1) [8] represents, at the atomic level, the charge density distribution related to polarity. Table 2.1 is to be used only as a guideline and is accurate to a degree, depending on material composition. The material at the top of the table, item 1, easily gives up electrons and therefore acquires a positive charge. The material at the bottom of the table, item 36, absorbs electrons, thus accumulating a negative charge.

Table 2.1 Triboelectric Series

Positive charge	
1. Air	19. Sealing wax
2. Human skin	20. Hard rubber
3. Asbestos	21. Mylar
4. Rabbit fur	22. Epoxy glass
5. Glass	23. Nickel, copper
6. Human hair	24. Brass, silver
7. Mica	25. Gold, platinum
8. Nylon	26. Polystyrene foam
9. Wool	27. Acrylic rayon
10. Fur	28. Orlon
11. Lead	29. Polyester
12. Silk	30. Celluloid
13. Aluminum	31. Polyurethane foam
14. Paper	32. Polyethylene
15. Cotton	33. Polypropylene
16. Wood	34. PVC (vinyl)
17. Steel	35. Silicon
18. Amber	36. Teflon
	Negative charge

2.4.3 Failure Modes from a Static Event

Static events must be considered in terms of current flow, not voltage. It's like a burst dam—it's the water flow that does the damage, not the pressure behind the dam before it burst. The voltage is merely a convenient metric of the electrostatic "pressure" before the electrostatic event occurs [9]. Most electrostatic problems fall within two categories—component damage and operational disruption:

- *Component Damage.* Happens whether or not the component is installed in a circuit. A semiconductor affected by an electrostatic event fails because of junction puncture burn-through or fusing. This type of damage is permanent and easy to detect. A subtle type of damage is weakening of the component. When subjected to a static event, the circuit is partially damaged but still fully functional. When stressed by a power supply, high temperature, or abnormal operating conditions, the damaged component can then fail permanently. This latent effect is very difficult to identify and solve.
- *Operational Disruption.* Caused by either direct or indirect injection of energy. Direct discharge occurs when electrostatic current finds its way to circuits through ports: power, ground, input, or output. When a sufficient amount of current is present, circuits will react. Permanent damage is possible and operational errors may be observed. For logic circuits, state changes can cause program halts and memory scramble.

Direct discharge and indirect discharge are defined as follows:

- Direct discharge is the discharge directly to the EUT. It may be by direct galvanic contact between source and circuit or it may be by a discharge through air to metallic items (e.g., traces) on the PCB.
- Indirect discharge is to an adjacent metallic surface by EM radiation. The radiated fields couple to the circuit.

Damage to a component is determined by the device's inability to dissipate the energy of the discharge or to withstand the voltage level involved. This concern is identified as device sensitivity to an electrostatic pulse. Some devices are prone to static damage while others may be robust. Damage can occur at any time. Most devices are susceptible to damage at relatively low voltage levels, such as 100 V. Many disk drive components are sensitive to discharges above 10 V. Component technology has progressed to the point where potential problems will develop under various environmental conditions and at low levels of energy disruption.

There are four basic failure modes from static (ESD-related) events to PCBs [9]. The first two failures are shown in Figure 2.16.

1. *Upset or Damage Caused by ESD Current Flowing Directly Through Vulnerable Circuit.* This relates to any current discharged directly into the pin of a component that causes permanent failure. This condition occurs when handling a PCB or

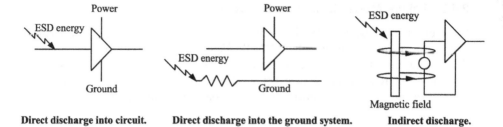

Direct discharge into circuit. Direct discharge into the ground system. Indirect discharge.

Figure 2.16 Failure modes caused by an ESD event.

digital component. Through this mode, direct discharge to the component from the outside environment (i.e., keyboard) can carry damaging ESD energy. Even a small amount of series resistance or shunt capacitance will limit the ESD current, although the acceptance value is specific to each component type.

Solutions to this problem include special ESD-handling requirements during installation or maintenance. Digital circuits have minimal ESD protection designed into the silicon wafer. For interconnects, damage typically occurs when a charge buildup is carried from the body of the installer to the connector backshell. The backshell then discharges directly to the signal pins. To prevent damage from a connector discharge, it is important to provide a filtered connector designed to dissipate ESD energy directly or to use a connector with mating pins recessed deeply inside the housing. Filtered pins do not dissipate the ESD energy but rather divert the energy to ground. Also, do not forget the straightforward solution, using a shielded cable and properly terminating both ends of the shield.

2. *Upset or Damage Caused by ESD Current Flowing in Ground Circuit.* This situation is typically observed in systems where chassis ground is directly connected to circuit ground. Most circuit designers assume that all circuit grounds have a low-impedance path to chassis ground. Once the ESD charge is injected into circuit ground, problems will be observed anywhere within the PCB and are nearly impossible to stop. A discharge to metal results in currents traveling in unpredictable paths, often upsetting circuits but not necessarily destroying them. Current distribution does not follow intended circuit paths with fast rise time transitions. A small amount of stray capacitance becomes a low-impedance path because wires are too inductive to pass the current. The current follows a ground path rather than a signal path.

With 1 ns rise time for an ESD event, ground impedance may not be low. Hence, the ground will "bounce." The usual result is an upset. Ground bounce or level shifting of the 0-V reference can drive complementary metal–oxide–semiconductor (CMOS) circuits into latchup. Latchup is a situation where the ESD does not actually do the damage—it just sets things up so that the power supply can destroy the part or, at best, the circuit becomes nonfunctional without a power cycle reset.

The ground discharge problems can be resolved by ensuring that all locations receiving the ESD energy have a low-impedance ground path to the remainder of the enclosure. A low-impedance connection must be by a direct bond or ground braid. The length of the braid cannot exceed a width-to-length ratio of 5 : 1. Width is better than length! Braid dissipates ESD energy quickly. Braids are good up to about 10 MHz. Beyond this frequency they become highly inductive. Braids are sufficient to carry the charge, eliminating slow charge buildup; however, a fast discharge cannot occur through a highly inductive strap. Direct bonding is preferred whenever possible.

If a plastic enclosure is used, grounding to a metal chassis cannot be performed. For this application, one must insulate or recess all metal parts and assemblies to prevent direct discharge damage. Metal assemblies need to be connected together by a 10–100-kΩ resistor to reduce peak currents from traveling between assemblies. These resistors do not affect the 0-V referencing between assemblies. If a lot of metal is located in close proximity, the resistor will not work as capacitive coupling will dominate.

3. *Upset Caused by Electromagnetic Field Coupling (Indirect Discharge).* This effect usually does not cause damage, although damage to very high impedance components can occur. Damage is rare because only a small fraction of the ESD energy is coupled directly into the vulnerable circuit. The induced voltage is usually not enough to do more than upset the logic. This failure mode depends heavily on the rise time of the discharge (dI/dt) and circuit loop area, regardless of shielding. This effect is often called the indirect coupling mode. Electromagnetic field sources do not have to be very close to cause disruption to sensitive circuits. Tests performed to ensure ESD protection levels are identified as the air or furniture discharge model.

The indirect discharge is produced by intense magnetic field coupling to adjacent current loops. Coupling increases with loop area. Loop areas of a PCB allow radiated emissions to exist in addition to permitting external magnetic fields to enter the circuit. This type of event is generally observed on products packaged in plastic enclosures, where an external discharge couples directly to internal circuits by EM means.

The solution to preventing a discharge from upsetting circuits is to minimize loop areas. Multilayer boards are preferred owing to their smaller loop sizes. Loops on double-sided boards can be minimized by careful layout. Routing of cables and interconnects must be as close to the chassis to minimize loops that may be present. Cables routed across cable seams are optimal for reception of intense magnetic fields.

4. *Upset Caused by Predischarged (Static) Electric Field.* This failure mode is not as common as the other modes. This mode appears in very sensitive, high-impedance circuits.

A predischarged (static) electric field is caused by stripping electrons from one object (resulting in a positive charge) and depositing these electrons on another object (resulting in a negative charge). In a conductor charges recombine almost instantly, whereas in an insulator the charges can remain separate. In an insulator it may be a long time before significant charge recombination happens and, consequently, a voltage builds up. If the voltage level becomes large enough, a rapid breakdown happens through the air or insulator, creating the familiar ESD arc or spark.

Because ESD is transient in nature, fast digital circuits are more prone to ESD upsets than slow analog or, for that matter, low-bandwidth digital device circuits. In fact, ESD rarely upsets the functionality of analog circuits. However, both analog and digital circuits are vulnerable to ESD damage from a direct hit. Digital circuits with edge rates faster than 3 ns are particularly vulnerable because phantom ESD pulses can fool them. As a result, digital circuits are more vulnerable than older circuits with slower edge rates.

To illustrate the intensity of an ESD pulse, rise times of 500 ps or faster have been measured. With this knowledge, an ESD pulse with a rise time of 500 ps and tens of amperes of peak current translates to an equivalent slew rate of gigaamperes per second across the circuit interface.

REFERENCES

1. Paul, C. R. 1992. *Introduction to Electromagnetic Compatibility.* New York: Wiley.
2. Hartal, O. 1994. *Electromagnetic Compatibility by Design.* West Conshohocken, PA: R&B Enterprises.
3. Montrose, M. I. 2000. *Printed Circuit Board Design Techniques for EMC Compliance— A Handbook for Designers.* Piscataway, NJ: IEEE.
4. Ott, H. 1988. *Noise Reduction Techniques in Electronic Systems,* 2nd ed. New York: Wiley Interscience.
5. Montrose, M. I. 1999. *EMC and the Printed Circuit Board Design—Design, Theory and Layout Made Simple.* Piscataway, NJ: IEEE.
6. Williams, T., and K. Armstrong. 2000. *EMC for Systems and Installations.* Oxford: Newnes.
7. Mardiguian, M. 1992. *Electrostatic Discharge—Understand, Simulate and Fix ESD Problems.* GA: Interference Control Technologies.
8. ANSI C63.16. 1993. *American National Standard, Guide for Electromagnetic Discharge Test Methodologies and Criteria for Electronic Equipment.*
9. Gerke, D., and W. Kimmel. 1994. The Designer's Guide to Electromagnetic Compatibility. *EDN,* January 20.
10. Williams, T. 1996. *EMC for Product Designers,* 2nd ed. Oxford: Newnes.
11. Mardiguian, M. 2000. *EMI Troubleshooting Techniques.* New York: McGraw-Hill.

CHAPTER 3

INSTRUMENTATION

This chapter introduces the most commonly used instrumentation available for performing measurements. Time-domain analysis is helpful during debugging and troubleshooting and for investigating effects of signal integrity. Frequency-domain analysis is used to measure RF energy, be it for certification testing or troubleshooting.

There is a vast amount of information available from vendor manufacturers and textbooks that provide more depth than the simple contents of this chapter. Therefore, we do not present everything there is to know about the instrumentation described herein or specifics on how to use the equipment. We examine only major features regarding the most common devices used. The reader is directed to manuals published by vendors. Photographs provided are a representative sample of products available and do not imply endorsement of any particular vendor or product.

3.1 TIME-DOMAIN ANALYZER (OSCILLOSCOPE)

Signals can be examined in many ways. One of the most common measurement techniques available is through a time-domain display. Time domain analysis is best used for signal integrity measurement and transient analysis. Reference 1 provides details on the relationship of signal integrity and EMI. A poorly terminated transmission line can cause development of common-mode RF currents. The source of a problem circuit can be difficult to locate using frequency-domain analysis; thus, the discussion on time-domain instrumentation is provided herein.

A sine wave is one of the most fundamental time-domain signals. Figure 3.1 illustrates this waveform where the vertical scale is at 1 V/div and the horizontal time base at 0.001 s/div. One cycle takes 0.004 s. The frequency is defined as $1/T$ or 250

Testing for EMC Compliance. By Mark I. Montrose and Edward M. Nakauchi
ISBN 0-471-43308-X © 2004 Institute of Electrical and Electronics Engineers

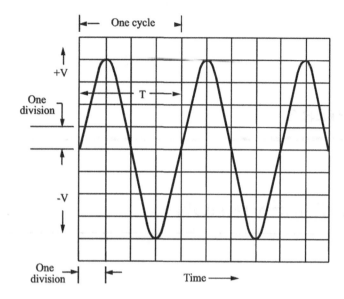

Figure 3.1 Time-domain display.

Hz with an amplitude of 4 V peak. The RF energy is identified as a signal with units of hertz, megahertz, and so on. The unit of hertz (cycles per second) describes a sine wave, which can easily be measured with a time-domain analyzer regardless of frequency.

A useful instrument with which to view time-domain waveforms is the oscilloscope ("scope"). A typical scope is shown in Figure 3.2. Other common wave shapes besides the sine wave are shown in Figure 3.3.

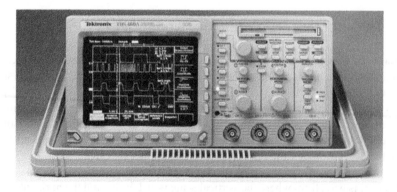

Figure 3.2 Typical digital storage oscilloscope. (Photograph courtesy of Tektronix.)

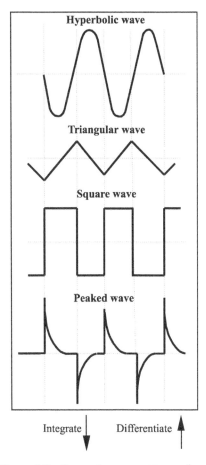

Figure 3.3 Some other common waveforms.

The vertical axis of the oscilloscope represents the amplitude or magnitude of the signal versus time on the horizontal axis (time period per division). For a display to be clearly presented, a trigger must be provided. The trigger starts a time-base generator that sweeps the trace across the screen from left to right. The waveform or signal can be analyzed and its various parameters measured. A waveform shape reveals the following information about itself, information that deals mainly with signal integrity concerns:

- Transmission line characteristics
- Periodic or nonperiodic waveform analysis
- Signal distortion (includes ringing and overshoot)

- Glitches in the transmission line
- Crosstalk
- Timing concerns (skew, propagation delay, dielectric losses)
- Power or return (ground) bounce
- Noise spikes (AC mains power, lightning and radiated noise coupling into circuits)
- Electrostatic discharge events
- Can monitor multiple channels simultaneously but is triggered by one channel at a time

Oscilloscopes measure various parameters. Two of the more important parameters are frequency and period. Frequency is measured in hertz, which is the number of times the signal repeats itself in a second. The inverse of frequency is the time required to complete one cycle and is called the *period*. These are described as

$$f = \frac{1}{T} \qquad T = \frac{1}{f} \tag{3.1}$$

where f = frequency
T = period of the signal

The amplitude or voltage level of a signal can be measured as well as phase or phase shift. Phase shift is the difference in timing between two signals. Today, most modern oscilloscopes are digital (digital storage oscilloscope, or DSO). Digital scopes can measure many more parameters than analog units. Since a DSO stores digital data, it can perform numerous calculations, such as Fast Fourier Transform (FFT), and calculate the signal's frequency spectrum. One may purchase a combined oscilloscope and spectrum analyzer; however, there is a limitation on measuring higher frequency components of a signal. There are front-panel buttons or screen-based soft buttons that one can select to perform any number of fully automated measurements.

An important parameter in choosing a DSO is sampling rate. Generally, the sampling rate must be reduced by some factor to yield actual digitized bandwidth. As an example, a 100-Mbit-sampling-rate DSO should only be used to display a 50-MHz signal. The rationale for this gets into *aliasing* and *Nyquist sampling rates*. Aliasing is a distortion of the signal being measured shifted to a lower frequency from the original frequency. This occurs due to an insufficient sampling of the measured signal. The Nyquist criterion requires two or more samples per measured signal cycle to prevent this aliasing phenomenon.

Other characteristics of an oscilloscope to consider in choosing the right test equipment include the following:

1. *Bandwidth.* The bandwidth of an oscilloscope must be sufficiently wide to reproduce the signal faithfully. A rough approximation is to use the "rule-of-

thumb equation" $f = 1/(\pi t_r) \approx 0.32/t_r$ of the signal being measured. For some signal integrity applications, the numerator of 0.35 is used. Above this bandwidth frequency, the spectrum envelope rolls off by 40 dB/decade. Therefore, in measuring a 1-ns-rise time pulse, the oscilloscope must have an absolute minimum bandwidth of 350 MHz. To see a faster rise time or to view any ringing or overshoot present within the transmission line requires at least 1 GHz bandwidth.

2. *Input Impedance.* For accuracy in signal acquisition, the impedance of the scope must match that of the coax and probe. Using a 75-Ω coax with a 50-Ω scope input will provide significant error in measurement data. Sometimes, a high-input impedance input is selected (typically 1 MΩ). This is typically done to avoid loading of the measured circuit or for use with certain high-impedance probes.

3. *Triggering Capability.* This is one area where digital oscilloscopes have an advantage over analog oscilloscopes. One can actually select a trigger point either before or after the signal of interest.

The greatest advantage of a DSO is its ability to capture and hold a single transient event. This type of signal is quite common in EMI/EMC testing and troubleshooting. Other parameters that can be measured with a modern digital oscilloscope include the following:

- Rise time
- Fall time
- Pulse width
- Duty cycle
- Delay
- Peak-to-peak
- Mean
- Root-mean-square (RMS)
- Overshoot
- Undershoot
- Minimum voltage level
- Maximum voltage level

The two basic measurement features of an oscilloscope are voltage and time. An oscilloscope is primarily a voltage-measuring device. Other quantities or parameters are only a calculation away. As an example, current can be calculated from the measured voltage value by knowing the resistance or impedance of the circuit, generally based a 50-Ω input. Voltage is read off the vertical axis, based on the number of divisions on the scale. Any signal being measured with a DC will be shown as a straight line either above or below the reference line, depending upon its voltage value. When one needs to measure only an AC waveform, the unit is switched to the AC coupling mode. This places a DC blocking capacitor in the measurement path and removes all DC elements of the signal. However, when measuring low-frequency signals, DC coupling is preferred.

Time measurements are recorded using the horizontal axis of the oscilloscope. Time measurement includes both period and pulse width of the signal. Since frequency is the reciprocal of period, by knowing the period, we also know the frequency.

When dealing with pulses, width and rise time are important and critical parameters for describing the signal. Pulse width is the amount of time it takes the pulse to go from low to high and back to low in amplitude. Typically, it is defined as the time measured between the 50% amplitude points. Edge rate transition, commonly called rise or fall time, is the period it takes for a pulse to transition logic states. Typically, it is the amount of time measured between 10 and 90% of the signal amplitude. Signal integrity engineers are generally concerned with the 10–90% or 20–80% edge transition period within a particular time frame, whereas EMC suppression is concerned with the 0–100% portion of the waveform.

Other measurements features include phase shift and capturing fast transient signals, depending upon the application. For more insight on other aspects of DSOs related to digital design engineering not commonly used in the field of EMC, the References in this chapter provide a selection of recommended reading.

Scopes with FFT analysis functions can be useful. It is important to remember that higher rates of change of voltage or current per unit time (dV/dt or dI/dt) mean greater threats of radiated emissions. Spurious ringing on a waveform identifies the resonant frequencies of a circuit and provides information on frequencies likely to be observed in the radiated spectrum.

A useful technique to locate a functionality or signal integrity problem is to follow a digital signal (e.g., clock) from its source to the final load. Look to see how badly the waveform degenerates. A correct PCB design and layout will maintain a quality signal integrity waveform along the length of routed trace. When waveforms degrade significantly, their ringing frequency identifies likely emissions or functionality problems.

One of the few benefits of scopes over spectrum analyzers is that scopes can be triggered from a clock or other waveform. This triggering is usually performed using a standard voltage probe and a spare channel. If it is impossible to look at a signal due to multiple waveforms mixing together, one can "freeze" parts of it by triggering from a different clock or other signals to ascertain which of them is contributing to the unwanted noise.

3.1.1 Oscilloscope Probes

Connection to an oscilloscope requires a probe that is appropriate for intended use. The probe must be well matched to the front end of the oscilloscope in order to ensure measurement integrity. When measuring a signal, two connection points are required. One is the probe tip to the actual signal point and the second is usually a reference (ground) location. In making measurements, one must be careful on how the ground connection is made. An incorrect use of the reference lead can produce errors in measured results.

Radio-frequency current probes used for conducted emission measurements, for instance in MIL-STD EMC testing, and current probes used for oscilloscopes are not the same. If one uses the word *probe,* a clear distinction must be made to correctly identify the proper device.

The two primary types of probes commonly used with oscilloscopes are passive and active. Both probe types comprise of either high- or low-impedance tips. Most probes are of the high-impedance version, identified as the 10X version. This type of probe has a trimmer adjustment to compensate for both high- and low-frequency elements of the measured signal in order to prevent measurement inaccuracies. Adjustment should be made to achieve a perfect square-wave pattern that is displayed on the screen with no overshoot or round-off (Figure 3.4). These 10X probes have an input impedance of 1 MΩ with a parallel capacitance of about 10 pF. This input impedance can create a problem during measurement of a signal high-frequency analysis due to capacitive loading. The total impedance of this parallel input circuit at 100 MHz is about 150 Ω. This is very different from the 1 MΩ impedance of an active probe. This high impedance is enough to cause inaccuracies in the measurement, depending upon circuit impedance.

The ground connection lead wire is a potential source of error. The inductance of the ground lead can cause resonant effects to appear as ringing on the waveform (Figure 3.5). The ground connection must be as short as possible to avoid this problem. A poor ground connection can also create a noise ground loop between the circuit board or system being measured. The reference (ground) terminal of the probe is essentially at the same reference potential as the chassis of the scope. Radio-frequency currents can flow in this loop circuit, creating a voltage drop in the probe's coaxial cable (usually the shield of the coax). This voltage drop will add to the amplitude of the signal being measured. Differential, high-impedance active probes will solve this problem. However, active probes have problems of their own as they are expensive, fragile, and sometimes limited in their bandwidth availability. For high-frequency measurements, a low-impedance

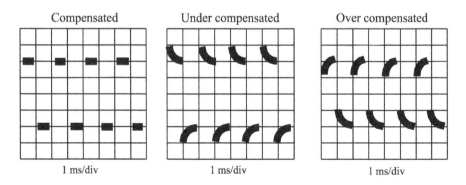

| Compensated | Under compensated | Over compensated |
| 1 ms/div | 1 ms/div | 1 ms/div |

Figure 3.4 Properly compensated scope probe.

passive differential probe will be sufficient [2]. The frequency response of a probe must be known before detailed measurements are made. Equations (3.2) and (3.3) illustrate how to calculate both the frequency response and bandwidth of a probe based on the edge rate transition of a digital signal. From this value of frequency or bandwidth, it becomes a simple task to calculate the actual (real) edge rate transition of a digital signal, which is generally distorted by an excessively long ground lead wire or by capacitive loading:

$$\text{Edge transition} = \frac{1}{\pi \text{BW}} \quad \text{or} \quad \text{Frequency} = \frac{1}{\pi t} \qquad (3.2)$$

$$\text{Actual } t = \sqrt{(\text{measured } t)^2 + (\text{probe } t)^2 + (\text{scope } t)^2} \qquad (3.3)$$

where f = frequency
$\quad t$ = edge rate transition (the faster of either the rising or falling edge)
$\quad$ BW = bandwidth

As an example, if one measures a signal with a rise time of 2.2 ns with a 350-MHz-bandwidth scope and probe, we have an actual rise time of the signal at 1.6 ns. With a higher frequency bandwidth value for both scope and probe, the equivalent edge rate transition value becomes inversely proportional, making them smaller with less effect per Eq. (3.3). Therefore, the measured rise time approaches the actual rise time.

No matter which probe type is used, it is important to perform a "null experiment" [2] before taking any measurements. Essentially, this null test connects the probe tip to the ground wire. If the voltage measured is much smaller than the measured signal level, then the reading can be believed. If the null reading is higher than the measured signal, the reading cannot be trusted. Reducing the length of the ground lead wire, changing to a different type of probe, and performing a differential measurement are some techniques to reduce potential error.

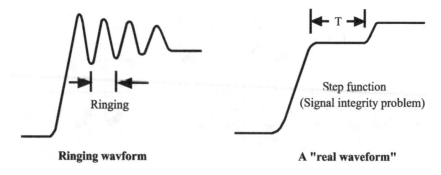

Ringing

Step function
(Signal integrity problem)

Ringing wavform   **A "real waveform"**

Figure 3.5   Ringing induced by a long ground wire.

When one performs voltage probing with an oscilloscope, problems can arise:

1. The loop formed by the probe's signal pin and ground wire behaves as an unshielded loop antenna, picking up noise from its local environment. Most oscilloscopes have very poor common-mode rejection. It becomes difficult to determine if we are measuring the signal, the probe's ground wire, induced coupling from a nearby circuit, or RF common-mode potential difference between the EUT and the scope's chassis. It is important to know the common-mode voltage of the circuit under investigation. When mixed with the waveform of a signal being measured, difficulties occur in making accurate measurements. A couple of 32-mm-long split-ferrite suppressors clipped onto the scope lead can help improve measured data accuracy.

2. Transducers such as close-field probes, current probes, and antennas have no metallic connection to the EUT. These probes do not suffer from the problems of clip-on voltage probes. Unfortunately, these probes do not measure signal waveforms directly, which a voltage probe is capable of performing.

3. The EUT's common-mode voltage can be detected with an electric field probe or a pin probe connected to the chassis or 0-V reference point. By connecting a ÷10 probe to the reference point (chassis or 0-V) without connecting a ground lead, one can detect common-mode voltage. The problem with AC mains frequency hum can be solved with use of a high-pass filter.

Using scopes and probes is the lowest cost manner to perform development or diagnostic emissions testing. It requires considerable skill in reading waveforms to achieve any *quantitative* correlation with standard EMC tests. First testing a known good system, commonly called the "golden product," and then comparing other systems to this particular unit allow one to make a rough judgment of performance criteria.

3.2 FREQUENCY-DOMAIN ANALYZERS

There are many analyzers available to engineers. Each instrument serves a unique purpose and application. Some analyzers examine the frequency domain in a visual manner while others investigate wave shape, phase, and other characteristics of a transmitted signal. Knowledge of test instrumentation is becoming mandatory for high-speed and high-technology products. This section identifies the most commonly used equipment with a brief overview of their function along with basic operating procedures. As with the section on oscilloscopes, specific details on how to use the tool is not provided. The reader is directed to vendor manuals and application notes.

A problem with using oscilloscopes is with systems that have a large bandwidth of operation. The measured signal will contain various types of noise, thus making it difficult for analysis. There is also a problem of converting waveform information into the frequency domain to compare with specification limits. Fast Fourier transform functions give different results depending on the window selected for the conversion. Scopes do not have quasi-peak or average responding detectors required for proper emissions testing.

Another manner in which to measure a signal, this time in the frequency domain, is to use a spectrum analyzer (or receiver, which is another type of analyzer). For spectrum analyzers, the horizontal axis is in units of frequency, or hertz, instead of time, as for oscilloscopes (in seconds). The vertical axis on analyzers represents the magnitude of the signal and is usually in decibels (dB) or impedance (ohms).

Figure 3.6 shows a typical digital pulse train and its frequency domain counterpart as would be displayed in the frequency domain.

A frequency-domain analyzer or receiver can measure the following parameters among other elements not identified below:

- Power level of a signal (amplitude)
- Frequency of operation
- Spectral density throughout a range of frequencies
- Modulation [AM, FM, continuous wave (CW), pulse width modulated (PWM), etc.]
- Intermodulation distortion

Low-cost analyzers have minimal features and functionality. A more expensive analyzer that provides automatic compensation for cable and antenna factors, dis-

Time domain representation of a signal pulse train

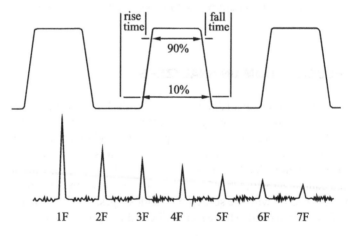

Frequency domain representation of a signal
xF = harmonic number of the primary frequency

Figure 3.6 Time- and frequency-domain display analyzers.

plays limit lines, and has accurate quasi-peak and average detectors along with data storage is one of the most useful tools available for EMC work. There are a number of manufacturers of analyzers and receivers, some aiming directly at either the low-cost or high-end EMC testing market.

3.2.1 Spectrum Analyzers

A spectrum analyzer displays a spectral distribution of RF energy. Essentially, a spectrum analyzer is a swept-tuned heterodyne receiver. The signal being measured is received from a transducer, such as an antenna or probe. From here, the signal travels to the analyzer's mixer stage and is mixed with a voltage-tuned local oscillator swept at the same rate as the horizontal deflection of the cathode ray tube (CRT) by means of a sawtooth generator. This gives a frequency-related horizontal display on the screen. The output of the mixer is amplified and sent to the vertical deflection stage. Thus, a spectrum analyzer sweeps frequency while an oscilloscope sweeps time. The widest frequency range that can be measured in a single sweep is called the *spectrum width* or *span*.

Spurious responses generated in the mixer can generate errors. This is typically caused by excessive signal input amplitude. A high-amplitude signal can cause the mixer to go into a nonlinear region of operation. Responses are then produced that are not linearly related to the input and can easily be mistaken for part of the real signal. An attenuator, either internal or external, is used to prevent signals higher than the dynamic range of the analyzer from being measured. There is an easy method to identify this potential problem: Decrease the input signal level by 10 dB with the attenuator installed. If the displayed signal decreases by 10 dB, then the signal is probably a real RF component. If it is a spurious response, the magnitude will decrease a great deal more than 10 dB.

A spectrum analyzer is a complicated test and measurement tool (Figure 3.7). Recent technical innovations have brought forth lightweight, palm-sized, battery-operated personal-size portable spectrum analyzers. It is ideal for signal characterization, identification of unknown signals, harmonic and spurious measurements, signal monitoring, field strength measurements, and EMC precompliance. Data can be displayed in either dBm or dBμV. Being battery operated gives it complete versatility to be used anywhere in a field environment. Some units has a built-in preamplifier; therefore, use of just about any type of passive probe is possible for troubleshooting purposes. Either electric or magnetic field probes can be accommodated. Finally, some analyzers have a serial RS-232 or IEEE 488 bus for allowing interfacing and transferring of data to a PC for further or more extensive analysis.

The primary advantage of using a spectrum analyzer for development, diagnostic, and qualification verification is that it is much easier to correlate measured results from prequalification analysis with those of a formal EMC test. Making the frequency domain visible enhances the ability of a design and development engineer to understand what is happening and where it is occurring.

Low-cost spectrum analyzers are readily available. These units use a time-based display in *XY* mode. In addition, low-cost analyzers contain few features

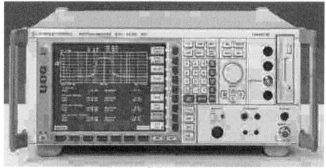

Full Size

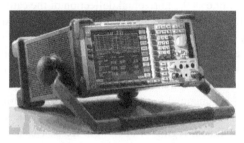

Portable

Handheld

Figure 3.7 Typical spectrum analyzers. (Photographs courtesy of Rohde & Schwarz GmbH and Bantum Instruments.)

and functionality. A more expensive unit, especially one that provides the features listed below, can save a great deal of time and money during its first few months of use.

There are a number of manufacturers of spectrum analyzers, each design for different target users, from prequalification to full compliance. Some useful features built into spectrum analyzers are as follows:

- Preamplifier
- Automatic compensation for cable and antenna factors
- Allows for different units of measurement to be displayed (dBm, dBμV, dBmV, etc.)
- Allows display of limit lines
- Has reasonably accurate quasi-peak and average detectors
- A demodulator to determine if a signal is an ambient or broadcast station
- Variable bandwidths for resolution bandwidth and video average
- Memory storage capability
- A floppy disk storage unit
- Computer control and peripheral interface circuitry

Spectrum analyzers and receivers can be used with closed-field probes, current probes, voltage probes, and antennas. When making a direct voltage measurement with a spectrum analyzer, ensure that attenuation is provided for every DC or low-frequency signals. A DC block prevents harmful voltage levels from entering the unit. Low-frequency signals or DC voltage can damage the very sensitive (and expensive to replace) analyzer input circuitry.

Spectrum analyzers are prone to overload from strong signals, even if they are outside the desired frequency span under investigation. For this reason, it is recommended to use a preselector. For precompliance testing, a preselector is not required; however, formal test facilities may use this expensive add-on module. A preselector increases the probability that the signal being measured is a real signal and not ambient. The noise floor of the analyzer becomes lower, allowing weaker signals to be observed. A preselector is more important for performing conducted emissions tests than radiated emissions.

An attenuator, either internal or external, can be provided to prevent signals higher than the dynamic range of the analyzer. There is an easy method to identify this potential problem. Decrease the input signal level by 10 dB with the attenuator installed. If the displayed signal decreases by 10 dB, then the signal is probably a real RF component. If, however, it is a spurious response, the magnitude will decrease a great deal more than 10 dB.

Low-cost analyzers can have a high noise floor, making measurements of weak signals difficult. An RF preamplifier may improve the signal-to-noise ratio. Ensure that the amplifier does not provide a DC signal to the spectrum analyzer's input. An RF preamplifier is also vulnerable to signal overload. Verify functionality with a 10-dB through-line attenuator. The preamplifier should also have a noise figure of less than 4 dB to prevent adding noise to the signal as well as amplifing the signal.

Important parameters to consider when choosing or selecting a spectrum analyzer include the following:

1. *Frequency Range.* The frequency range is the spectral distribution of frequencies that one desires to measure and analyze. The three primary areas re-

lated to frequency range is; Start Frequency, Stop Frequency and Center Frequency.

2. *Span.* Frequency span is how small or large a desired frequency range is to be displayed on the screen in hertz on the horizontal scale. Having the correct SPAN value is important when attempting to resolve signals that may be very close to each other in frequency.

3. *Resolution Bandwidth.* Resolution bandwidth is the passband of the internal filters that aids in distinguishing two signals that are very close together. A smaller resolution bandwidth means a better ability to "resolve" adjacent signals. The ideal is having the lowest resolution bandwidth possible. With a smaller bandwidth the sweep time increases.

4. *Dynamic Range.* The dynamic range is a value that determines the lowest level signal that can be measured and is related to the internal noise floor of the analyzer. Signals being measured must be higher in amplitude than the minimal value capability of the analyzer in order to be a valid measurement. Ideally, one should choose an instrument with the lowest noise floor possible. To improve the ability of the spectrum analyzer in measuring very low level amplitude signals, use of a preselector and/or a low noise figure preamplifier is added to the input of the unit. An increase of 26 dB amplitude range is a typical value for supplemental front-end equipment.

5. *Video Bandwdith.* Video bandwidth refers to a monitor's ability to refresh the screen. High bandwidths allow more information to be painted across the display in a given amount of time, which translates into support for higher resolutions and higher refresh rates. Lower bandwidths result in flickering, ringing artifacts, and ghosting.

A spectrum analyzer displays the amplitude of a signal (vertical axis) versus frequency (horizontal axis). The vertical axis is calibrated in either linear or logarithmic mode. In the linear mode, the scale is in units per division (i.e., μV/division). In the logarithmic mode, the axis is calibrated in dB/div (e.g., usually 10 dB/div). Because EMI standards are generally specified in dB's, most engineers find using the logarithmic scale more convenient.

A signal is measured by observing its vertical height on the display screen. A spectrum analyzer provides the relative amplitude difference between signals at various frequencies. In order to obtain absolute amplitude levels, the spectrum analyzer must be calibrated with a known input level. Once calibrated to a known amplitude, any signal above or below this reference level indicates the absolute value or the received signal. The top graticule of the vertical scale is the reference or calibration line. Actual signal amplitude is then determined by counting the number of divisions *down* from this reference level based on the scale on the vertical axis. Spectrum analyzers display data in dBm (decibels relative to one milliwatt of power). Most, if not all, spectrum analyzers allow for easy conversion from dBm to dBμV or to other units of measurement for easier comparison to EMI test standards.

Determining the frequency of the signal is similar to measuring amplitude. The frequency is read off the horizontal scale, usually in MHz. The reference for the horizontal frequency axis is the center frequency point located on the middle or center graticule on the horizontal axis. The horizontal axis can be displayed in the linear or logarithmic mode.

Figure 3.8 shows two signals. The signal on the left side of the display is 16 dB down from the top reference line (graticule). A second signal (right side of the display) is 28 dB down from the reference line. Measurement is always made from the reference line. In Figure 3.8, each vertical division is 10 dB.

Spectrum analyzers as well as EMI receivers convert an input signal to a video signal using an envelope detector. These signals have a frequency range from DC (0 Hz) to a specified upper frequency determined by the detection circuit elements. An envelope detector consists of a diode followed by a parallel *RC* combination. The output of these components is applied to a specific detector. The time constants of the detector are chosen such that the voltage across the capacitor equals the peak value of the signal. This requires a fast-charge and slow-discharge time constant.

If there is more than one signal within the measurement bandwidth of the instrument, their interaction can create a beat note. The envelope of the IF (intermediate-frequency) signal will vary according to the phase change between the two sine wave input signals. The maximum rate at which the envelope of the IF signal changes is determined by the value of the resolution bandwidth.

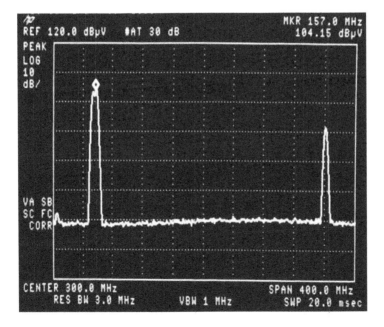

Figure 3.8 Typical spectrum analyzer display screen.

One peculiarity of EMC emission measurements, both radiated and conducted, is the need for CISPR16 quasi-peak (QP) and average (AVG) detectors to detect voltage, current, and power. Most commercial compliance measurements require QP detection and sometimes AVG detection. Peak readings, if under the quasi-peak or average level specification limit, can be considered compliant. Also, if the QP emissions are below the average limit, this is also acceptable. Detectors analyze broadband noise differently from narrowband emissions. Only high-end EMC measuring instruments intended for formal compliance certification requires accurate QP and AVG detectors.

Originally, QP detectors were designed to obtain a receiver reading proportional to the annoyance effect of interference found on commercial broadcast radio station reception signals. The design of the detector circuit, with various charge and discharge constants, is dependent on the frequency range resulting in weighting of the broadband signal as a function of its repetition rate. For those signals with a low repetition rate, less annoyance occurs.

Emissions with a high repetition rate get more emphasis related to regulatory compliance because they create more harmful interference to the ear. As the repetition rate approaches that of a CW signal (i.e., 100% duty cycle), the signal is at maximum annoyance. No weighting is applied to the amplitude level during the measurement cycle. The time constant of the detector has a smoothing effect on the output signal so that a steady amplitude reading is obtained. Quasi-peak amplitudes of signals will always be less than or equal to the amplitude measured with a peak detector. Any appearance of modulation, broadband noise or data spikes, will cause the peak readings to become smaller in amplitude with these two detectors.

When a narrowband signal exists with a secondary impulsive signal, the QP detector can completely mask the lower level narrowband signal. This is due to the detector's response to the predominant peaks of the broadband or impulse signal. On the other hand, average detection is used to suppress broadband signals and is therefore well suited to recover the true amplitude of narrowband signals. The recovery of narrowband signals in the presence of broadband emissions is the primary purpose of the average detector.

The choice of detector type in a given application depends on which characteristic is likely to produce maximum interference or performance degradation. Average and/or RMS detectors are useful in measuring broadband interference that is random in nature along with certain types of narrowband interference. One type of detector, QP, has a very high ratio of discharge time constant to charging time constant. A variation of the peak detector is the slide-back type, in which a bias voltage is applied to the measuring diode. This bias voltage provides a threshold cutoff for the output of the detector. Table 3.1 lists various types of detectors and their use.

International EMI regulations mandate use of average detection, either in addition to or as a replacement of QP for conducted emissions. Average detection, at the time of writing, is the only acceptable measurement detector for signals above 1 GHz. Average detection was introduced to measure low-level narrowband signals in the presence of broadband noise. In general, the interference potential of narrowband signals is higher than that of broadband signal.

Table 3.1 Comparison and Application of Selected Detectors

Detector	Output Response	Typical Application[a]
Root-mean-square	Proportional to the square root of the bandwidth	Broadband interference; atmospheric noise; random noise
Quasi-peak	Large ratio of discharge time constant to charge time constant	AM receiver interference; industrial, scientific, and medical (ISM) equipment; overhead transmission line noise; lighting devices; home entertainment and personal computing equipment
Average	Average value of signal envelope	Modulated radio carriers; atmospheric noise; narrowband sources; communication networks; ISM equipment
Peak	Direct reading of peak value above a threshold level	Military standards; impulsive interference; low-repetition-rate impulses

[a]Not comprehensive

Emissions from microprocessor clocks and similar signals that are unmodulated radiate at approximately the same amplitude level on peak detectors found in low-cost EMC analyzers within a few decibels. When more randomized emissions from data buses, DC motors, relay contacts, and low-rate pulse emissions occur, measurement accuracy can be in error by as much as ±20 dB if proper QP and AVG detectors are not used.

Spectrum analyzers read true peak values. Results obtained by a true CISPR receiver should match those of a spectrum analyzer only for narrowband signals. For impulse measurements, the spectrum analyzer gives true peak value while the CISPR receiver will give QP values. To compensate for the difference in broadband measurements, the spectrum analyzer must be corrected with internal software. One part of the correction takes into account the minor frequency measurement bandwidth difference. As an example, the CISPR receiver bandwidth of 9 kHz specified by test standards is not available as a default bandwidth function for most spectrum analyzers. The standard value of bandwidth for a spectrum analyzer is 10 kHz. A correction factor of $20 \log(10 \text{ kHz}/9 \text{ kHz}) \approx 1$ dB is required. The other portion of the correction takes into account the detector function difference from peak to QP. Fortunately, spectrum analyzers are used so often for EMI measurements that most analyzers have adapters or built-in features to handle these correction factors automatically.

Spectrum analyzers can be used with close-field probes, current probes and a variety of antennas. Spectrum analyzers can also be used with voltage probes, although the typical oscilloscope probe may not be ideal owing to the need of the spectrum analyzer's 50 Ω input impedance.

Low-cost spectrum analyzers can have a high-noise floor. This means the analyzer is not capable of measuring low-level signals. Everything observed on the screen may be background noise or ambient. An external low-noise RF preamplifier may improve the signal-to-noise ratio. Ensure that no DC level signal is applied to the analyzer's input. Low noise radio-frequency preamplifiers are also vulnerable to overload. Validate functionality with a 10-dB through-line attenuator.

3.2.2 Receivers

An alternative to the spectrum analyzer is the measuring receiver. Receivers have been around since the early 1900s and were the first EMI measurement tool. With the advent of swept frequency techniques, spectrum analyzers began to replace the point-to-point measurement method that was unfortunately necessary with receivers. Modern receivers now incorporate a swept-tuned superheterodyne design much like spectrum analyzers. The main difference between a receiver and spectrum analyzer is that the RF amplifier of receivers has a narrow bandwidth of operation, unlike spectrum analyzers which must be wide open to allow all the signal through. The front-end RF amplifier within a receiver is tuned to a narrow bandwidth around a single frequency of interest. What this achieves is less sensitivity to overload problems if a large-amplitude signal is nearby to a desired signal that is to be measured. There is also enhanced noise figure performance (less internal system noise) and hence more dynamic range. This enhanced dynamic range is due to the presence of tuned front-end amplifiers that limit the total amount of power to the input mixer. Figures 3.9*a* and *b* show several types of EMI test receivers.

Modern receivers have monitors associated with them to display the frequency spectrum, much like a spectrum analyzer, and/or have the capability of connecting an external monitor to display information being measured. Receivers are also becoming microprocessor based. This allows a great deal of flexibility for the receiver. Going from peak to AVG, peak to QP, or QP to AVG is done by simply pushing a button; adding in correction factors for various accessories is an example of the flexibility associated with modern receivers.

Receivers have the ability to be computer controlled through a bus interface. Therefore, the measurement process can be automated (i.e., input and output) through a keyboard with data recorded by a printer/plotter. Automated control is reserved to shielded enclosures and not for use at Open-Area Test Sites.

There is very little difference between a spectrum analyzer and a receiver with today's technology. The main difference lies in the capability of the receiver to handle overload or large-input-signal conditions over that of a spectrum analyzer. However, if proper input attenuation and bandwidth filtering (preselection) are available for the spectrum analyzer, differences are minimized.

Summarizing, a receiver has the following basic functions and characteristics:

- Variety of weighting functions (e.g., peak, average)
- Audio-frequency demodulation
- Measures modulation depth and frequency deviation

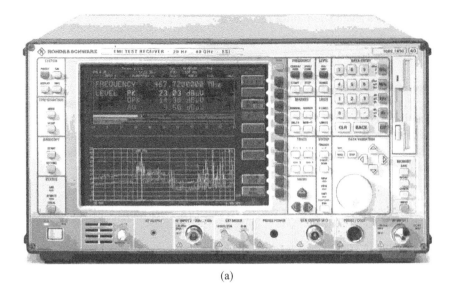

(a)

(b)

Figure 3.9 (*a*) Typical EMI test receiver with graphical display. (Photograph courtesy of Rohde & Schwarz GmbH.) (*b*) Typical high-end EMI test receiver with preselector. (Photograph courtesy of Agilent Technologies, Inc.)

- Provides analog output for recorders
- Low noise figure
- High overload capacity

Receivers not specifically designed for EMC will not have QP or AVG detectors. Their meters will also be linear, not logarithmetic. The most useful signal strength meters are those that respond to the peak of the RF signal and have a dwell time of under 0.1 ms. A known compliant product can be used to roughly calibrate the signal strength meter on a radio receiver for detecting emissions from different types of sources.

3.3 PRECOMPLIANCE VERSUS COMPLIANCE ANALYZERS

Electomagnetic compliance measurement equipment, especially spectrum analyzers if used for formal conformity testing, must comply with the provisions of CISPR 16-1. Most test instrumentation used is "precompliance," which cost less than "full compliant," appealing to the majority of test engineers and their companies. Many qualification tests are performed with noncompliant equipment.

Significant differences exist between precompliance and full-CISPR 16-1 requirements [3]. The primary differences include the following (in no particular order) and are equally applicable to both radiated and conducted measurements [4].

1. Bandwidth
2. Detectors
3. Input VSWR and sensitivity, overload performance, and pulse accuracy

Bandwidth. Bandwidth refers to the size of the signal spectrum being measured. All analyzers have the ability to display a specific bandwidth using default values or by manually punching in numbers on a keypad. This is shown in Figure 3.10.

If the interfering signal is wider than the bandwidth of the analyzer, the signal is classified as *broadband.* Broadband signals will generally escape detection as a valid emission. If the measured signal width is smaller than the preset bandwidth of the analyzer, the signal is called *narrowband,* and is the signal of interest that defines EMI.

When performing emission measurements, one must select a proper bandwidth to determine if the signal is broadband or narrowband. For known signals, like radio and television, this is easy to achieve. For unknown signals, the characteristics of the interfering signal are not known in advance. CISPR 16-1 defines the bandwidth required to properly measure a propagating signal, either radiated or conducted. In addition, the shape of the bandwidth filter internal to the analyzer must conform to specific requirements detailed in CISPR 16-1, illustrated in Figure 3.10 [4].

For conformity testing, only receivers that comply with CISPR 16-1 bandwidth

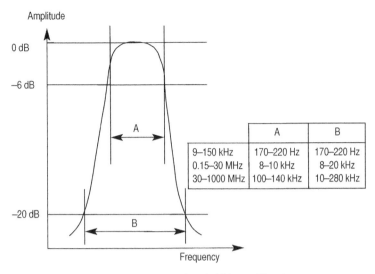

Figure 3.10   CISPR 16-1 bandwidth specifications.

requirements are permitted. Most signals measured have some broadband noise. For narrowband interference, such as a clock signal and its harmonics, the performance of the bandwidth filter will have little effect and a strong peak reading will be measured. If all signals present are smaller than the bandwidth specification outlined by CISPR 16-1, these analyzers may be qualified for conformity testing; however, one does not always know if every signal observed from a device under test is narrowband.

Detectors. CISPR 16-1 specifies three principal detector types; peak, QP, and AVG. The majority of signals that are measured use the peak detector. The peak detector responds almost instantly to the peak value of the interference signal. Regulatory standards define emission compliance levels using the QP detector, although some standards still require use of the peak detector. The emissions limits for AC mains interference voltage (Line Conducted Interference) is specified for both QP and AVG detectors. When performing full-compliance measurements, use only the correct detector. The difference between detectors is in how they respond to pulsed or modulated signals. An illustration of these detector features is provided in Figure 3.11 [3]. All three types of detectors provide the same response to unmodulated continuous signals, such as clock.

The QP detector weights the measured value in terms of its perceived "annoyance factor." This annoyance factor is defined for low-pulse-repetition frequencies. These low-pulse-repetition frequencies are less annoying when experienced on broadcast radio and television than higher pulse repetition frequencies. The QP detector is specified in terms of its attack and decay time constants. The average detector simply returns the average value rather than the peak of the interference sig-

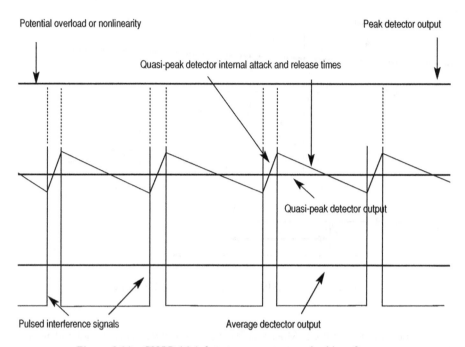

Potential overload or nonlinearity

Peak detector output

Quasi-peak detector internal attack and release times

Quasi-peak detector output

Pulsed interference signals

Average dectector output

Figure 3.11   CISPR 16-1 detector response to pulsed interference.

nal as measured. The average filter is simply a low-pass filter whose time constant is slower than the slowest pulse repetition frequency of the input.

Input VSWR and Sensitivity, Overload Performance, and Pulse Accuracy. Another concern regarding CISPR 16-1 analyzer requirements is input VSWR. The VSWR is specified to be 2 : 1 with no input attenuation, dropping to 1.2 : 1 with 10 dB attenuation. Impedance matches in the measurement instrumentation directly relate to VSWR. For broad-spectrum receivers, a VSWR of 2 : 1 without attenuation is difficult to achieve. Most analyzers cannot achieve this value. If input attenuation is added, receiver sensitivity is degraded, although VSWR is lowered. The sensitivity requirement of CISPR 16-1 is expressed in the form that the noise component should not degrade the measurement accuracy by more than 1 dB. This implies that the system noise floor must be at least 6 dB below the lowest level of the signal being measured. Most low-cost receivers often have inadequate sensitivity at frequencies throughout the majority of the frequency spectrum.

3.4   CORRELATION ANALYZER

An ordinary spectrum analyzer has a single-input channel capability and only displays scalar results. No phase information is kept. The analyzer will display what is at the input regardless of whether the input signal is intentional or not. Bandpass and band-reject filters can be used to attenuate noise that is concurrent with the desired signal of interest. This presents two problems: (1) the characteristics of the interfering frequency must be known and (2) filtering should not affect the desired signal. Achieving both of these is almost impossible, especially when the desired signal and interfering signal are both at the exact same frequency. An analyzer that is independent of the interfering frequency is required. What makes a noise-canceling system unique is the addition of a second input channel that is frequency and time synchronized together between the two inputs and the capability of performing calculations in complex form. Retaining the complex form means keeping not only the amplitude information but the phase information as well. This type of spectrum analyzer is called a correlation analyzer.

The basic concept of a correlation analyzer system is that there is a single signal composed of the sum of the desired signal S_j and an uncorrelated undesired signal I_j as one input to the system. A second input reference signal I_j contains no desired signal S_j or is uncorrelated to it but is correlated to the undesired signal, I_j, in some manner. The reference should be proportional in order for an exact replica of the desired signal S_j to become possible. The output from the measurement system is the original signal minus the reference signal. Figure 3.12 shows schematically the concept of a basic cancellation system where n_{0j} and n_{1j} are from a common origin. They are correlated with each other but uncorrelated with the EUT emissions signal S_j.

Using Eqs. (3.4)–(3.6), it is observed that, if the right proportionality constant W is selected, the output becomes the desired signal S_i and the undesired noise signal is gone. This is based on the least-mean-square (LMS) algorithm. It is possible to perform complex calculations owing to tremendous advances made in digital signal processing technology and PCI-based DSP plug-in cards. Therefore, a successful correlation analyzer requires two inputs with (1) a reference sensor and (2) a reference input that are correlated with the ambient:

$$y = W \times (Ij_j + n_{ij}) \tag{3.4}$$

$$d = S_j + I_j + n_{0i} \tag{3.5}$$

$$\text{Output} = d - y = S_j + I_j + n_{0i} - y = S_j + I_j + n_{0i} - [W \times (I_j + n_{1i})] \approx S_j \tag{3.6}$$

The LMS algorithm contains two processes. One process is filtering, which computes the output of the filter produced by the weighting factors and provides an estimation error to the desired response. This is part of the closed-loop portion of the correlation analyzer. Since filtering is involved in the feedback loop, stability is essential. The mean-square error produced by the algorithm at time t and its final value are constant. This satisfies the requirement called convergence. This is all computed in complex form. In other words, Eqs. (3.4)–(3.6) are still valid with $S_j = A_0 \cos(\omega_0 n + \phi)$.

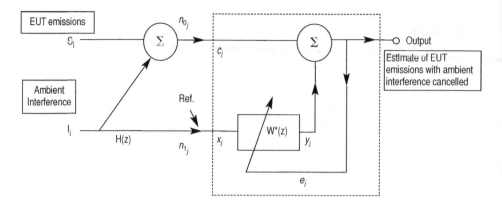

Figure 3.12 Block diagram of a basic cancellation system.

3.4.1 Characteristics of Correlation Analyzer

Since real-time sampling is performed, the operation of a correlation analyzer is theoretical independent of the signature profile of the undesired signal. One primary characteristic of this system is continuous operation regardless of the amplitude, frequency, or waveform shape of the undesired signal(s). Both inputs, desired and undesired, have identical propagation paths. Since the process is continuous, cancellation occurs even for transient or short-time-interval occurrences. In addition, the response time of the system and amount of cancellation derived are highly dependent upon the gain of the closed-loop controller. Like many closed-loop designs, the response time is dependent not only on the loop bandwidth but also on the initial or previous conditions. For example, if the undesired signal is either a short time duration or one that appears and disappears on a regular periodic basis, the response time could be zero. This is because cancellation parameters are held for a finite time period. If the undesired signal reappears during this period, cancellation is maintained. A nonzero response time will indicate if the disappearance of the undesired signal is such that the cancellation parameters have changed since the disappearance of the undesired signal.

In summary, good cancellation can be characterized as follows:

1. Cancellation quality should be independent of the characteristics of the undesired signal.
2. Cancellation should require no a priori knowledge of the undesired signals characteristics.
3. Amount of cancellation is dependent upon the characteristics of the closed-loop response.
4. Response time is not only dependent upon the closed-loop response but also on the initial or previous conditions.
5. The cancellation system must be a linear process so as not to generate any intermodulation products between the desired and undesired signals.

3.4.2 Coherence Factor

Normally, a spectrum analyzer is used for tracking down potential EMI noise sources. A spectrum analyzer is limited in this capability of identifying noise sources because of its single-input channel and because it is a scalar device designed to measure only the frequency and amplitude of a signal. Knowing both phase and amplitude becomes a tremendous help in isolating noise sources. Thus, a cancellation system with the unique capabilities of a correlation analyzer allows one to use correlation algorithms and two *simultaneous* measurement signal inputs to gain another vital bit of information: the *coherence factor.*

Coherence is a measure of the relationship between two signals. It requires two signals to be simultaneous in time. The coherence function is a measure of how or if the signals are related to each or if they are from the same source. If the signals are perfectly correlated, the coherence factor will be unity. If the signals are uncorrelated, the coherence factor will be zero. The correlation function is calculated based upon using signal cross-correlation techniques that compare spectral power densities. Equation (3.7) mathematically expresses how this coherence function is obtained between two signals. Remember that the two signals must be measured simultaneously and therefore frequency and time synchronized:

$$\text{Fourier Transform of } x(t) \ X(\omega) = \int_{-\infty}^{\infty} x(t)e^{j\omega t}dt$$

$$\text{Fourier Transform of } y(t) \ Y(\omega) = \int_{-\infty}^{\infty} y(t)e^{j\omega t}dt$$

$$\text{Coherence } (\omega) = \sqrt{\frac{|\Sigma[Y(\omega)X^*(\omega)]|^2}{(\Sigma|Y(\omega)|^2)(\Sigma|X(\omega)|^2)}} \quad \begin{array}{l} \boxed{\begin{array}{c} \text{cross-spectral density} \\ + \\ \text{coherent averageing} \end{array}} \\ \boxed{\begin{array}{c} \text{power spectral density} \\ + \\ \text{noncoherent averaging} \end{array}} \end{array} \quad (3.7)$$

The usefulness of the coherence factor can be seen in an illustration of a multiple noise source measured at the same frequency, 72 MHz (third harmonic of 24 MHz and ninth harmonic of 8 MHz). A test PCB was constructed with three crystal oscillators sources. Two of the oscillators are of the same frequency (8 MHz) with the third having its fundamental at the third-harmonic frequency (24 MHz) of the other two oscillators. With an ordinary spectrum analyzer, one would observe a common 72-MHz signal from all three oscillators. The oscillators drive a loop antenna capacitively coupled to a second etched loop, which in turn drives a simulated data cable that radiates the RF field. The second 8-MHz oscillator is reduced slightly in level by an attenuator to simulate a lesser source amplitude. Refer to Figure 3.13 for the test setup. An antenna serves as the reference input, picking up radiated noise, while a small loop probe was used to provide the second input to the correlation an-

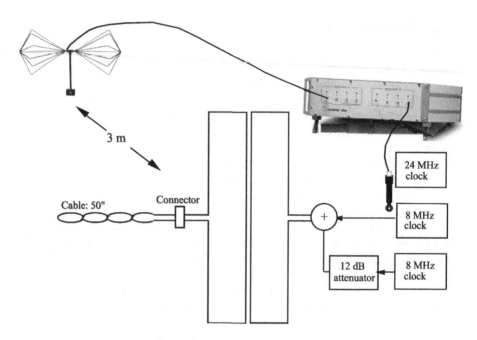

Figure 3.13 Coherence test setup.

alyzer. With the analyzer examining the 8- and 24-MHz oscillators directly to determine the primary source of the 72-MHz radiated noise, it would be difficult to determine which oscillator is the source of the problem. It would be like viewing only the top line of each chart individually. Looking at two channels simultaneously, it would still be difficult to determine the source of the 72-MHz radiation. Generally, one might guess the 24-MHz oscillator to be the main culprit, or perhaps one of the two 8-MHz oscillators due to relative amplitude differences. However, by using a correlation analyzer and looking at the coherence factor, which is the bottom half of each waveform chart in Figure 3.14, it is observed that the 8-MHz oscillator without the attenuator is the main source since it has the highest coherence factor. Therefore, with the coherance factor, one can determine which source is the main source of noise from a multitude of identical frequency sources.

Another advantage in utilizing a correlation analyzer is that measurements are taken in real time without requiring components or circuits to be powered down or disconnected. Significant time and labor are sometimes involved in shutting equipment on and off, if it can be done at all. Mechanically, it may be easier to disconnect components. Equipment may be significantly altered to not operate at all, in which case software changes may be required. This scenario is repeated as many times as necessary. The advantage of being able to make comparative measurements with a coherence parameter can save hours of troubleshooting time by not having to disconnect cables, wires, or circuits, tear down setups, or reprogram the equipment.

This technique can be applied to other testing situations, such as measuring

24 MHz clock

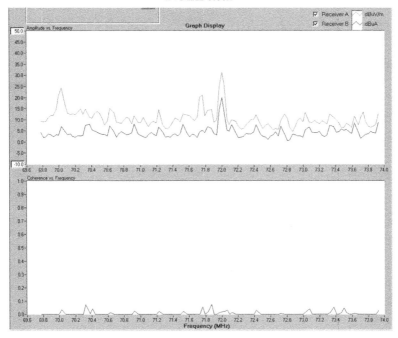

8 MHz clock with attenuator

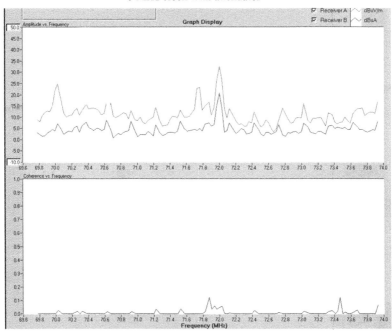

Figure 3.14   Results from test PCB (*continued*).

8 MHz clock without attenuator

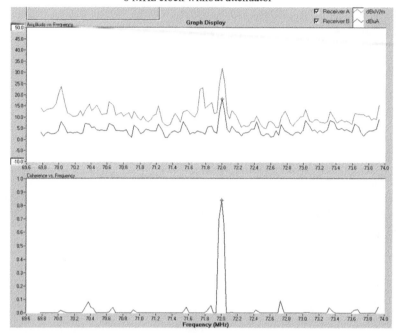

Figure 3.14   (*cont.*).

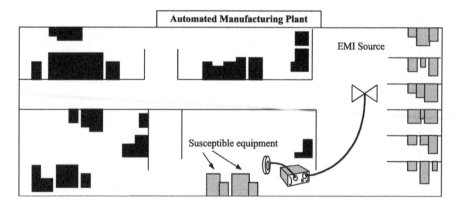

Figure 3.15   Plantwide EMI isolation.

shielding effectiveness in a noisy environment or using ambient noise as a source when measuring in a particularly sensitive environment. One channel can monitor the ambient outside RF noise while the second channel monitors the levels inside the shielded area. The coherence function can then be used to determine which data points are usable. If a data point has high coherence (e.g., >0.75) between the two channels, then it is measuring the same noise. Another application utilizing the usefulness of the coherence parameter is the ability to locate a noise source within a manufacturing environment. The susceptible equipment, or noise-impacted area, can be monitored with one channel and the antenna moved about the facility near the suspected or potential noise source while monitoring the coherence function (Figure 3.15). When the source of the noise correlates with the measurement of the noise in the impacted area or equipment, the coherence function will be high.

REFERENCES

1. Montrose, M. I. 1999. *EMC and the Printed Circuit Board Design—Design, Theory and Layout Made Simple.* Piscataway, NJ: IEEE.
2. Smith, D. C. 1993. *High Frequency Measurements and Noise in Electronic Circuits.* New York: Van Nostrand Reinhold.
3. CISPR 16-1. Specification for radio disturbance and immunity measuring apparatus and methods - Part 1: Radio disturbance and immunity measuring apparatus and methods— Part 1: Radio disturbance and immunity measuring apparatus.
4. Williams, T., and K. Armstrong. 2000. *EMC for Systems and Installations.* Oxford: Newnes.
5. Prentiss, S. 1992. *The Complete Book of Oscilloscope.* 2nd ed. Blue Ridge Summit, PA: TAB Books.

CHAPTER 4

TEST FACILITIES

Electromagnetic measurements must be conducted under specialized conditions for both emission and immunity testing. Whether the purpose of the test is for certification to a regulatory standard, customer requirement, or troubleshooting, testing must be performed in an environment that allows for measurement to occur without disruption from external ambients. This chapter provides insight into various types of facilities available to the test engineer, regardless of the type of measurement to be taken. It is difficult to troubleshoot a problem area when undesired signals from broadcast stations (AM/FM radio, televisions, personal communication equipment, etc.) occur at the same frequency as the signal of interest being measured in an open or uncontrolled environment.

It is not expected that those using this book for the purpose of testing and troubleshooting for EMC compliance will build the facilities detailed herein. Material is provided that allows one to understand the test environment that products are placed within and necessary steps toward achieving a test environment that provides correlated test results.

For EMC compliance, we are concerned with attempting to measure the amount of electromagnetic field strength being radiated by a device under test. For immunity, we attempt to determine if a specified amount of electromagnetic energy will couple into the unit and cause degradation of performance.

For both emissions and immunity, it is important to have a reflection-free area as well as eliminate all external ambients from the test environment. Each proscriptive standard (i.e., military or commercial) specifies in detail requirements and test procedures to be followed. Included are open-area test sites (OATSs), screened rooms (anechoic or semianechoic), and specialized test cells [e.g., Transverse Electromagnetic (TEM)/Gigahertz TEM (GTEM)].

Testing for EMC Compliance. By Mark I. Montrose and Edward M. Nakauchi
ISBN 0-471-43308-X © 2004 Institute of Electrical and Electronics Engineers

This chapter provides an overview of different facilities, both indoor and outdoor, used for both emission and immunity testing. For each facility, a basic review on how to perform tests is presented. The reader is referred to the vendor of each product for specific details along with appropriate application notes. Not every possible item that can be used for EMC testing is detailed herein. Only the more commonly used facilities are presented.

Appendix B provides guidance on how one conducts specific EMC tests for the most common emissions and immunity requirements specified in compliance standards. This appendix is a simplified guide on how to set up equipment and operate instrumentation for those not comfortable or familiar with performing EMC testing and troubleshooting. Knowledge on how a test facility actually performs certification testing may prove valuable as one attempts to correlate testing (troubleshooting data) with in-house measurements.

4.1   OPEN-AREA TEST SITES

An OATS is the prescribed facility for performing the majority of radiated emission testing. It provides the most direct and universally acceptable approach. An OATS requires a calibrated receive antenna, a proper ground plane, and quality coaxial cables and must be located a significant distance from metallic objects and high-ambient electromagnetic fields such as broadcast towers and power lines. This allows accurate radiated emissions from a EUT to be measured. Similarly, using a calibrated transmitting antenna for susceptibility tests on specialized equipment can be investigated under specific field and test conditions, especially if the frequency to be transmitted does not fall within a frequency range designated for communication purposes. If susceptibility measurements are to be performed for formal certification purposes, use of test chambers or a test cell is mandatory, with certain exceptions for in situ tests detailed in Chapter 7.

The primary disadvantage of using an OATS is the need to search the entire frequency spectrum for unintentional radiated emissions within an electromagnetic environment that may have ambient noise that far exceeds a propagated signal from the EUT being measured. An example is trying to measure a weak 200-MHz clock harmonic in the presence of a television signal at 199.25 MHz (channel 11, video, United States and Canada) or a clock source in the middle of the FM radio band, especially if the radio station has a strong signal. In addition, scattered signals from metallic objects near the EUT may also result in inaccurate measurements.

4.1.1   Requirements for an OATS

The test site shall be flat, free of overhead wires, and away from reflecting structures. The EUT and measurement antenna form the foci of an ellipse where the major diameter is twice the distance d between the EUT and antenna. The minor diameter is defined by 1.73R, or $\sqrt{3}d$. The distance between source and receive antenna is usually 3, 10, or 30 m depending on the test specification used and physical size of the EUT. This is illustrated in Figure 4.1 [1].

For repeatability between different test sites, a substantially larger surrounding area free from reflecting objects is recommended. This includes the control room containing test instrumentation. As an alternative to a remotely located control room, this facility can be installed directly below the OATS, separated by a solid ground plane or using an available room within a building that is a reasonable walking distance from the range.

Reflecting Ground Plane. The minimal size of a metallic ground plane is specified in the standard applicable to the product being evaluated [1]. This plane is usually rectangular with a width twice the maximum test unit dimension and extending the plane 1 m beyond the perimeter of the test unit and 1 m beyond the measurement antenna on both sides. The ground plane must not have gaps or voids that are a significant fraction of a wavelength at 1 GHz. The recommended mesh size of a perforated metal ground plane is $\frac{1}{20}$ of a wavelength at 1 GHz, or about 30 mm. For installations on a paved area or rooftop, the metal ground plane should be the size of the ellipse. Acceptable performance may be achieved without a metal ground plane

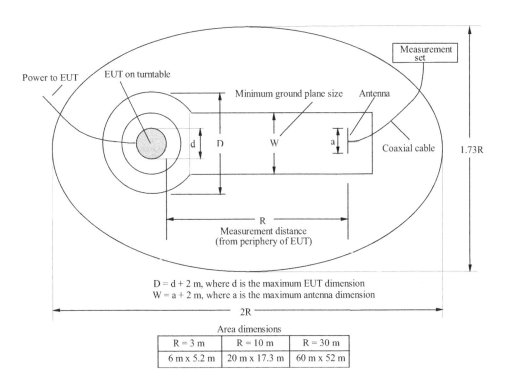

Figure 4.1 Site configuration—OATS.

provided the ground is composed of homogenous good soil (not sand or rock) and free of buried metal.

A larger ground plane, extended beyond the minimal dimensions, will bring site attenuation (calibration) closer to theoretical values. Radio-frequency scattering from the edges of the ground plane can contribute significantly to inaccuracies in measured data. These inaccuracies can be minimized by termination of the ground plane, at its edges, into the surrounding soil. Solid metal sheets welded together work best. For many locations, this may not be achievable.

The measuring equipment and test personnel should be in the plane of the antenna orthogonal to the site axis and at least 3 m minimum from the antenna. It is preferred to be at the greatest distance possible, such as 30 m.

Weather protection may be required as long as the enclosure does not have any metallic material, such as support beams, rods, nail, and door hinges. Within certain parts of the world, environmental conditions will not permit use of an OATS in its pure form. In addition, cost of land can be prohibitive. For example, a 30-m-test site requires an area 60 m × 52 m. This large area of land may be too expensive to procure in a densely populated industrial area. In addition, for industrial or urban areas, the ambient electromagnetic environment is likely to be too high to ensure meaningful measurements. For locations where climate presents practical difficulties, viability of an OATS must be considered. The cost to build a covered site with material transparent to EM waves requires a considerable investment.

The preferred antenna is the half-wave long dipole. This simple antenna is telescopic in nature. For frequencies to be measured, the antenna must be extended or shortened using a ruler or measuring tape. Although highly accurate, scanning a product from 30 to 1000 MHz could take many hours, if not days. Detailed description of antenna types commonly used for both emission and immunity testing is provided in Chapter 5.

EUT Turntable. A continuously rotatable, remotely controlled turntable is required. The turntable must be able to support the weight of the EUT and accessory equipment. Most turntables can handle a large amount of weight. Do not forget to add the weight of multiple test personnel to the support weight of both table and EUT. A turntable facilitates determination of direction of maximum radiation at each EUT emission frequency. For floor-standing units, the turntable shall be metal in construction and installed to be flush with the metal ground plane. Bonding of the turntable to the ground plane must occur. This connection is generally provided by flexible gaskets or shims. For table-top systems, the turntable should be nonmetallic, located on top of the reference ground plane. The height of the support table for table-top equipment *must* be nonmetallic and located at a height of 0.8 m above the ground plane.

Antenna Positioner. A continuously variable height, remotely controlled antenna positioner is recommended to change the distance of the antenna above the ground plane. By changing the height of the antenna from 1 to 4 m, reflected signals from the ground plane can be discovered. If a remotely controlled positioner is not available, the test engineer must manually change the height of the antenna between different positions. This means that a simple test has now turned into a very time consuming task, as the entire frequency spectrum must be investigated at least four times (1, 2, 3 and 4 m height). The turntable is rotated to determine the maximum radiation pattern of a propagated signal. After locating the angle on the turntable that produces the highest levels of RF energy, the antenna height is then varied.

Measuring Distance. Each test standard indicates the distance between the EUT and receive antenna. Typical values are 3, 10 and 30 m. Alternate antenna distances may be used with a correction factor applied to the specification limit using the equation

$$\text{New limit (dB}\mu\text{V/m)} = \text{published limit (dB}\mu\text{V/m)} + 20 \log (d)/[\text{new distance (m)}]$$

$$(4.1)$$

where d is the distance identified within the test standard.

EXAMPLE

For a reduction from 10 to 3 m, an increase in the limit of 10.5 dB is added to the specified limit.

For a reduction from 30 to 5 m, an increase in the limit of 15.5 dB is added to the specified limit.

For a reduction from 30 to 2 m, an increase in the limit of 23.5 dB is added to the specified limit.

Distance separation is defined between the boundary of the EUT and a reference point on the antenna, usually the centerline of a dipole or biconical antenna, or halfway in the middle of a log period. Trying to extrapolate accurate radiated emissions from 3 to 10 m may not provide value due to site design and the performance characteristic, physical size, and RF characteristic profile of the EUT. At 3 m the antenna may be located in the near field and thus not be able to measure a signal, or if it does measure a signal, the amplitude may not be accurate. Equation (4.1) provides a way to determine the theoretical antenna distance for extrapolation purposes; however, one should be cautious when extrapolating data from one antenna distance to another. *Studies have shown that, in reality, emissions do not fall off as 1/r. In fact, emissions can remain at the same amplitude or actually increase in amplitude with increased distance from the EUT.* This is due to the effects of the ground plane as well as nearby reflections creating multipath transmissions. If one does extrapolate data, then it is preferred to use distances relatively close to each other to minimize the potential for errors. An example of extrapolation is 3 m to 1 m, or 3 m to 10 m but not 10 m to 1 m. A better extrapolation would be from 5 m for a distance spacing between 3 m and 10 m. If a test is performed at 3 m, extrapolated from the recommended 10 m distance and certified as compliant, a customer or regulatory agency may retest at the specified 10 m distance. If the unit is significantly out of specification, the company is liable for nonconformity. Verification testing is always performed at the distance specified for the product in its respective test standard.

Site Attenuation. All OATSs must be calibrated to ensure accuracy of measured data. The process of calibrating an open-area test range is called *site attenuation.* This is defined as the ratio of source antenna input power to the power induced into a load connected to a receiving antenna. Methods for site attenuation are described in Ref. 1. This document is used worldwide.

An antenna transmits a known frequency at a specific power level located at where the EUT would be on a turntable or floor mounted. The transmitted RF energy is measured by a receive antenna along with proper instrumentation (Figure 4.2). Cable and instrumentation errors are corrected by repeating the measurement with the two antenna feeders first connected together (null test). The measured site attenuation must be within ±4 dB of the theoretical normalized site attenuation (NSA) specification for an ideal test site. Tuned dipoles are the preferred antenna for measurement purposes. This is due to the accuracy of dipoles at a specific frequency versus broadband antennas that can have significant errors or attenuation across the frequency spectrum. Validation of acceptability should be reviewed every year but could be extended based on environmental conditions (e.g., indoor vs. outdoor site).

Although measurement of NSA appears to be simple, it is extremely complex in application. The loss in the system from a source generator through both transmit and receive coax to a receiver is measured for two conditions. One is the null test mentioned above and the other is when the antennas are coupled to each other. The NSA is the difference between these measurements less the antenna factors of both

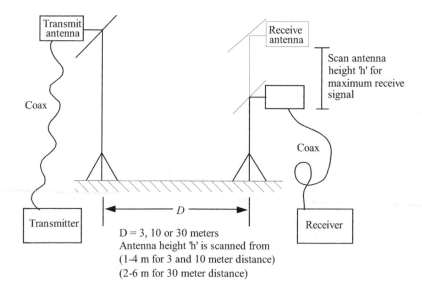

Figure 4.2   Schematic of site attenuation measurement.

antennas. The CISPR criterion allows for a 3-dB measurement uncertainty for instrumentation and only 1 dB for variations in the site itself. A well-constructed site can meet this requirement; however, very little margin for inadequacies in measurement exists. A particular problem is that the antennas used must be calibrated for the specific geometry of the NSA measurement and not for free space. Differences in antenna factors between these two conditions can be large enough to invalidate results for an otherwise satisfactory site. The theoretical NSA curves for an OATS are provided in Figure 4.3 [2, 3].

The measured NSA cannot be used as a "correction factor" during actual test conditions. Normalized site attenuation relates only to the artificial attenuation between two antennas at specific locations. The attenuation that exist between the EUT and measurement antenna, even at the same location, may be quite different because of electromagnetic coupling between the antennas and the radiation characteristics of the EUT, which is not the same as that of a particular antenna type.

Experience has shown that two antennas at 3 m distance will couple very well to each other. Effects of the environment, discontinuities in the ground plane, and reflective structures will not affect the propagated signal to a significant degree. Although the site may appear to have no problems at 3 m, imperfections at the site can make a significant difference in measured data taken at 10 m. At this distance, both transmit and receive antennas are not well coupled and ground reflections start to play a major role in changing the signature characteristics of the transmitted signal.

One method for performing site attenuation is use of a comb generator. This method is discouraged for formal assessment of the OATS. A comb generator is a small battery-operated device that transmits a particular signal and all harmonics throughout the frequency spectrum at a known amplitude value. The problem with using a comb generator is that the generated spectrum is not continuous. Therefore,

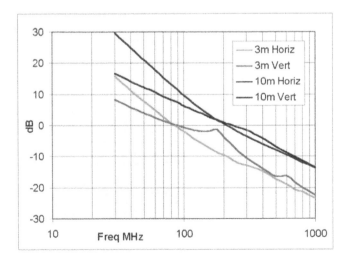

Figure 4.3 Theoretical NSA curves.

discrepancies due to resonance or antenna characteristics may not be identified. After an OATS has been calibrated using the prescribed method in Ref. 1, using a comb generator is a quick-an-easy test method to check if any significant problems have occurred to the test site, such as a damaged ground plane or defective coaxial cable. This type of testing is called verification. An example of a field strength (comb) generator is provided in Figure 4.4.

Electromagnetic Scattering. Another concern related to the performance capabilities of an OATS is electromagnetic scattering. Scattering refers to RF reflections from buildings or other metal structures (control room, houses, concrete driveways, water tanks, auxiliary generators), electrical power lines and fences, vegetation (trees and bushes), underground water and utility cables buried close to the surface, and the like. The easiest manner to avoiding interference from underground scattering is to use a metallic ground plane to prevent strong reflections from underground sources. In addition, there are specifications regarding the quality of the ground plane as well as its installation, bonding to earth ground, and continuity with the turntable. These are detailed in Ref. 1. Objects outside the elliptical boundary, such as buildings and parked automobiles, may affect measurements. Care must be taken to choose a location as far as possible from large or metallic objects of any sort.

Measurements Above 1 GHz. Studies are being completed to provide guidance for measurements above 1 GHz. Use of low dielectric materials appear to be necessary. It has been discovered that wood tables at frequencies greater than 1 GHz can cause discrepancy of several decibels. With directionality becoming more and more

Figure 4.4 Field strength generator for calibration of radiated emission facilities. (Photograph courtesy of ETS-Lindgren.)

critical, bore sighting of the measurement antenna may be required. Also, the speed of the turntable will need to be slower and with smaller increments used to locate the radial lobes. It may also be required to change from a ground plane to no ground plane and of course, change the procedure for NSA at these higher frequencies.

4.1.2 Test Configuration—System, Power, and Cable Interconnects

To ensure quality performance not only for the EUT but also for support equipment, it is desired that all remote connections to the EUT be made with high-quality, low-loss shielded cable. It is recommended that these cables be provided in metal conduits with the conduits properly grounded at both ends for EMC protection. All AC mains cable must be properly filtered to ensure that undesirable RF energy from the EUT or support/auxiliary equipment or to test instrumentation (e.g., receiver) is not disturbed.

All interconnect cables should be typical of normal use and be the specified type marketed with the EUT. Where cables of variable length are used during testing, the cables most typical for use with the product should be used throughout the test. When the cable length is unknown, a recommended length of 1 m nominal should be provided. The same type of cable (i.e., unshielded, braided, foil shield) specified for the EUT in the user manual should also be used.

A basic configuration for table-top equipment setup is detailed in Figure 4.5 for both conducted and radiated emissions for table-top equipment, while Figure 4.6 illustrates floor-standing and table-top arrangements [1]. These illustrations represent the CISPR-recommended test setup.

Conducted Emissions. For table-top equipment, excess cable length shall be draped over the back edge of the table. If any cable extends to a distance closer than 40 cm to the reference ground plane, the excess cable shall be bundled in the center in a serpentine fashion at 30–40 cm in length and located to maintain 40 cm height above the reference plane. If the cables cannot be bundled because of physical limitations (bulk, length, or stiffness), they shall be draped over the back edge of the table unbundled in a manner that all portions of the cable remain at least 40 cm from the reference ground plane. Interconnect cables between EUT and peripherals shall be bundled in the center to maintain a 40 cm height above the ground plane. The end of the cable may be terminated. The overall length of each bundled cable shall not exceed 1 m (Figure 4.7) [1].

For AC mains conducted emission testing, detailed in Chapter 6, power cords of equipment other than the EUT do not require bundling. The power cords of non-EUT equipment can drape over the rear edge of the table, routed to the floor and over to a LISN. For radiated measurements, all power cords must drape to the floor and are routed to the AC main receptacle.

For floor-standing equipment, excess interconnect cable lengths shall be folded back and forth in the center to form a bundle between 30 and 40 cm in length. If the cables cannot be bundled because of physical limitations (bulk, length or stiffness), they shall be arranged in a serpentine fashion. Interconnect cables that are not con-

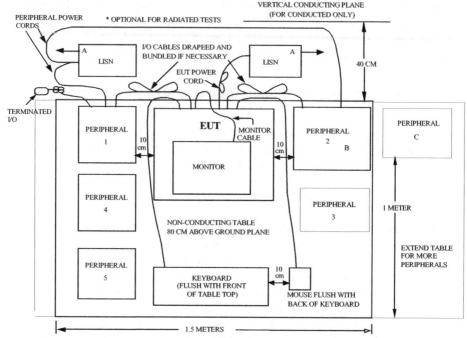

Figure 4.5 Test configuration for table-top equipment (radiated and conducted emissions).

LEGEND:

A - LISN(s) may have to be positioned to the side of the table to meet the criterion that the LISN receptacle shall be 80 cm away from the EUT. LISN(s) may be above ground plane only for conducted emission measurements.

B - Accessories, such as ac power adapter, if typically table-mounted, shall occupy peripheral positions as is appliable. Accessories, which are typically floor-mounted, shall occupy a floor position directly below the power cord of the EUT.

C - Table length may be extended beyond 1.5 m with all peripherals aligned with the back edge. Otherwise, the additional peripherals may be placed as shown and the table depth extended beyond 1 m. The 40 cm distance to the vertical conducting plane shall be maintained for conducted testing.

nected to peripherals may be terminated, if required, using correct terminating impedance. Cables that are normally grounded shall be grounded to the ground plane for all tests. Cable normally insulated from ground shall be insulated from the ground plane by up to 12 mm of insulating material. For combined floor-standing and table-top equipment, the interconnect cable to the floor-standing unit must drape to the reference ground plane with the excess bundled. Cables not reaching the reference ground plane are draped to the height of the connector or 40 cm, whichever is lower (Figure 4.8) [1].

Radiated Emissions. Figures 4.9–4.11 illustrate a simple test setup for radiated emissions based on the type of product being investigated. Details on performing radiated emissions are found in Chapter 6. Test personnel are not permitted within the perimeter of the area during testing. For this setup, the EUT is turned on. The re-

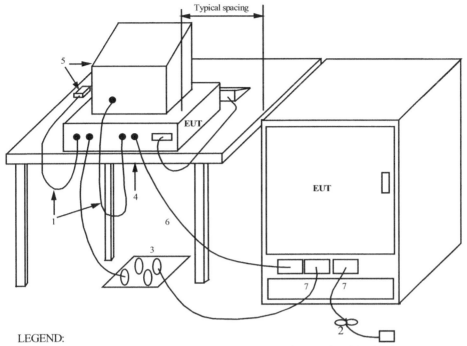

LEGEND:
1. Interconnecting cables that hang closer than 40 cm to the ground plane shall be folded back and forth forming a bundle 30 to 40 cm long.
2. I/O cables that are connected to a peripheral shall be bundled in the center. The end of the cable may be terminated, if required, using proper terminating impedance.
3. If LISNs are kept in the test setup for radiated emissions, it is prefered that they be installed under the ground plane with the receptacle flush with the ground plane.
4. Cables of hand-operated devices, such as keyboards, mouses, etc. shall be placed as for normal use.
5. Non-EUT component of EUT system being tested.
6. I/O cable to floor-standing unit drapes to the ground plane and shortened or excess bundled. Cables not reaching the metal ground plane are draped to the height of the connector or 40 cm, whichever is lower.
7. The floor-standing unit can be placed under the table if its height permits.

Figure 4.6 Test configuration for combination, floor-standing and table-top equipment (radiated and conducted emissions).

ceiver (spectrum analyzer) is then scanned over a specified frequency range to measure all radiated emissions from the EUT in an effort to determine compliance with mandated specifications [1].

4.1.3 Operating Conditions

The best equipment and facilities provide no value if testing is not performed correctly or there are deficiencies in the test environment. The most important part of performing emission tests are the skills of the test engineer. Use of automated

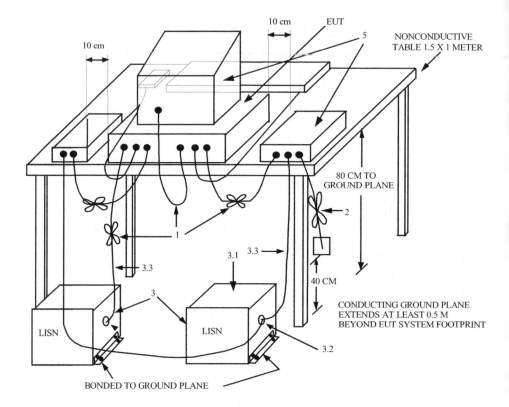

LEGEND:
1. Interconnecting cables that hang closer than 40 cm to the ground plane shall be folded back and forth forming a bundle 30 to 40 cm long.
2. I/O cables that are connected to a peripheral shall be bundled in the center. The end of the cable may be terminated, if required, using proper terminating impedance. The overall length shall not exceed 1 meter.
3. EUT connected to one LISN. Unused LISN measuring port connectors shall be terminated in 50 ohms. LISN can be placed on top of, or immediately beneath, reference ground plane.
   3.1 All other equipment powered from additionas LISN(s).
   3.2 Multiple outlet strip can be used for multiple power cords of non-EUT equipment.
   3.3 LISN at least 80 cm from the nearest part of EUT chassis.
4. Cables of hand-operated devices, such as keyboards, mouses, etc. shall be placed as for normal use.
5. Non-EUT component system being tested.
6. Rear of EUT, including peripherals, shall be aligned and flush with rear of tabletop.
7. Rear of the tabletop shall be 40 cm removed from a vertical conductive plane that is bonded to the ground plane.

Figure 4.7 Test arrangement for conducted emissions (table-top configuration).

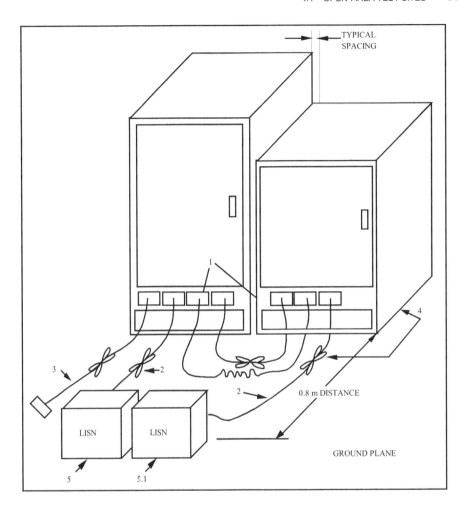

LEGEND:
1. Excess I/O cables shall be bundled in the center. If bundling is not possible, the cables shall be arranged in a serpentine fashion not exceeding 40 cm in length.
2. Excess power cords shall be bundled in the center or shortened to appropriate length. I/O cables that are not connected to a peripheral shall be bundled in the center. The end of
3. the cable may be terminated, if required, using proper terminating impedance. If bundling is not possible, the cable shall be arranged in serpentine fashion.
4. The EUT and all cables shall be insulated from the ground plane by up to 12 mm of insulating material.
5. EUT connected to one LISN. LISN can be placed on top of, or immediately beneath the ground plane.
   5.1 All other equipment shall be powered from a second LISN or additional LISN(s)
   5.2 Multiple outlet strip can be used for multiple power cords of non-EUT equipment.

Figure 4.8 Test arrangement for conducted emissions (floor-standing configuration).

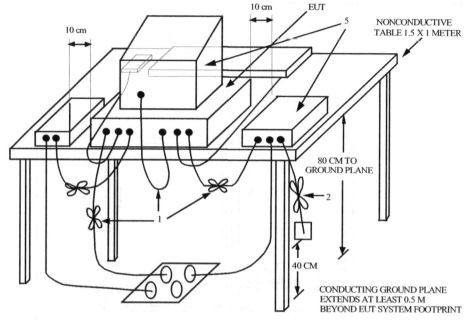

LEGEND:

1. Interconnecting cables that hang closer than 40 cm to the ground plane shall be folded back and forth forming a bundle 30 to 40 cm long.
2. I/O cables that are connected to a peripheral shall be bundled in the center. The end of the cable may be terminated, if required, using proper terminating impedance. The overall length shall not exceed 1 meter.
3. If LISNs are kept in the test setup for radiated emissions, it is preferred that they be installed under the ground plane with the receptacle flush with the ground plane.
4. Cables of hand-operated devices, such as keyboards, mouses, etc. shall be placed as for normal use.
5. Non-EUT component system being tested.
6. Rear of EUT, including peripherals, shall be aligned and flush with rear of tabletop.
7. No verticle conducting plane used.
8. Power cords drape to the floor and are routed over to receptacle.

Figure 4.9 Test arrangement for radiated emissions, table-top equipment.

software for conducted emissions is acceptable, but for radiated emissions, use of automated software is impossible at an OATS. This is due to the number of ambients present and the ability to distinguish between a clock signal and an FM signal. Automated software is, however, appropriate for radiated emissions only when used within a shielded facility.

It is imperative that the worst-case emission level be determined. This is achieved by placing cables in worst-case position through experimentation, displaying a pattern on the video monitor that stresses the video amplifiers to their limits, and that each subassembly or I/O interconnect is exercised in a full data-

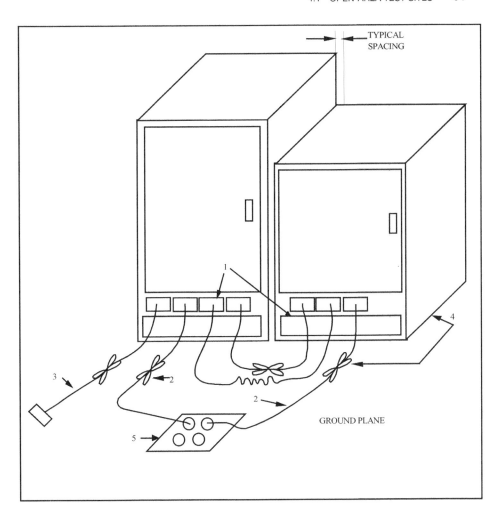

LEGEND:
1. Excess I/O cables shall be bundled in the center. If bundling is not possible, the cables shall be arranged in a serpentine fashion not exceeding 40 cm in length.
2. Excess power cords shall be bundled in the center or shortened to appropriate length.
3. I/O cables that are not connected to a peripheral shall be bundled in the center. The end of the cable may be terminated, if required, using proper terminating impedance. If bundling is not possible, the cable shall be arranged in serpentine fashion.
4. The EUT and all cables shall be insulated from the ground plane by up to 12 mm of insulating material.
5. If LISNs are kept in the test setup for radiated emissions, it is preferred that they be installed under the ground plane with the receptacle flush with the ground plane.

Figure 4.10 Test arrangement for radiated emissions, floor-standing equipment.

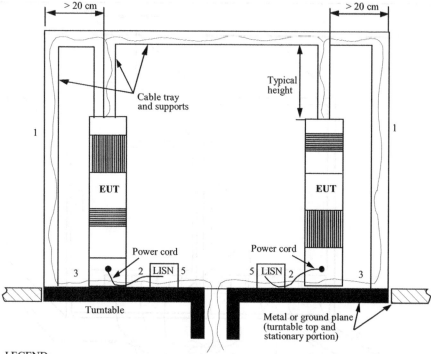

LEGEND:
1) Only one vertical riser may be used where typical of system under test.
2) Excess power cord shall be bundled in the center or shortened to apropriate length.
3) EUT and cables shall be insulated from ground plane by up to 12 mm. Where the manual has specified or there exists a code of practice for installation of the EUT, the test arrangement shall allow the use of this practice for the tests.
4) Power cords being measured connected to one LISN. All other system power cords powered through other LISN(s). A multiple receptacle strip may be used for other power cords.
5) For conducted tests, the LISN(s), if used, should be installed under, with the receptacle flush with the ground plane.

Figure 4.11 Test arrangement for radiated emissions (overhead cables), side view.

rate manner. For example, a typical personal computer should be performing the following operations all at the same time, assuming the operating system is able to handle this, such as Unix:

- Full-speed DMA transfers on all floppy drives, hard disks, and compact disks, read and write, (CD/CDW/CDR) simultaneously
- Sending continuous data or performing I/O on the parallel and all serial ports
- Ethernet connection must have LINK enabled and communicating to a remote hub with another computer passing data (not just pinging each other)
- Audio that can be heard on speakers

- USB ports accepting data from a remote video camera or another device
- External peripherals transferring data via synchronous/asynchronous protocols

To ensure that worst-case emission level from the EUT has been detected, one must deal with a number of variables [3]:

- EUT related: physical radiation pattern, operating configuration, build state (engineering level, preproduction, production), various revision levels of PCB layout, connected cables, periodic and cyclic effects
- Measurement related: frequency range, antenna polarization, antenna height, detector time constant

The vast majority of electronic products are not designed to be intentional radiators. Digital logic has as a byproduct the development of unintentional RF energy. Because of this situation, knowing the direction of maximum radiation will always be unknown. This is especially true if cables or external interconnects are provided, even user interfaces, including mice and keyboards. Movement of interconnect cables a small distance can be the difference between pass and failure. For this reason, one must maximize cables to ascertain maximum radiating levels "throughout" the entire frequency spectrum. At one frequency, the signal amplitude may drop; however, at another frequency the signal increases. It is imperative that one does not rely on the results of cable maximums at only "one" frequency but that the spectral distributions of energy from all cables are taken into consideration.

The layout of both the EUT's internal and external components and assemblies along with remote peripherals, especially their interconnect cables, will modify the radiation pattern. It is virtually impossible to predict unintentional radiated patterns in advance. It can be assumed that matching polarization of cables to the polarization of the antenna will be significant.

Maximum emissions are also likely to change in time if there are periodic operations or functions within the EUT. These emissions may change depending on the revision level of the software. Precompliance testing prepares the system for formal conformity assessment by determining worst-case configuration and functional parameters. This saves considerable time exploring these parameters during final testing.

CISPR measurement procedures include the requirement to determine worst-case emissions from both horizontal and vertical antenna polarizations. The antenna is changed in height from 1 to 3 m (for 3 and 10 m test distances) or from 2 to 4 m (for 30-m testing) in order to eliminate the cancellation effect of ground plane reflections. Contrary to appearances, the height scan is not intended to pick up emissions from the EUT in the vertical polarity.

The full frequency range of 30–1000 MHz is required for the majority of emissions requirements of commercial products. Certain standards require testing above 1000 MHz, especially for the FCC, if the highest generated operating frequency is above 108 MHz. For these tests, specialized antennas are required. Site attenuation specifications above 1 GHz do not exist at time of writing. In addition, procedures and methodology to perform testing above 1 GHz is under consideration as mentioned in Section 4.1.1.

When using a spectrum analyzer or receiver, proper instrument settings must occur for accuracy of measurement. This includes a fixed resolution bandwidth of 120 kHz. The frequency step size, dwell time at each frequency, measurement bandwidth, detector response time, and EUT emission cycle time are all interrelated. Their interaction with each other determines how long any given test will take.

With all these concerns, a typical full-compliance measurement procedure is shown in Figure 4.12 [3]. These procedures have largely become standard practice at most test laboratories, and optimize the overall measurement time by performing an initial series of prescans with a fast peak detector. Using the peak detector, instead of a quasi-peak or average detector, will always provide a higher amplitude reading on the signal. During preliminary scans, a table of frequencies is generated that identifies which signals are close to the specification limit. These signals are then investigated using the slower quasi-peak detector. Use of this prescan procedure provides the greatest amount of benefit when performed in an anechoic chamber or in a facility that guarantees the ambient noise (RF energy) does not skew

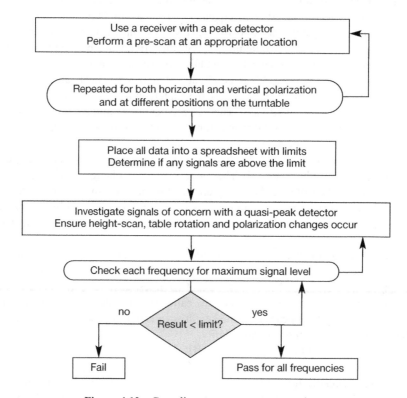

Figure 4.12 Compliance measurement procedure.

measured results. Ensure that turntable rotation and antenna height scan along with both vertical and horizontal polarization are performed. If any signals are missed because the scan was performed too fast or over an inadequate range of orientations, the method identified in Figure 4.12 becomes worthless. For this reason, a competent test engineer is required.

4.1.4 Measurement Precautions

Performing radiated emissions is simple and straightforward. Several areas of concern may affect the accuracy of test data. In addition, it may be impossible to achieve repeatable results at another OATS should the product be retested.

Electromagnetic Environment. The electromagnetic environment for an OATS needs to be relatively quiet and free from ambient or strong signals. These signals include broadcast radio or television transmitters, aeronautical communication, cell-phone towers, man-made radiators such as automobile ignition systems and arc-welding equipment, along with a host of other products. It is desirable that both conducted and radiated ambient radio noise and signal levels, measured without any EUT, are equal to or greater than 6 dB below the allowable specification limit of the applicable specification or standard. Most OATSs must maintain a NSA characteristic of ±4 dB. Below the lower tolerance limit, ambient noise is expected. If ambient or man-made noise does occur within the frequency range that contains signals related to the product, it may be impossible to acquire an accurate reading. The use of a correlation analyzer may be necessary to obtain a reading in a noisy ambient condition.

Magnetic Field Strength Measurements. If an OATS is to be used for magnetic field measurements, the test site should be modified. The ground plane is not required. If a reference ground plane is present, the measured level of radiated emissions may be higher than if measurement was made without the ground plane. Magnetic field measurements are referenced without the plane should verification testing be required by a third party. For this reason, this test should be performed in a shielded enclosure at frequencies below the chamber's resonant frequency. Most shielded enclosures are designed to operate from 30 MHz and above. Magnetic field measurements are generally below 30 MHz.

4.1.5 Alternate Test Sites

Measurements may be performed at facilities that differ from the standard test site described above. These sites include RF absorber-lined or metal test chambers, office or factory buildings, and weather-protected OATSs. A spare room or area within a facility can suffice for preliminary investigation of radiated emissions. Alternative sites should comply with volumetric NSA requirements over the volume of the EUT. Metal enclosure and other sites that do not comply with volumetric requirements may only be used for exploratory radiated emission measurements, unless it can be demonstrated that results achieved are equivalent to those obtained at a standard or alternative site that does comply with volumetric requirement.

4.2 CHAMBERS

The main advantage of testing in a chamber environment is having a clean (RF-free) ambient environment to work within. For radiated emission, this can save considerable time as no effort is wasted in attempting to resolve ambient signals from real signals, especially if frequencies happen to be coincidental to those of the EUT. In the case of immunity testing, it is essential to test in a chamber to prevent potential interference to the general electromagnetic spectrum.

A screened room or chamber is nothing more than an all-metal enclosure with special precautions for proper electrical, mechanical, and personnel access. The limitations of a bare-wall enclosure lies in having resonance. Any electromagnetic energy from the EUT or test antenna will interact with specific resonant frequencies of the chamber in a complex manner. Since specific dimensions of the room determine resonance, each chamber will react differently. Resonances produce minima and maxima owing to the electromagnetic propagating standing waves produced. This makes it difficult to have a high degree of confidence in the measured results. Field strength variabilities have been recorded to be as much as 10–20 dB. These factors of resonance are in addition to effects of reflections

Following is a brief overview of test facilities described here.

Anechoic and Semianechoic. To resolve problems with the electromagnetic environment, absorbing material can be added to reduce RF reflections from the walls, thus making the anechoic chamber perform similar to an OATS environment. In common practice, it is desirable to leave the floor reflective to act like the ground plane of an OATS but then cover the floor with removable absorber material to reduce resonant effects and ensure a uniform field required for radiated immunity testing. The material used for absorber-lined chambers are usually carbon-loaded polyurethane foam in the shape of pyramidal cones in combination with ferrite tiles. Ferrite tiles are effective for low frequencies from about 30 MHz to several hundred megahertz. Carbon-loaded foam is efficient at higher frequencies. Cost is a significant concern associated with designing, operating, and maintaining this type of chamber.

Reverberation Chamber. Another alternative and less expensive facility is the reverberation chamber. This facility maximizes both reflection- and resonant-mode effects. It does this mode conversion by having tuners or stirrers rotating inside the chamber. These tuners create an ever-changing resonant frequency pattern. The advantage of this technique is that the stirrers are capable of providing power efficiency. Because of the resonant multiplication effect due to an increased Q-factor, this chamber requires less RF input power to create a corresponding equivalent field than an anechoic-type chamber, resulting in a randomly varied wave polarization within an isotropic homogenous field. What this implies is that there is no need to reorient the EUT or repeat the test for all polarizations.

TEM and Other Specialized Test Cells. Another type of test chamber is called a TEM (transverse electromagnetic) or sometimes the Crawford cell. This cell is essentially an expanded coaxial design with the inner conductor called the septum. The TEM cell provides a uniform RF field as long as the higher order modes are not excited (i.e., *H* distance from the septum to the outer wall is $<\lambda/2$). The maximum height of the EUT should be less than *H*/3. This type of chamber is usually relegated to testing small devices only. It can be used for both emission and immunity testing. Its primary advantage is low cost.

A close relative of the TEM cell is the GTEM (Gigahertz Transverse Electro-Magnetic). This cell is gaining in popularity because of its cost advantage. With its larger and unique design, this cell can operate to higher frequencies without exciting higher order modes of field propagation. It is designed as a tapered transmission line that is terminated with absorber material to reduce end-wall reflections. The septum is offset to allow testing of larger devices. The GTEM handles personal computer size devices. Other than these basic features, both GTEM and TEM are similar and can be used for both emission and immunity testing.

4.2.1 Anechoic Chamber

A typical anechoic chamber (Figure 4.13) is the most common shielded facility in use. This chamber contains carbon-filled absorber cones, ferrite tiles, or a combination of both. A full anechoic chamber has shielding material on the floor while a semianechoic facility has a solid metal ground plane, simulating the effects of an OATS. The size of the absorbers should be suitable for the size of the EUT to be tested in order to achieve adequate control over the tolerance of the required field strength for radiated immunity tests.

Test standards state that the antenna-to-EUT distance should be greater than 1 m; however, 3 m is commonly used in an anechoic chamber due to limitation in size. For official testing at an OATS, according to the American National Standards Institute (ANSI) the test limit is 3 or 10 m depending on the category of product. For CISPR, the distance should be performed at 10 or 30 m. The test report must indicate if the test distance is other than that recommended by the test standard. In case of disputes, the measurement distance will be that specified in the appropriate standard being used.

The primary criterion for a chamber is the uniformity of the field strength that illuminates the EUT. Larger or smaller screened enclosures may be used if it can be demonstrated that the radiated field is not altered in such a way as to affect test results. Anechoic chambers are not effective at lower frequencies because the absorbers are less efficient. This usually is observed at frequencies below 80 MHz. If

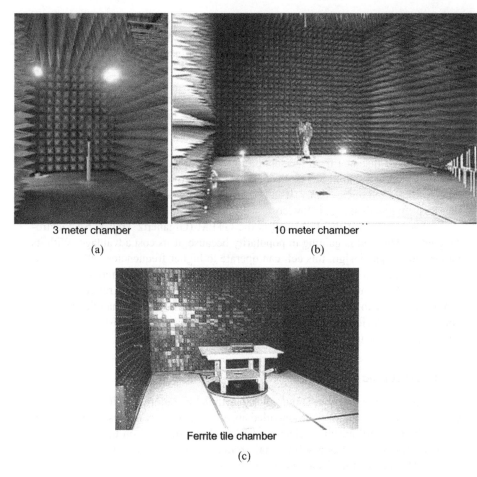

3 meter chamber   10 meter chamber

(a)   (b)

Ferrite tile chamber

(c)

Figure 4.13 Sample anechoic chamber configurations. [Photographs courtesy of (*a*) EMC Test Systems and (*b, c*) Lehman Chambers.]

a 10-m or greater size chamber is available, achieving accurate field uniformity becomes easier.

The following items are of concern when using an anechoic chamber [7].

Field Uniformity. In order to ensure that the proper field strength is presented to the EUT, a series of calibration tests need to be performed. A uniform field of RF energy must be present along the front plane of the device under test. The uniformity of the field is measured by placing field-sensing probes at the EUT position and adjusting the carrier signal level of the transmitting system until the required field strength is obtained. This procedure is repeated for a total of 16 positions (Figure 4.14). Within any chamber, be it anechoic or not, resonances cause standing waves. At higher frequencies, standing waves can result in significant variation in the field

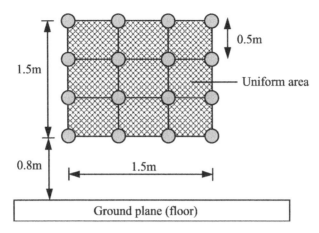

Figure 4.14 Field uniformity measurement.

strength over a small volume, generally less than that occupied by the EUT. The International Electrotechnical Commission (IEC) test standard IEC 61000-4-3 describes this calibration procedure in detail.

When performing field uniformity measurements, 16 points are required over a grid covering a plane area. Calibration is performed without the EUT. The grid plane represents the front plane of the EUT. A tolerance level of –0 to +6 dB is permitted for each location. If the tolerance of the field strength data is nonuniform, changes to the test chamber must be made to ensure uniformity. In addition, placement of the antenna plays a significant role in the accuracy of this calibration procedure.

Signal Source. Any RF signal generator that covers the frequency range of 80–1000 MHz is acceptable for radiated immunity tests. For conducted immunity, the signal generator needs to operate between 150 kHz and 80 MHz, minimum. For conducted immunity measurements, a second, separate amplifier can be used to minimize the cost of a single-wideband RF amplifier. The output level of the signal generator must match the input requirements of the amplifier to within a few decibels.

The signal generator must be able to provide a modulated AM carrier of 1 kHz to a depth of 80%. A spectrum analyzer with a tracking generator can be optionally used to monitor the frequency, amplitude, and amount of modulation. A computer controller will determine the step frequencies across the band to be tested. The required frequency accuracy depends on whether the EUT exhibits any response to narrowband interference. A manual frequency-setting ability allows one to investigate the response around a particular frequency with appropriate debugging equipment (e.g., sensors, probes, software).

Power Amplifier and Field Strength. Due to the high level of power needed for radiated immunity testing, an RF amplifier is required. An RF signal at a particular frequency is provided to the input of an RF amplifier. That same frequency is then amplified and sent to the antenna. Depending on the type of chamber used and the power level involved, amplifiers from 25 to 100 W are usually required. The larger the output power capability of the amplifier, the more expensive the test becomes. It is common to find two or more amplifiers at a test facility. One amplifier is typically used for conducted immunity (150 kHz–80 MHz) and one for radiated immunity (80–1000 MHz).

The power required is dependent on the transducer (antenna) and varies with frequency. In addition, the antenna factor (gain) plays a significant part in the calibration process. The lower the frequency, the greater the drive strength and cost. The relationship between antenna gain and power provided to the antenna related to the field strength in the far field is identified in effective radiated power (ERP):

$$\text{ERP} = \frac{E^2 r^2}{30} \tag{4.2}$$

where E = field strength (V/m)
$\quad r$ = distance from transmitting source in meters

This translates into

$$E = \frac{\sqrt{30(\text{ERP})}}{r} \tag{4.3}$$

The Friis field equation (4.2) illustrates a troubling development within EMC, expressing in very simple mathematical terms what is typically a complex situation. This equation is true if the observation point is in the *true far field*. However, very few EMI scenarios qualify as true far field. For example, practically none of the standardized EMC test setups emulate a true far-field condition because in most situations it is not practical to do so. Equation (4.2) provides at best a rough first estimate. It should never be used to decide if a given measurement result makes the EUT pass or tail. When used by experienced EMC test engineers, such simple expressions are helpful tools. The proliferation of this simple equation by inexperienced or junior engineers can result in significant errors when determining the magnitude of emissions from a radiating source.

The gain of a broadband antenna varies with frequency; hence, the required power for a given field strength will also vary with frequency. Less power is required due to the higher gain of a log-periodic antenna. Antenna calibration factors are likely to vary for values of r between 1 and 3 m due to proximity effects.

For example, for a 50-kW transmitter, the field strength at 5 km is 0.2 V/m. With a 500-kW transmitter at the same distance, the corresponding field is less than 1 V/m. The directivity of the antenna, type of terrain, weather, and other environmental factors can affect this amplitude level.

Based on actual measurements, the field strength from hand-held transceivers is statistically averaged to be

$$E = \frac{3\sqrt{P}}{d} = 7\,\frac{V}{m} \tag{4.4}$$

The difference between this level and the theoretical level (16 V/m), derived by assumption of a uniform area of illumination from a dipole antenna, is due mostly to absorption of energy by the human operator and other relative inefficiencies in the antenna, such as the absence of a solid ground plane.

The power output versus bandwidth is the most important parameter of an amplifier. It largely determines the cost of the unit. Most broadband amplifiers (1–1000 MHz) are available with powers of a few watts. This particular amplifier may be insufficient to generate required fields strengths with biconical antennas in the lower frequency range. A higher power amplifier with a bandwidth restricted to 30–300 MHz will also be required. This is a primary reason why more than one amplifier is sometimes required.

The power delivered to the antenna (net power) is not the same as the power provided by the amplifier. Losses occur in the coax as well as the antenna. Reflected power is sent back to the amplifier as a VSWR.

If any significant amount of reflected energy is sent back to the amplifier, damage could occur to the output circuits. An attenuator, typically 3–6 dB, is usually located at the output terminal of the amplifier to prevent damage due to high VSWR.

Factors to take into consideration when specifying a power amplifier include the following:

1. *Linearity.* When RF power is on, distortion of the signal should not be excessive. If distortion exists, it will be observed as harmonics of the test frequency, giving rise to spurious responses within the EUT. The amplifier should be designed to have distortion products at least –20 dB below the carrier frequency.
2. *Power Gain.* The amplifier must be able to provide full power levels across the frequency spectrum with a safety margin of gain to allow for losses within the test environment.
3. *Ruggedness.* The amplifier may be subjected to high levels of VSWR or power levels in excess of maximum ratings. Under abnormal operating conditions, the amplifier should not shut down but should fall back to a predetermined set of maximum values. Due to changes and losses in the coax to the antenna, in addition to the facility's environment, the amplifier must remain stable at all times.
4. *Availability.* To ensure constant uptime, a spare amplifier should be available. If not possible to have a second amplifier, the unit must be capable of being repaired quickly at minimal cost.

Field Strength Monitor and Leveling. It is essential that the correct field strength be provided to the EUT at all times. Too often, a higher level field strength occurs, causing system failure. Monitoring the field strength with a closed-loop control system that regulates the output of the RF amplifier in accordance with the antenna gain, attenuation loss of the coaxial cable and chamber configuration is required. This area of concern is significant when determining the measurement uncertainty of the test environment.

The RF fields are monitored through field strength sensors. Typical sensors are shown in Figure 4.15. These sensors contain a small internal antenna and detector replicated in three orthogonal planes. This allows three orthogonal planes to be monitored at all times and is insensitive to polarity. The units can be battery powered or remotely controlled. Units with a fiber-optic link are optimal for use in a chamber as fiber-optic cable is immune to radiated fields.

There are two primary means of controlling field strength intensity: closed-loop leveling and substitution. In non-anechoic rooms, closed-loop monitoring is pre-

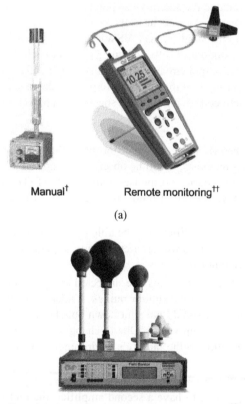

Manual[†]   Remote monitoring[††]

(a)

Complete test system (field probes and monitor)[††]

(b)

Figure 4.15 Typical field strength sensor. [Photograph of (*a*) Instruments for Industry, Inc. and (*b*) Amplifier Research.]

ferred. The sensor is placed next to the EUT. Disadvantages of using this method include the following:

1. The sensor monitors only one position along the front plane of the EUT. The field strength intensity at other locations can be significantly different.
2. If the sensor is located at a null position for a particular frequency, the amplifier will increase its output power to damaging levels. This stresses not only the EUT but also the amplifier.
3. Attempts to determine the correct field strength at a particular step frequency may result in overcorrection of the applied power. As a result, excessive field strengths may occur for a brief moment of time.

When an anechoic chamber is used, the substitution method is preferred. This involves precalibrating an empty chamber with the 16-point field uniformity test configuration. For each frequency, the amount of amplifier power is determined and recorded. The EUT is then placed in the chamber. At each frequency, the amplifier provides the predetermined value of power. This test method is viable when the field uniformity is closely defined. When a change in power is required, it is best to vary the output of the amplifier (forward power) instead of the net power to the antenna. Antenna characteristics can be significantly altered when a EUT is placed within its environment.

Transducers. The same antenna used for radiated emissions can be used for radiated immunity. The greatest area of concern lies in the maximum amount of power that the antenna can handle before damage occurs. A balun limits the power-handling ability of the amplifier. The balun (balanced–unbalanced) transformer matches the wideband impedance of the antenna to that of the amplifier. All antenna calibrations must include coax losses. Losses in the antenna and coax can be substantial while losses in the balun are generally small. Some of the power delivered to the antenna will be reflected back to the amplifier. This reflected power causes heating in the balun core and windings in addition to possible damage to the amplifier. The balun determines the maximum amount of power that can be delivered to the antenna.

Biconical antennas have high losses at lower frequencies. This is a primary reason why most of the power from the amplifier is reflected back rather than radiated. As with radiated emission testing, the polarity of the antenna (horizontal and vertical) must be considered. This involves two test runs throughout the frequency spectrum.

Sweep Rate. Sweep rate refers to how fast the frequency band is scanned (swept) during testing and is critical to the acceptance criteria specified. A signal generator is manually or automatically swept across the frequency range of interest at 1.5×10^{-3} decades/s or slower, depending on the speed and response of the EUT. For simplicity, as an alternative, the signal from the generator can be increased (decreased) in steps typically at 1% of the frequency applied. Slower dwell times translate into longer test times. For example, the frequency range 80–1000 MHz with a step size of 1% and dwell time of 3 s takes 12.7 min. Both horizontal and vertical polarities must be investigated and for all four sides of the EUT, making this a total of eight scans to perform.

Many systems are sensitive to the speed of the sweep rate when modulation of an RF signal is present. Systems that pass the immunity test using a fast sweep rate may fail the test when a slower sweep rate is used. Sweep rate is so important that MIL-STD-461E specifies a minimum sweep rate—hence the need for determining the slowest sweep rate necessary for maximum excitation of the system.

Modulated signals have a very broad bandwidth signature and are typically several megahertz wide. Nonmodulated signals are narrowband in nature and may not be detected by the EUT within the period that the RF signal is applied. For EUTs with very narrow bandwidth detectors, system failure may not be noted. This is especially true for analog-to-digital converters operating at a fixed clock frequency when interfering frequencies are aliased down to the baseband frequency. If the sweep rate is too fast, a response from the EUT may be missed. Narrowband susceptibility can be up to 25–30 dB worse than broadband. It therefore becomes important for the test engineer to understand the EUT's internal circuitry in order to assess the magnitude of narrowband versus broadband interference potentials. Probably more important than knowledge of the internal circuitry is the knowledge of the EUT functionality, bit cycles, operation modes, failure modes, and so on. For other types of systems, especially monitors, modulated RF can be extremely disruptive.

Biological Safety Requirements. The presence of high-power RF fields can cause biological harm. For this reason, test personnel are not permitted inside a test facility with the amplifier turned on. Because of this safety requirement, closed-circuit cameras with a zoom function are generally used inside the chamber. The advantage of having a camera in the chamber with a monitor in the control room allows one to view disruptive errors that may occur on video screens in addition to noting changes in the EUT and antenna configuration. This camera should be capable of operating in high-induced RF fields without disruption.

4.2.2 Screen/Shield Rooms

A screen or shield room is a facility that is used for diagnostic testing or debugging known EMI problems. In addition, this room can be used to perform conducted emission tests for regulatory compliance (LCI, line-conducted interference). Use of this room for radiated emission compliance is not permitted. This is because multiple reflections of RF waves are bouncing around the chamber, distorting the real signature profile of the EUT.

A screen/shield room that meets the CISPR NSA requirement without including absorber material on the walls and ceiling would be extremely large and very expensive. This is because RF energy reflection from the six surfaces (four sides, top, and bottom) will severely distort NSA figures. A typical small unlined screened room just big enough to encompass a 3-meter test range will have so many reflections that the actual NSA could vary by more than ±30 dB at certain frequencies. Making a measurement in such a room is impossible. The best one can achieve is to identify emissions frequencies at which there *might* be problems and measure these individually at an open-field test site. The alternative is to line the walls and ceiling with absorber (ferrite tiles or pyramidal carbon-loaded foam), which damps reflections and makes the room usable for measurements. This room is now physically smaller.

Most screen/shield rooms are built from modular steel and wood sandwich panels welded or clamped together. Honeycomb assemblies provide ventilation. A shield room will have no windows. A screen room may contain a copper mesh screen for the walls of the room. This mesh not only allows for air ventilation but also provides a scenic view of the test laboratory. All electrical services that enter the chamber must be filtered. This includes AC mains, I/O, and lightning protection. Lights must be incandescent as fluorescent bulbs emit broadband interference. The most critical part of a screen/shield room is the door assembly. A double-wiping action knife-edge gasket must contact the door frame on all four sides with beryllium copper finger strips. Figure 4.16a shows various screen room assemblies. Figure 4.16b illustrates a low-cost shield room made of conductive fabric and expanded metal, which is easy to build within one's facility if a limited budget exist.

A shield room isolates the EUT and support equipment from the external environment, which may cause erroneous operation or false readings. Interconnect cables leaving the room must be filtered and shielded. A removable bulkhead panel is generally provided to allow multiple cable assemblies to enter the room. Depending on the application, certain interconnect cables will require filtering.

Regarding immunity testing, it is recommended that the facility be capable of having a minimum of 3 m distance between EUT and antenna. Immunity tests are performed in the same manner as with an anechoic chamber, with the exception that anechoic material must be placed on the floor between the EUT and antenna in addition to the rear wall, which is optional (Figure 4.17) [3]. Anechoic material is required to minimize reflections off the floor back to the EUT, increasing the amplitude of the injected field. It is recommended to perform immunity testing in a chamber that meets required test specifications for radiated electric fields. One good advantage of using this particular type of shield room is that a solid ground reference plane is always provided.

Chambers that do not have anechoic material exhibit field peaks and nulls at various frequencies. These peaks and nulls are determined by the chamber's physical dimensions. The larger the room, the lower the resonant frequency. For most rooms, signals below 80 MHz are not easily measured. To minimize resonances, use of anechoic material is required. The easiest material to install or remove is that manufactured with pyramidal carbon-loaded foam sections. These cones are expensive and subject to damage. Even if cones are used, the lower frequency limit of efficiency is 200 MHz. Below 200 MHz, measurement uncertainty becomes a concern. Above 200 MHz, significant benefit is realized.

Figure 4.16 Typical screen/shield rooms. (Photographs courtesy of (a) ETS-Lindgren and (b) The Expanded Metal Co. and *EMC Compliance Journal* [3].)

Measurement within the frequency range of 30–200 MHz is subject to large variations of field intensity. This intensity can be on the order of 30–40 dB at resonant frequencies. A small change in antenna position will also change the measurement uncertainty to a significant degree. As a result, performing RF immunity testing in a screen/shield room is not repeatable. Regardless of configuration, anechoic cones will help the performance of screen/shield rooms to a limited degree.

Another material available is ferrite tile. This tile is best used when permanently installed. Due to the brittleness of the tile, damage will occur if mishandled. When placed on the floor, a piece of plywood is generally placed on top of the tile to pre-

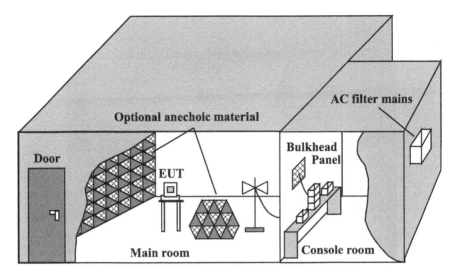

Figure 4.17 Screen/shield room configuration with anechoic material.

vent damage. This wood does not affect the anechoic performance of the tile to any degree.

4.2.3 Reverberation Chamber

A reverberating chamber consists of a rectangular chamber with walls and a mode-stir paddle that disrupts RF fields internal to the enclosure. This chamber can be used as an alternate facility for conducting radiated immunity tests. The principal of operation is based on the existence of multimode resonance mixing. A typical chamber is shown in Figure 4.18.

The reverberation chamber is simple in design; however, testing products is complex. This is due to the need for comprehensive theoretical analysis to describe field behavior inside the chamber and a means to correlate test results with actual operating conditions. Reverberating chambers emulate free-space conditions using the process of mode stirring in an enclosed area. Data from testing can be extensive during the monitoring process, which requires computer support.

Use of a reverberating chamber for radiated susceptibility (RS) testing has advantages over other chambers. This is because the chamber provides excellent isolation from the external electromagnetic environment. In addition, these chambers are relatively inexpensive to build and are capable of yielding efficient field conversion, thus making it possible to conduct RS testing at high field levels. One disadvantage is relating the measurements made in the chamber to actual operating conditions. Polarization properties of the EUT are unknown.

A reverberating chamber produces an environment where the radiated field is uniform throughout the volume of the enclosure walls. As a result, several modes of

Figure 4.18 Typical reverberating chamber. (Photograph courtesy of Defence Science and Technology Organisation, Australia.)

electromagnetic fields are present at the same time. Modes include Transverse Electric (TE), Transverse Magnetic (TM), and Transverse Electromagnetic (TEM). Details on various modes and how they affect field propagation may be found in vendor application notes. A simple rectangular chamber does not produce a uniform field throughout its volume. Wherever the EUT is physically located in the chamber, a different field strength exists, which leads to uncertainty of measurement. Within the chamber are metallic vanes that are inserted from adjacent walls and rotated at different speeds around an axis perpendicular to the wall. The time variation of the chamber geometry along with the rotation of the vanes produces a continuous variation of mode mixing with the same statistical distribution of field modes. Regardless of where the EUT is located within the chamber, the field intensity is uniform.

Reverberation chambers are mainly used for military system testing such as Hazards of Electromagnetic Radiation to Ordnance (HERO), automotive applications, and large-EUT hardness testing at high field strengths. They are also used for commercial standards such as the volumetric uniformity requirements of IEC 61000-4-3 (EN 61000-4-3) and commercial avionics test standard RTCA/DO-160D. These chambers provide random, complex, real world conditions similar to the environments found in avionics bays and automobile engine compartments.

Figure 4.19 illustrates the basic setup when performing tests within a reverberation chamber [2]. Both emissions and immunity can be performed with the same setup. When performing tests, additional connections must be made to the EUT to monitor its performance level or malfunction.

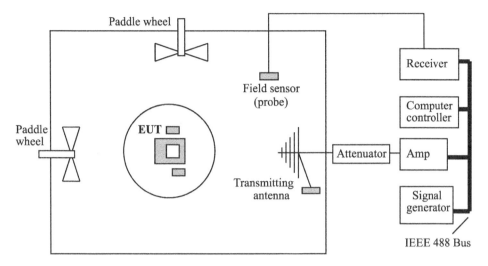

Figure 4.19 Basic setup for measurement of radiated emissions and immunity in a reverberation chamber.

The desired field intensity for immunity tests is established with a signal generator, attenuator, and amplifier. The stirrer vanes are constantly kept in rotation. Monitoring the EUT for malfunction must be performed at each field intensity level. It is important to ensure that sufficient time is provided for both the field level and performance of the EUT to stabilize. Testing should be performed throughout the entire frequency spectrum.

4.3 CELLS

There are two basic types of cells, TEM and GTEM. Both have unique applications and use. The advantages of using cells are for products that are physically small, especially components.

4.3.1 TEM Cell

The TEM (Transverse ElectroMagnetic) cell is a small enclosure used in normal laboratory environments for both emissions analysis (non–product qualification) and radiated immunity (product qualification). An example of various types of TEM cells is shown in Figure 4.20.

The TEM cell is a rectangular coaxial transmission line resembling a stripline with outer conductors closed and joined together. Both ends of the rectangular section are tapered to where a 50-Ω coax cable is connected. The center conductor of the cell and the outer shield (top and bottom plates plus sides) facilitate propagation

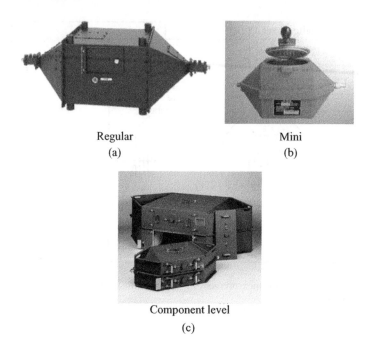

Regular
(a)

Mini
(b)

Component level
(c)

Figure 4.20 Various configurations of typical TEM cells. [Photographs courtesy of (*a*) Instrument for Industry, Inc., (*b*) CPR Technology, and (*c*) Amplifier Research.]

of electromagnetic energy from one end of the cell to the other in the TEM mode. The center conductor is supported by dielectric supports. The EUT is located on the rectangular portion of the transmission line between the top or bottom plate and center conductor. A dielectric spacer is used to electrically isolate the EUT from the outer and inner conductors of the transmission line. The presence of a closed outer shell provides an effective shield to isolate electromagnetic fields from both entering and leaving the cell. Transverse electromagnetic cells can also be designed with other cross sections, such as square or asymmetric rectangular.

An advantage of a TEM cell is small size, low cost, and lack of need for a high-power amplifier. This chamber can be located almost anywhere. In addition, no additional shielding is required to attenuate external radiated fields. One disadvantage is that a window is required if one is to view operation of the EUT, such as a video screen. The cell is optimal for small EUT dimensions (up to one-third the volume of the cell). A drawback is the ability to operate at higher frequencies. If a larger size chamber is used, the lower frequency limit can be extended.

The size of the EUT should be small relative to the test volume inside the TEM cell. If the EUT is not small, the RF field inside the cell is essentially shorted out in certain locations, causing a higher intensity of field strengths to occur elsewhere in the chamber. Tests may have to be repeated after unnecessary modifications to the EUT to increase the immunity levels that were unfortunately unrealistically high.

4.3.2 GTEM Cell

The GTEM (Gigahertz Transverse ElectroMagnetic) cell has certain advantages over a regular TEM cell. The restriction on upper frequency limit is eliminated by tapering the transmission line continuously outward from the feed point to a termination system. The tapered point and anechoic absorbers at the larger end of the chamber are what allow this chamber to operate well into the gigahertz range. Overall, the unit looks like a pyramid lying on its side (Figure 4.21).

One advantage of a GTEM cell is that is allows the full frequency range to be applied without the need for a shielded facility. In addition, a small output power capability amplifier is required, saving cost. As with typical TEM operations, the

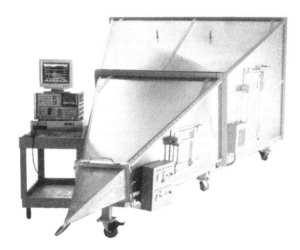

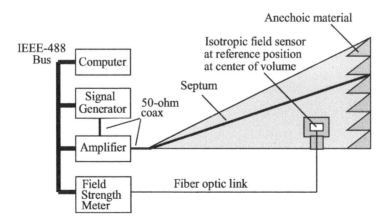

Figure 4.21 Photograph and diagram of a typical GTEM cell. (Photograph courtesy ETS-Lindgren.)

EUT must be tested in three orthogonal orientations (x, y, and z axis). This requires a special rotational system that the device under test is to be mounted on that will not affect the radiated fields internal to the chamber. Cable routing thus needs to be carefully considered when installing a system for testing.

REFERENCES

1. ANSII C63.4. *American National Standard for Methods of Measurement of Radio-Noise Emissions from Low-Voltage Electrical and Electronic Equipment in the Range of 9 kHz to 40 GHz.*

2. Williams, T. 1996. *EMC for Product Designers,* 2nd ed. Oxford: Newnes.

3. Williams, T., and K. Armstrong. 2001. EMC Testing Part 2—Conducted Emissions. *EMC Compliance Journal,* April. Available: www.compliance-club.com.

4. CISPR 16-1. *Specification for Radio Disturbance and Immunity Measuring Apparatus and Methods.*

5. Kodali, P. 2001. *Engineering Electromagnetic Compatibility,* 2nd ed. New York: IEEE.

6. Williams, T., and K. Armstrong. 2000. *EMC for Systems and Installations.* Oxford: Newnes.

7. Williams, T., and K. Armstrong. 2001. EMC Testing Part 4—Radiated Immunity. *EMC Compliance Journal,* August/October. Available: www.compliance-club.com.

CHAPTER 5

PROBES, ANTENNAS, AND SUPPORT EQUIPMENT

The purpose of this chapter is to highlight various tools used to measure RF currents or electromagnetic fields. It is important to recognize that each type of transmission is different depending on the application or functional use. Formal certification testing requires many of the probes and antennas described in this chapter; however, for purposes of troubleshooting, one may need to use multiple sensors to provide an indication of where RF currents are created and how they are propagated.

When using many of the devices detailed in this chapter, remember that (for troubleshooting only) calibration and accuracy of measurement are not required. If a product fails a particular test, it may be easy to locate the problem area with a particular probe or debug tool. Changes can them be made to the circuit. If the relative amplitude of the problem signal is significantly reduced in the near field, then probably a significant reduction in overall amplitude will be observed in the far field. This is the ultimate goal for engineers: Get the product to pass the test or elevate the quality level to a higher standard.

5.1 NEED FOR PROBES, ANTENNAS, AND SUPPORT EQUIPMENT

In order to determine if RF energy is present, instrumentation is required. A spectrum analyzer, receiver, oscilloscope, or any measurement tool has to interface with the outside world. The method of inputting voltage, current, or electromagnetic fields is through a device called a transducer. A transducer provides a means whereby energy flows from one or more transmission systems or media to another trans-

Testing for EMC Compliance. By Mark I. Montrose and Edward M. Nakauchi
ISBN 0-471-43308-X © 2004 Institute of Electrical and Electronics Engineers

mission system or medium. The energy transmitted may be of any form (electric, mechanical, or acoustical) and may be of the same or different form in various input and output systems or media. Antennas or test/diagnostic probes are types of transducers.

An ideal transducer is a hypothetical passive transducer that transfers maximum available power from the source to the load (instrumentation). In linear transducers with only one input and one output and for which impedance is a concern, an ideal transducer will

- dissipate no energy and
- when connected to a specified source and load, present a conjugate impedance.

There are different types of transducers, depending on what is to be recorded:

1. *Active Transducer.* A device whose output and detection are dependent upon the source of power, apart from that supplied by any of the actuating input. The power level is controlled by one or more of the input sources. Essentially, the output is one in which there is an impressed driving force.
2. *Passive Transducer.* A transducer that has no source of power other than the input signal(s) and whose output signal power cannot exceed that of the input. Most transducers used for EMC measurements are passive.

A concern in dealing with transducers has to do with gain and loss, where loss is negative gain. Called the *transfer function,* this can be defined as follows:

1. *Gain.* The ratio of power delivered to a specified load under specific operating conditions to the available power of the specified source.
2. *Loss.* The ratio of the available power of the specified source to the power that the transducer delivers to the specified load under specified operating conditions.

If the input and/or output power consists of more than one element, such as multi-frequency signals or noise, then the particular components used and their weighting must be specified. The gain achieved is usually expressed in decibels. Gain is only developed by powered devices.

(Support equipment involving measurement instrumentation was described in Chapter 3. The use of probes and antennas with measurement instrumentation will be provided in Chapter 7).

When performing measurements, one must keep in mind that reasonable results are desired. Sometimes the test environment is poor or the item to measure may be in a prototype stage. The cost of test instrumentation and transducers will not guarantee accuracy of any measurement results. All measurement results must be verified by actual testing on a sample of the product during the design-and-evaluation stage of product development or by performing a null test experiment.

5.2 VOLTAGE PROBES

A voltage probe is a transducer that measures an RF voltage level. Normally, for diagnostic purposes, the primary component to measure is common-mode voltage. This probe consists of a series resistor, an AC blocking capacitor, and an inductor to provide a low-impedance input to a receiver, and is used for direct connection to a conductor. It is unaffected by the current level in the transmission line.

One generally uses a voltage probe for in situ testing as an optional method of measuring mains interference voltage defined within CISPR 16-1. The advantage of using a voltage probe is that the full mains supply current is not impressed across the probe.

The basic configuration of the voltage probe is shown in Figure 5.1. A series resistor, a 50-Hz AC blocking capacitor, and an inductor provide a low-impedance connection at 50 Hz to the line requiring measurement. The resistor provides an insertion loss of approximately 25 dB, which must be corrected for by using the calibration table provided by the manufacturer. This loss must be added to the measured reading to maintain accuracy of the data recorded. The center conductor of the probe connects to a measuring instrument, while the outer connector is referenced to a local 0-V location (generally earth ground).

A disadvantage of using the voltage probe is stabilizing the RF impedance across a wide frequency range. This is because the probe is inserted across the mains con-

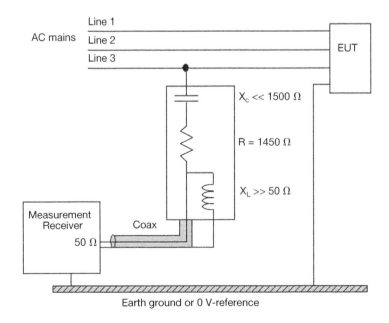

Figure 5.1 Voltage probe configuration.

nection instead of in series. The impedance of the probe must be high enough to not affect the measured energy or disrupt operation. Use should be restricted to in situ measurements where results are relevant to only the particular EUT or installation. Measured data may be different for the same EUT in different locations, although the systems are exact copies of each other. This difference is due to the impedance of the mains distribution network, the length of the routed cables, and other RF noise present within the facility's power distribution system.

Voltage probes with carefully selected *RC* values may be used to couple to data signal conductors. The *RC* values are set to mimic the appropriate impedance values. Also, voltage probes are sensitive to radiated interference signals due to their high input impedance and may exhibit false responses in high-impedance-fields environment.

5.3 CURRENT PROBES*

A current probe is a valuable tool (transducer) for measuring current levels within transmission lines. These probes contain a magnetic core material that detects the magnitude of magnetic flux present and delivers this field measurement to a receiver. It can be used as a substitution device to the voltage probe. Certain versions of current probes can be clamped around a conductor for ease of measurement. Cutting or removing a line for insertion of a sensor is not required. Examples of various types of probes are shown in Figure 5.2. These include, but are not limited to, clamp-on, surface, donut, simple pick-up, and flat cable.

Current probes can be used to measure common-mode currents from cable as-

Figure 5.2 Selected types of current probes available. (Photograph courtesy of Fischer Custom Communications, Inc.)

*Material in Sections 5.3 and 5.3.1, including Figures 5.3 to 5.6, has been adapted from Ref. [1]: *High Frequency Measurements and Noise in Electronic Circuits* by Doug Smith.

semblies and current on individual wires. Close-field loop probes measure differential and common modes at the same time. This dual-measurement feature depends on where they are placed in relation to the conductors in the cable.

Since most high-frequency circuits are relatively low in impedance, placing a resistor in the transmission line and then placing the probe around that resistor is one manner of taking data. Another form of measurement allows one to determine the magnitude of balance in circuits (common-mode rejection), net current within a wire harness or bundle, and phase measurement between two current sources.

A current probe is placed around a transmission line to measure various forms of conducted EMI. When placed around a cable, only common-mode EMI is observed as differential-mode currents cancel out. Depending on cable configuration, with respect to the probe, it becomes possible to ascertain which mode of current is present. The construction of the probe must ensure that it is isolated from external radiated fields. It is supposed to respond only to the magnetic flux (current) in the line.

Current probes generate an output voltage by one of two mechanisms. One is inductive coupling to generate a voltage and the other uses a Hall effect sensor. For probes with inductive coupling, consider a transformer whose primary winding contains the transmission line passing through the current probe and the secondary of the transformer is a pickup coil inside the assembly. The pickup coil senses the magnetic field linked to the primary. Typical designs use a ferrite core around the line to be measured. The pickup coil is wound around a ferrite core. This type of coil is shown in Figure 5.3.

A Hall effect sensor produces an output voltage in response to the magnetic field of the circuit being measured. A Hall effect sensor appears similar to an inductively coupled transformer probe; however, operation is quite different. A semiconductor circuit inside the probe is exposed to the magnetic field of the transmission line to be measured. An output voltage is generated in response to the field strength present. An amplifier within the probe increases the internal voltage level to a measurable value and buffers it. An advantage of this probe is the ability to measure down to DC levels. Most Hall effect sensors operate up to hundreds of kilohertz.

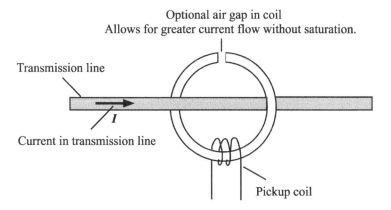

Figure 5.3 Typical configuration of an inductively coupled current probe.

Both types of probes maintain isolation of measurement equipment from potential high levels of damaging voltage and current levels. Isolation is a significant advantage of the clamp-on or Hall effect sensor.

Oscilloscope probes can create problems when measuring RF energy with false data recorded. The ground wire associated with the probe can result in substantial ground currents flowing in the circuit. These ground currents will affect measured data and provide incorrect information. When this happens, we refer to this as "ground loading." Current probes prevent this source of measurement error from being developed.

5.3.1 Specifying a Current Probe

Before specifying a current probe, we examine how these probes work [1]. Figure 5.4 illustrates a Thevenin equivalent circuit. The open-circuit voltage is $M(dI/dt)$ [$j\omega MI(t)$ when viewed in the frequency domain]. The Thevenin impedance is the self-inductance portion of the probe coil wound around a core of magnetic material. The DC resistance of the wire used for the winding and core loss of the magnetic material are assumed negligible for purposes of discussion herein. The output is connected by a coaxial cable to an oscilloscope or spectrum analyzer terminated in its characteristic impedance. For most probes, this impedance value is 50 Ω. The 50-Ω load from the output of the probe together with the probe self-inductance forms a single-pole L/R low-pass filter. When connection to instrumentation is made, it is generally assumed that the current probes do not significantly affect the current flowing in the transmission line or wire. Normally, this assumption is valid as long as the probe is terminated in 50 Ω or less.

If a frequency sweep of constant current is injected through the probe, the Thevenin circuit (Figure 5.4) produces two responses (Figures 5.5 and 5.6). The first response shows the voltage source increasing at 20 dB/dec as dI/dt increases linearly with frequency, and only occurs for a constant current source. At the frequency where the reactance of the probe coil inductance equals the load resistance (50 Ω), the L/R divider response will decrease at 20 dB/dec. The second response is transfer impedance, discussed below. To obtain the value of output voltage delivered to the load, the open-circuit voltage must be multiplied by the L/R divider. On a log scale that means adding the two curves together.

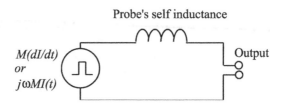

Figure 5.4   Equivalent circuit of a current probe.

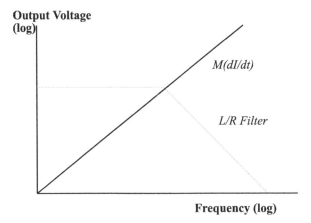

Figure 5.5 Frequency response to a constant current source.

Usually, it is desired to use the probe in the flat region of the frequency response labeled "desired frequency range of operation" (Figure 5.6). In this region, the probe exhibits flat transfer impedance: the ratio of output voltage to current flowing through the probe. Below this break point, the transfer impedance changes with frequency, making the probe inconvenient to use in this frequency range and difficult for a time-domain measurement with an oscilloscope.

For transfer impedance, the current probe output voltage is divided by the sensed current flow in the transmission line. This output is defined with units of volts per ampere or ohms ($V = IZ$). Current probes are usually specified to have a transfer im-

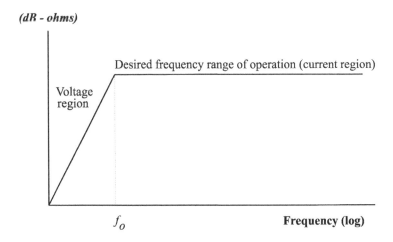

Figure 5.6 Typical transfer impedance for a current probe.

pedance value when their output is terminated in a load, usually 50 Ω. If the transfer impedance is provided in dBΩ and not ohms, then this specification is identified as $20 \log(|Z_t|)$ (in decibels) relative to 1 Ω. Transfer admittance is the reciprocal of transfer impedance. Note that the transfer impedance of the probe is actually additive to the conductive impedance of what is being measured, which increases the circuit impedance and reduces the current flowing in the circuit being measured.

The output voltage of a current probe is related to the current present and the transfer impedance Z_t. A probe having 1 Ω transfer impedance will have a 1-V output when passing 1 A of current, again illustrated in Figure 5.6.

Multiplying the L/R voltage divider by $j\omega MI(t)$ to find the output voltage delivered to the load, rearranging, and combining terms, the equation for transfer impedance is

$$Z_t = \left(\frac{M}{L}\right) R_L \qquad (5.1)$$

where Z_t = transfer impedance
M = mutual inductance between probe coil and current-carrying wire
L = inductance of coil wound on the magnetic core of the probe
R_L = load impedance of probe, usually 50 Ω

The corner frequency in Figures 5.5 and 5.6 is now calculated from

$$|X_L| = |2\pi f L| = R_L \qquad (5.2)$$

where: X_L = reactance of the probe coil
f = frequency
L = inductance of the probe coil
R_L = load impedance of probe, usually 50 Ω

Combining Eqs. (5.1) and (5.2) and solving for M result in

$$M = \frac{Z_t}{2\pi f_c} \qquad (5.3)$$

From Eq. (5.3), one can calculate M since Z_t and f_c are known from the probe calibration data sheet provided by the vendor.

5.3.1.1 Current Probe Applications.
Numerous applications exist for current probes. These include, as an overview:

1. Measuring very small current on the order of microamperes
2. Measuring common-mode currents on cables to predict radiated emissions for regulatory compliance
3. Measuring balance between wire pairs to ensure optimal signal integrity

Common-mode current measurement is the most frequent use for current probes. The radiated field from cable assemblies and harnesses allows electric field measurement to exist. In addition, near-field magnetic fields can be characterized. The magnitude of radiated field strength is proportional to frequency, length, and current within the transmission line. Equation (5.4) details the magnitude of the radiated common-mode electric field for a transmission line in free space (Chapter 2 provides additional detail on this topic):

$$E \propto (fI_{CM}L) \qquad (5.4)$$

where E = electric field strength
f = frequency (MHz)
I_{CM} = common-mode current (A)
L = antenna length (m)

To illustrate how a small amount of common-mode current in a transmission line can cause a product to fail FCC Class A limits for Information Technology Equipment at 75 MHz on a 1-m-long cable, 15 μA of common-mode current is required. The specification limit is 40 dBμV at 10 meters. Current probes are very efficient at low-amperage measurements and are an effective precompliance tool.

With a constant current source and fixed cable length, the electric field at a prescribed distance is proportional to frequency. Unlike differential-mode radiation that is easy to minimize using proper design techniques, common-mode radiation is a more difficult problem to find and solve. The only variable available to the designer, if it can be determined, is the common path impedance for the return current. In order to eliminate or reduce common-mode radiation, common-mode currents in the conductor must approach zero. This is achieved using a sensible grounding scheme.

Another use for current probes is for isolated voltage measurements. This technique allows connection between two nodes of a circuit for which voltage is to be measured, such as a circuit return. The probe is placed over a known impedance source and the value is recorded. Once the measured current through the probe's transfer impedance is known, the voltage level can be calculated. For optimal operation of the probe, insert series impedance, usually a precision resistor with a value that is large compared to the transfer impedance of the current probe between the two nodes. The impedance of the resistor should be low enough to allow current to flow between the nodes.

For relative phase measurement, two matched current probes are required. Each probe is placed on a separate transmission line. The voltage levels in the two probes are compared against each other. Any imbalance is easily noticed and corrective action can be taken using a dual-channel oscilloscope or spectrum analyzer with memory to store data from one probe to compare with the second probe (Figure 5.7). For EMI debugging purposes, this is a valuable technique and easy to implement. The data measured can be determined to be common or differential mode based on the direction in which wires are inserted into the probes.

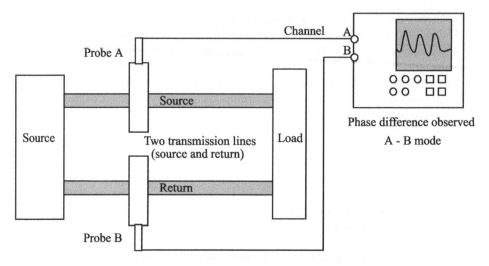

Figure 5.7 Measuring current imbalance using two matched probes.

To measure standing waves on a single transmission line, the simple technique illustrated in Figure 5.8 exists. If there are any standing waves, reflections may be present due to a poorly matched transmission line. These standing waves can propagate significant radiated electric fields if the cable is not shielded.

5.3.2 Limitations When Using Current Probes

Primary areas of concern include the following:

1. *Capacitive Coupling to Circuit Being Evaluated.* Capacitive coupling plays a significant role in assuring accuracy of measurement. Most probes have ca-

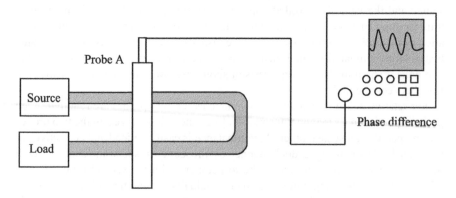

Figure 5.8 Measuring standing waves on a single cable.

pacitance of a few picofarads. Any capacitance within a circuit can cause potential problems at higher frequencies. For probes that have 1 pF capacitance or less, signal integrity concerns are minimized. Most probes have a bandwidth far lower than 1 GHz. The reactance of 1 pF at 1 GHz is 160 Ω ($X_c =$ $1/2\pi fC$), which is on the same order of many circuits and cable assemblies. A substantial fraction of current measured with the probe may be increased through this parasitic capacitance internal to the probe. To prevent loading down a transmission line and for accurate measurement, the probe should be placed where a low voltage level with respect to ground is present. By Ampere's law, the sum of the current must be equal everywhere in the transmission line. Placing the current probe on the return terminal and not the source line may provide a more accurate measured value of RF energy, or magnetic flux; however, measurement on the source line must also be taken to ascertain the magnitude of common-mode current present.

2. *Inserting Series Impedance into Circuit Being Evaluated.* With any probe, series impedance will always be present. Most vendors provide this information in their data sheets and identify if built-in termination is provided, such as a 50-Ω resistor. Some probes have series impedance that is too small to be of concern. The transfer impedance of the probe is usually the series impedance added to the conductor being measured. Inductance within any transmission line causes voltage potentials as well as total flux to vary; thus inaccurate measurement can occur.

3. *Effects of Transfer Impedance on Equipment Operation.* Transfer impedance allows one to know how much coupling is present between the transmission line and the probe. A significant amount of error can occur if the transfer impedance plot from the vendor of the probe is not taken into consideration during measurement. All probes have a loss curve across their frequency range of operation. The value of the transfer impedance must be added to the measured value to provide accuracy. The problem with achieving extreme accuracy lies with how accurate the probe has been characterized.

4. *Magnetic Core Saturation.* Magnetic core saturation plays an important part in large energy circuits, especially AC mains where LCI measurements are taken. Circuits that are capable of causing saturation may be caused by either high- or low-frequency currents that are outside the range of desired operation. High energy levels may be generated by pulsed electromagnetic fields in addition to AC mains. The probe must be able to handle the power level present as saturation can affect measurement accuracy. This is especially true with smaller sized probes. Most clamp-on versions are able to handle 50–100 A without saturation. When measuring two wires at the same time, in differential-mode, common-mode current present will not typically saturate the probe since differential flux in the wire pair essentially cancels out the magnetic field present.

5. *Stray Response.* The stray response of current probes is another concern. Current probes may be sensitive to stray electric fields as they are designed to

measure magnetic fields. Stray electric fields will cause measurement errors. One should hope for a spurious response from undesired electric fields to be at least 60 dB or better below the current response at a particular frequency. For 50/60-Hz measurements, this may not be possible. Under this condition, a voltage probe is a better choice.

5.4 LISN/AMN (AC MAINS)

A Line Impedance Stabilization Network (LISN), also identified as an Artificial Mains Network (AMN), is a device inserted into the AC supply mains of an apparatus to be tested. For a frequency range of interest, the network maintains a specific load impedance defined by international standards. This impedance allows measurement of disturbance noise voltages present within the AC mains distribution system. This type of measurement is known as Line Conducted Interference (LCI). Various LISNs are shown in Figure 5.9.

The AC mains generally provide power to multiple devices plugged into the wall outlet (or hard wired). Excessive RF noise from one device may cause harmful interference to other systems through the AC mains plug. This undesired noise is from switching energy developed within the unit. In addition, undesired RF energy from periodic switching circuits such as clocks or high-bandwidth components can inductive or capacitive couple into the power supply system through radiated or conductive means. Once undesired energy gets into a power supply, along with its own inherent switching noise, an excessive amount of harmful interference may be observed on the mains cable leaving the product.

Measurement of AC mains voltage disturbance is generally in the frequency range of 150 kHz–30 MHz for most product specifications; however, switching energy can be of sufficient magnitude down to 9 kHz. Between commercial and military applications, different specification limits exists along with the frequency range of interest to be measured.

A LISN/AMN offers a direct means toward measuring conducted disturbance from devices on the same branch circuit. The RF noise on the interconnect cable can be conducted, radiated, or both.

Certain test specifications require measurement on both AC mains and data/signal lines. Figures 5.10 and 5.11 illustrate various standard specifications for most commercial products based on international regulatory requirements for AC mains disturbance. Class A equipment is targeted for commercial installations (United States, Canada) or heavy industrial environments (Europe, worldwide), while Class B is marketed toward residential (United States, Canada) or light industrial (Europe, worldwide) environments.

The limits in Figures 5.10 and 5.11 assume the RF impedance at the EUT is well defined. In actual applications, the impedance of the mains cables can vary over a very large range. This large variance is frequency dependent in addition to the gauge of the wire and length of installation. What is being accomplished is

PMM
L2-16

Rohde & Schwarz
ESH3-Z5

Thurlby-Thandar
LISN 1600

Schaffner-Chase EMC
MN2050C

EMCO
482/5/2

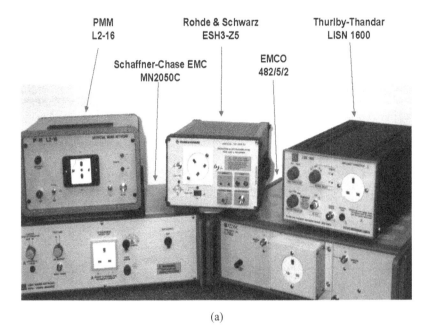

(a)

(b)

Figure 5.9 Various configurations of LISNs. Photographs courtesy of (*a*) *EMC Compliance Journal* [2] and (*b*) Fischer Custom Communications, Inc.

measuring the voltage level of RF energy. The source impedance of the EUT is usually unknown and may in fact be extremely high. A high impedance within the AC mains cables allows for improper measured voltages within the LISN/AMN. In order to make tests repeatable between laboratories, the LISN/AMN is to be normalized to international standards and must be stable across a wide range of frequencies. Above 30 MHz the impedance is not defined as most regulatory compliance limits do not exceed this frequency limit. Automotive and aerospace standards usually exceed 30 MHz. For these tests, a special measurement transducer is required as component parasitic reactance makes it difficult to achieve accurate results.

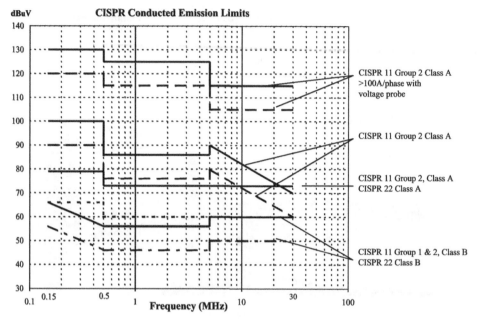

Figure 5.10 CISPR limits for mains interference voltage (conducted emissions). (Reference: CISPR 11+A1:1999 and CISPR 22:1993/A1:1995.) *Note:* Test limits subject to change. Refer to the latest issue of the standard for current values.

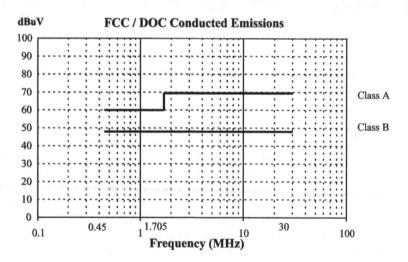

Figure 5.11 FCC limits for mains interference voltage (conducted emissions). (Reference: CFR 47, Part 15, Subpart B.) *Note:* Test limits subject to change. Refer to the latest issue of the standard for current values.

CISPR 16 provides guidance on the design and specification of a LISN/AMN. The value of impedance is the primary item of importance. The output for the measuring instrumentation input impedance is specified as a 50-μH inductor and a 5-Ω resistor. The remaining components are provided to decouple the incoming supply voltage. A high-pass filter is commonly added between the LISN/AMN output and receiver, cutting off below 9 kHz. This high-pass filter prevents the receiver from being affected by high levels of harmonics present within the mains supply. This high-pass filter must be maintained at 50 Ω impedance and have a defined insertion loss, preferably 0 dB at all measured frequencies.

The specification on the design of a LISN/AMN used for commercial EMC tests is shown in Figure 5.12, while the international standard for impedance versus frequency is in Figure 5.13, as defined by CISPR 16-1.

When performing diagnostics with a typical LISN/AMN, it is impossible to determine if the noise present is common or differential mode. The device measures total amplitude of interference voltage. Comparative currents measured with a current probe on each line individually, and then in combination, can provide indications of the mode structure. Certain vendors of LISNs/AMNs provide an option that allows one to configure the unit to detect either the sum or difference of the line and neutral voltages, which corresponds to common- and differential-mode voltages, respectively. Use of this special configuration is not required for compliance measurement. It does, however, make diagnostic analysis easier to achieve.

Note: *LISNs* may elevate the differential-mode source impedance above what is typically found as utility source values. When this situation is noted, increased EMI will be measured as a proportion of the artificial line impedance.

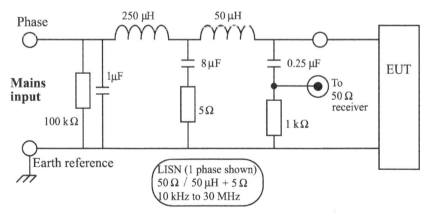

Note: The 5 ohm resistor is required for the range of 9 kHz to 150 kHz.
The network can be constructed such that it meets the requirements
of 150 kHz to 30 MHz.

Figure 5.12   Configuration of a typical LISN/AMN, per phase.

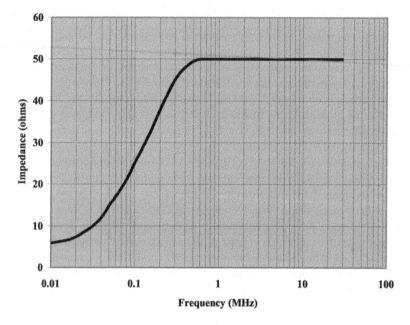

Figure 5.13   Impedance chart for a typical LISN/AMN per Figure 8 of CISPR 16-1.

5.5   CDNs (DATA AND SIGNAL LINES)

Certain regulatory standards require RF measurement on interface cables identical to AC mains disturbance. In other words, conducted emissions must be ascertained for all I/O. The propagated fields from these cables may be of sufficient magnitude to cause harm to other electrical products. If the cable is shielded, the possibility of harmful energy being present is minimized. In many applications, if the cable is shielded, this test is not required. The types of signals are all encompassing, including typical slow speed (RS-232) as well as telecommunication, including Ethernet and high-speed networking.

Use of CDNs (Coupling and Decoupling Networks) are not only used for compliance purposes but can be used for diagnosing potential problems in the laboratory long before sending the product out for formal testing. If a system has numerous interconnect cables (serial, parallel, video, Ethernet, telecommunication, etc.), determining which particular cable is causing problems may be difficult, if not impossible, after the product has gone into production. Testing all cables in a laboratory environment is extremely cost effective.

A sample CDN with unique interface adapter plugs for networking and signal/data cables is shown in Figure 5.14. Most vendor units are similar in appearance and functionality. The reason why there are many variations lies in the type of communication protocols and cabling in common use.

Figure 5.14 Various configuration of a CDN with adapter plugs for application-specific needs. (Photograph courtesy of Fischer Custom Communications, Inc.)

Coupling and Decoupling Networks work by measuring the common-mode current disturbance that may be present on a cable. In order to measure common-mode current, a CDN with a particular configuration must be used. In many cases, the I/O cable must be cut open and the CDN inserted in series. This is a destructive test for the cable.

For emissions testing, a coupling network is used. The coupling network operates identical to the LISN/AMN with different values for components.

For immunity tests (IEC 61000-4-6/EN 61000-4-6), decoupling networks are used to ensure that the disturbance signal does not significantly influence the auxiliary equipment (AE). This network is placed between the EUT and AE. Most decoupling networks are inductive and use a high-impedance choke. The second decoupling network combines resistive and inductive technique by using ferrite torroids around the cables connecting the EUT and AE. Both coupling and decoupling networks can be separate units or combined in the same instrument.

A current probe is sometimes placed adjacent to the CDN in order to measure or validate the magnitude of RF energy being injected. Injection of the signal must be performed on unscreened cables, shielded cables, balanced cables, coaxial cables, and power mains. A CDN acts as a low-pass filter, preventing the susceptibility test signals from interfering with the auxiliary equipment.

Most CDNs contain a high-pass filter to ensure that only the desired test RF current is injected into the EUT. Some current, small as it may be, may enter the AE when immunity testing is being performed. Coupling and decoupling networks use less power from a power amplifier than bulk current injection (BCI). The disadvantage of having to test products using CDNs is that a number of different units must be used, according to the number of different interfaces that must be tested. For cables with more than two or four wire pairs, a CDN test is impractical. Another method to perform this immunity test is to use an absorption clamp, which is much easier to use, detailed in the next section. Discussion of BCI probes follows the section on absorbing clamps.

5.6 ABSORBING CLAMP

The absorbing clamp is used for the measurement of radiated RF power on cable assemblies in the frequency range of 30–1000 MHz. For certain applications, current probes are preferred due to ease of use as the absorbing clamp can be quite cumbersome. The basic unit of measurement is dBpW, which can easily be converted to acceptable field strength levels. One advantage of using this probe is that a large open-area test site is not required. It is however recommended that tests be conducted inside a chamber. Another name for the absorbing clamp is the ferrite clamp. Typical absorbing clamps are shown in Figure 5.15.

International standards that make use of this clamp include EN 55013, EN 55014, and EN 55020. The specification limits for these test standards are provided in Figure 5.16.

Signal or interference energy beyond 30 MHz is generally transmitted by way of radiation rather than conduction. However, when the mains cable is well shielded to

Figure 5.15 Typical absorption clamps. (Photograph courtesy of *EMC Compliance Journal* [2].)

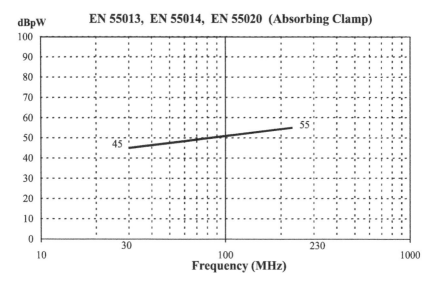

Figure 5.16   Absorbing clamp limits for select international specifications. *Note:* Test limits are subject to change. Refer to the latest issue of the standard for current values.

prevent radiated EMI, there is a possibility that the internal conductors will still contain interference voltage. The maximum intensity of conducted emissions exists near the connecting point of the power cable to the EUT. As a result, the clamp is inserted at this location to obtain maximum indication in the measuring receiver or spectrum analyzer.

An absorbing clamp contains current transformers using ferrite split rings. These rings are split to allow for cable insertion. A coupling loop current transformer is provided that connects to the spectrum analyzer. Outside of this current loop are additional ferrite split rings forming a power absorber and impedance stabilizer that also clamps around the cable. The device is calibrated in terms of output power versus input power. The ferrite absorbers attenuate reflections and extraneous signals that would otherwise appear at the current transformer. From the current transformer to the measuring instrument, lead wires are sheathed with ferrite rings to attenuate screen currents on the cable. Details on construction are shown in Figure 5.17.

Since the output of the clamp is proportional to the current that flows in common mode within the cable, direct measurement of noise power is possible. The clamp can be calibrated as a two-port network in terms of output power versus input power.

The specification, construction, and calibration of the clamp are described in CISPR-16. Advantages of this clamp are for use in certain compliance and certification tests as well as being used as a diagnostic tool. Repeatable measurements on a cable allows for determining the effects of a change within a source circuit, including improvement within cable shielding. For many situations, use of the clamp for

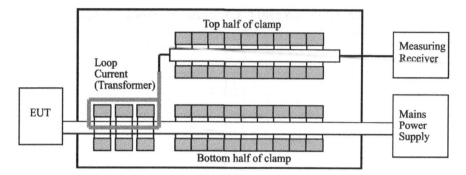

Figure 5.17 Absorbing clamp internal configuration.

radiated emissions may be preferred, especially when the EUT is small and the majority of the emissions from interconnect cables are substantial. The clamp is commonly used in pretests, especially when determining specific frequencies to be examined during formal testing.

5.6.1 Test Setup and Measurement Procedure

The EUT is placed on a nonmetallic table at least 70 cm from any other metallic object. The mains lead is placed horizontally in a straight line for a distance sufficient to permit movement of the clamp. The current transformer and the absorbing rings are placed around the mains lead. The output current transformer is connected to the measuring set by means of a 50-Ω coaxial cable (Figure 5.18). The ferrite clamp is moved along the mains cable to obtain the highest reading at each test frequency. The absorbing clamp is equipped with rollers so that movement along the mains

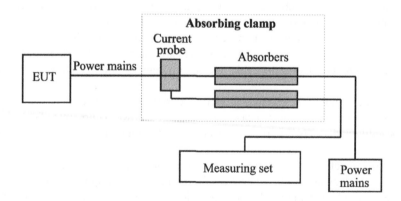

Figure 5.18 Test setup for use with an absorbing clamp.

line to the highest interference location is easily accomplished. The measured interference power is derived by reference to the calibration curve. The calibration curve is the insertion loss of the absorbing clamp. Each clamp is calibrated for insertion loss and correction factor.

Measurement to the CISPR/EN standards requires a measuring receiver having an input impedance of 50 Ω. The power measured using a 50-Ω load is equal to the square of the voltage measured divided by 50. Expressing the above in decibels gives

$$10 \log(P) = 10 \log\left(\frac{V^2}{50}\right) = 20 \log(V) - 10 \log(50)$$

$$= 20 \log(V) - 17 \text{ dB}$$

Taking into account the insertion loss of the clamp, power is computed by adding it to the voltage reading in decibels:

$$10 \log(P) = 20 \log(V) - 17 \text{ dB} + \text{insertion loss of clamp in dB}$$

If power P is expressed in picowatts, equivalent voltage V is in microvolts. The numerical value of power P is expressed in decibels (picowatts across 50 Ω) by subtracting 17 dB from the numerical value of V in decibels (microvolts across 50 Ω) plus the decibel value of the absorbing clamp that is supplied by the manufacturer.

The clamp can be positioned at more than one location. The location nearest the EUT gives the maximum reading on the receiver. The measurement procedure for using the clamp follows:

1. Initiate operation of the EUT.
2. Set the spectrum analyzer or measuring receiver to the desired frequency range of measurement.
3. Move the absorbing clamp until the largest interference level is measured.
4. Measure the interference power level. The interference power level is measured using a receiver having an impedance of 50 Ω, with the receiver set for dBμV.

The interference power level is given as:

$$\text{Interference} = \text{indication} + \text{clamp correction curve}$$

$$\text{dBpW} = \text{dB}\mu\text{V} + \text{dB}$$

An alternate method is to use the insertion loss of the clamp:

$$\text{dBpW} = \text{dB}\mu\text{V} - 17 \text{ dB} + \text{insertion loss of clamp in dB}$$

5.7 BULK CURRENT INJECTION—PROBE AND INSERTION CLAMP

Bulk current injection methods are used to evaluate the electromagnetic susceptibility (immunity) of a wide range of electronic devices, including automotive, avionics, computing, medical, and telecommunications. The BCI method is intended to simulate continuous-wave (CW) currents developed in electrical conductors of equipment during normal operation. These conductors include signal, control, and power circuits.

Bulk current injection uses RF transformers to inductively couple large amounts of RF currents into conductors linking parts of electronic systems. The injection probe acts as a multiple or single primary winding of a transformer and the transmission line or circuit under test acts as the secondary winding. Injection probes are used over a specific range of frequencies depending on design. Each probe is provided with an insertion loss curve plotted from the vendor.

There are several benefits for performing compliance testing using BCI. The primary benefit is that BCI test results sometimes correlate well with radiated susceptibility test results, relative to cable influences. Under certain situations, correlation is not possible. The design engineer should evaluate the effects of injected currents on the system or subsystem under development and the relative immunity level of different designs at the prototype stage of equipment development, saving significant redesign time and cost. Conducted immunity testing can function as an integral part of a production quality assurance program. Quality assurance engineers can use BCI to perform conducted immunity tests on 100% of all critical circuits, ensuring a high level of system compliance. Each injection probe must contain a calibrated list of attenuation loss over the operating frequency range provided by the manufacturer.

There is one advantage to testing multiple cables and wires simultaneously instead of individually. In real-life applications and installations, all cables will be exposed to the same field, not just one if the cable layout is identical between all assemblies.

The physical difference between a BCI probe and injection clamp is easy to see when compared against each other. Both require extensive calibration procedures prior to testing. Calibration refers to sending RF power from an RF amplifier over a range of frequencies to be tested to the BCI probe. The power level is measured by a dummy load (or calibration jig, Figure 5.19) using a power meter. Once the magnitude of power required for each frequency has been determined and recorded for future use, the calibration jig is removed and the probe/clamp is installed in the test setup. When a BCI probe is used, a current clamp monitors the amount of RF energy being injected and ensures that the drive level of the amplifier and probe do not exceed the specification limit. Since transmission lines have various impedance values, an excessive amount of RF energy can be injected, thus stressing the unit to levels beyond that required or desired.

For a bulk current clamp injection (Figure 5.20), the attenuation level of the injection probe is more stable throughout the frequency range of operation. When using the clamp, the cable sees simultaneously both electric and magnetic field coupling with a directional effect: The resulting power is sent to the EUT side with practically no power being sent toward the AE. This is an attractive method of mea-

Figure 5.19   Typical BCI probe with calibration fixture. (Photograph courtesy of Fischer Custom Communications, Inc.)

surement with high efficiency. This means that much less power from the RF amplifier is required, saving cost in instrumentation. The disadvantage of using this clamp is its size. It is difficult to use in the laboratory or test site.

Typical setup configurations are shown in Figures 5.21 and 5.22 for current injection testing.

5.7.1   Choosing a BCI Probe

When choosing a BCI probe, the following are of primary concern:

1. Physical dimensions
2. Bandwidth
3. Efficiency
4. Continuous-wave power rating

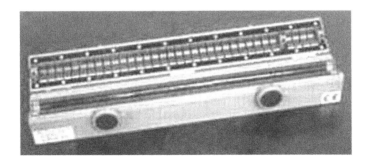

Figure 5.20   Typical BCI clamp. (Photograph courtesy of Fischer Custom Communications, Inc.)

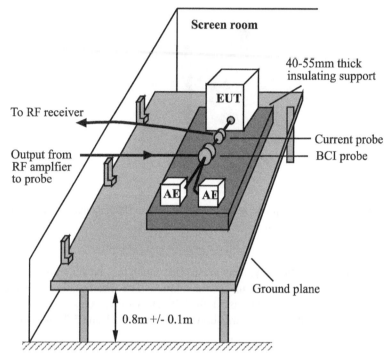

General requirements: ground plane and bench in a screen room

Figure 5.21 Test setup for BCI (probe).

The most important mechanical parameter is the size of the *internal* diameter or injecting window. The internal diameter must be able to accommodate the size of the circuit under test, including cable bundles as well as individual cables. Injection probes with an internal diameter of 40 mm can be used with signal and power cables for most commercial electronic equipment and systems. Injection probes with internal diameters of 66 mm are ideally suited for military and avionics equipment having large signal and power cable assemblies. Custom designs having larger than 66 mm internal diameter are available.

The *external* dimensions of the injection probe become important if the probe must be placed inside the EUT, such as a motor vehicle or aircraft.

The usable operational bandwidth of the injection probe must overlap the frequency range under evaluation. When conducting tests in accord with RTCA/DO-160 Section 20 (aircraft application), the usable bandwidth of the injection probe must be 10 kHz–400 MHz. Two injection probes are required to inject the specified current over the entire range due to the large bandwidth required. Most probes operate satisfactorily for EMC purposes within a fairly large frequency range.

The effectiveness of the injection probe over the operational bandwidth is determined by the probe's efficiency. The efficiency of an injection probe is a function

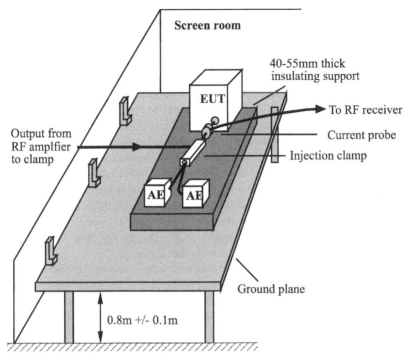

General requirements: ground plane and bench in a screen room

Figure 5.22 Test setup for BCI (clamp).

of the attenuation or insertion loss. The insertion loss is the ratio, in decibels, of the amplifier's forward power to the actual power delivered by the probe to a calibration fixture. To inject large CW currents with small amounts of amplifier power, the injection probe must have a low self-attenuation. If high attenuation is associated with the probe, a large amount of power from a more expensive amplifier is required.

The CW power rating of injection probes influences the effectiveness of the probe's operation. Each probe has limitations on power-handling capability. The probe's rating should be higher than the power required to inject the specified current. Most bulk injection probes can operate at a rated CW power level for a minimum of 30 min before damage occurs.

5.8 BASIC PROBE TYPES—NEAR FIELD AND CLOSED FIELD

Near-field or closed-field probes are used for detection of electric and/or magnetic fields. These probes are easy to build and are detailed in Appendix A. Near-field probes are commonly used during the development portion of a product design as well as diagnostic work and quality assurance verification. Most manufacturers of EMC test equipment sell variations of these probes, all at a significant cost. Homemade probes can be built for basically no cost.

There are three basic types of probes used for diagnosing EMI, as follows [3]:

1. *H-Field Near-Field Probe*. The *H*-field probe provides a voltage level to a measurement system (analyzer) proportional to the magnetic field strength observed at the probe location. With this probe, circuit RF sources may be localized in close proximity to each other. This effect is caused by interference sources, which in digital circuits are of low resistance (a relatively small change in voltage causes a large change in current). The source of radiated interference begins primarily as a magnetic field propagated from a current loop. At a specific distance away, based on wavelength of the signal, a transition from magnetic field (near field) to electric field (far field) occurs. The transition point is where the impedance of the transmission line approaches that of free space, or 377 Ω. The magnetic field intensity will decrease as the cube of the distance from the source.

When using *H*-field probes, one observes a rapid increase in the probe's output voltage as the probe is moved closer to the source of the radiating emissions. This allows one to locate a potential problem area, be it a component or a cable assembly. Using this type of probe helps isolate magnetic fields but will not provide information on how the undesired field was generated. In addition, this probe is useful for investigating leaks in the area of cables and wires for conducted interference.

2. *E-Field Far-Field Probe*. The *E*-field monoprobe is a very sensitive sniffing transducer. This sensitivity allows reception of almost any type of radiated field and is often used for radio and television reception. With this probe, the entire frequency spectrum from a circuit, cable, or system assembly can be examined.

The *E*-field probe is useful for determining the effectiveness of shielding materials and the effects of filters installed on cables that exit or enter any system. In addition, the *E*-field probe may be used to perform relative measurements for certification tests. This makes it possible to apply remedial suppression measures so that rework performed can be examined as to effectiveness. Using this probe, precompliance verification may be done in a quick and simple manner in the engineer's office or development laboratory. Use of sophisticated equipment is not required.

Another advantage of using this probe is the ability to make this sniffer extremely small in size, physically. With a very small tip, it is possible to determine the exact source of the emissions down to the pin of a digital component. Although at this distance one is in the near field, the fact that emissions are observed on a far-field probe is enough evidence to investigate this source of energy in greater depth.

3. *High-Impedance Probe*. A high-impedance probe is generally used with oscilloscopes to prevent loading of circuit traces. This probe, when used in the time domain, may allow for determination of RF interference on individual pins of components or PCB traces. This type of probe has a metal contact tip. If this probe is connected to a spectrum analyzer, damage to the analyzer will occur if a DC block is not provided. If the probe tip is isolated from contact with DC voltage (measured on AC signals), direct connection to a spectrum analyzer is possible.

A high-impedance probe contains a value near the insulation resistance of the PCB material and loads the circuit at the test point with usually 2 pF (80 Ω at 1 GHz). This high-impedance value permits measurements of a circuit directly without significantly influencing operation of the system (degrade signal integrity). One is able to measure the effectiveness of filters or other suppression techniques for PCB traces. Pins of components can be identified as sources of EMI, in addition to improving the signal integrity of an offending source.

Near-field probes require use of 50-Ω coax cable. This is because the input impedance of most RF-measuring instruments is 50 Ω. Should instrumentation not have 50 Ω input (i.e., 75 Ω or a very high impedance), a separate 50-Ω terminator built into the coaxial adapter is required at the input port of the test instrument. This terminator prevents cable resonance within the probe cable. Resonances are generally observed when the length of the cable exceeds one-tenth of the wavelength of the highest frequency of interest. For example, 50 Ω termination is required for cables longer than 300 mm if the highest frequency of 50 MHz is present. Keep in mind that the wavelength for a signal in a cable is roughly half that of an electromagnetic wave in free space.

The electric field probe is a valuable tool to inject fast transient pulses directly into conductors and component leads (Figure 5.23) [3]. In addition, this probe works extremely well as a pin probe to measure electric field radiation on cables and within circuits. In order to perform an immunity test with this probe, use of a high-voltage capacitor rated at around 10 pF will be required for DC isolation. This

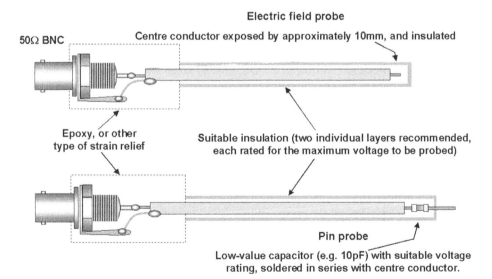

Figure 5.23 Various construction details of an electric field probe. (Artwork courtesy of *EMC Compliance Journal* [3].)

field probe must be carefully used so as to not short out electrical circuits. It is recommended to begin troubleshooting with a low voltage level and increase the voltage until the problem area is detected. Continue this procedure until all problem areas are located and solutions or fixes implemented.

Magnetic field probes are usually shielded. This shield prevents the pickup of stray electric fields. Stray electric fields can lead one to investigate an incorrect area of a product that is generating unacceptable levels of RF energy or prevent one from detecting the actual source. Generally, common-mode currents are radiated from cable assemblies that contain a defective shield. These defective shields cause the majority of radiated emissions. Sometimes, the electric field is generated by common-mode voltage present within the system's chassis. The magnetic field pin probe may be one of the easiest probes to use and is a valuable diagnostic tool in the engineering laboratory.

All loop probes are directional. When searching for sources of emissions, the probe should be positioned in two or three orientations to locate the strongest signal. Since radiated emissions are caused by both electric and magnetic fields, an unshielded loop probe may be desired. Unshielded loop probes are sensitive to both types of field: magnetic and electric. The resulting voltage from the probe is not accurate; however, their use can speed up the time it takes to locate a particular area from which an unwanted emission is coming.

The loop probe shown in Figure 5.24 works up to 1 GHz [3]. A smaller diameter loop (e.g., 10 mm) will have an improved high-frequency response with less sensitivity at lower frequencies. Probes with more than one turn are more sensitive at lower frequencies. There is inter-turn capacitance between the windings. This capacitance considerably reduces high-frequency response. Most designers use a variety of probe sizes for both low- and high-frequency investigations.

The magnetic field probes shown in Figure 5.25 are not well insulated against the hazard of electric shock. If these wires accidentally touch electrical circuits or components, permanent damage can occur rendering the system worthless. In some situations electrocution will happen. Care must be taken when using noninsulated probes. Use of heat-shrink tubing or a thick resin applied to the assembly is recommended. Shrink tubing will not affect the performance of the probe.

5.9 SNIFFER PROBES

Sniffer probes can be any small transducer used to isolate or locate radiating RF energy. Calibration is not a concern as their primary purpose is to help identify and locate sources of undesired RF energy. While there are commercial probes available, many probes can be home made for minimal cost. The key element when using a probe is making sure that the right tool is used for the task. Appendix A provides details on building home-made probes.

Probes are useful for measuring either electric or magnetic fields. Few probes measure both fields. Although a probe is specified as an E-field sensor, magnetic field components may be unintentionally measured, providing inaccurate data. Care

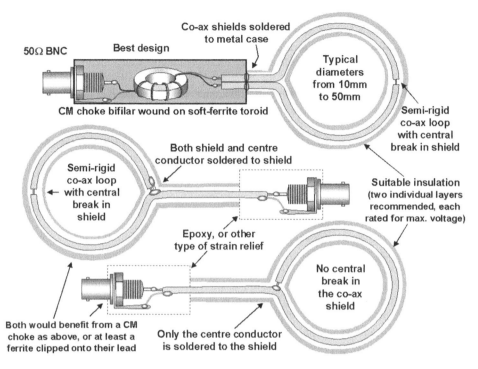

Figure 5.24 Construction details of a magnetic field probe. (Artwork courtesy of *EMC Compliance Journal* [3].)

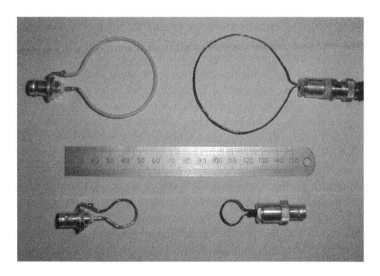

Figure 5.25 Various configurations for magnetic field probes. (Photograph courtesy of *EMC Compliance Journal* [3].)

must be taken to measure only the particular field desired: electric or magnetic. Probes are generally useful only in the near field. Once the distance spacing between source and probe approach approximately λ/6 at the frequency on interest, the field component becomes basically an electromagnetic field, identified as a Poynting vector. The magnitude of the electric (E) field to the magnetic (H) field becomes larger than 120π. At this point, it becomes nearly impossible to ascertain if the source emissions are common mode, differential mode, electric field, magnetic flux, and the like.

One should recognize that probes are useful only as a diagnostic tool. Commercial E- and H-field probes marketed for use as compliance sensors (with calibrated data) provides conditional results, as measurement uncertainty is high; however, their use as a diagnostic tool is valuable. On the other hand, commercial current probes are reasonably accurate and can be used for compliance testing when use of the calibration table is taken into consideration.

5.9.1   Near-Field Probes

When an EMI event is noted, examination must be made to ascertain the source of the radiating field. Near-field (sniffer) probes are therefore commonly used to locate problem areas. Two types of probes are usually required, one for the electric field (rod construction) and one for the magnetic field (loop construction). Probes are connected to a spectrum analyzer (frequency-domain measurement) or oscilloscope (time-domain measurement).

When using a near-field probe, a trade-off must be made between sensitivity and spatial accuracy. The smaller the probe, the more specific it becomes in locating a particular source of RF fields; however, it will not be very sensitive. Also, when using a differential-mode radiating loop with a rod-type antenna, false information may be observed. To increase sensitivity, an amplifier can be used, especially with low-power circuits. A well-designed magnetic field probe should be insensitive to electric fields if manufactured with care. Conversely, an electric field probe will detect nodes of high dV/dt but not the specific local voltage point.

Near-field probes are calibrated in terms of output voltage versus field strength. Calibration is generally found on probes that are purchased, not home made. Homemade probes generally do not need to be calibrated since their purpose is for diagnosing problem areas. It is nearly impossible to extrapolate data from a probe and relate it to far-field emission levels, such as an open-field or anechoic chamber). This is due to the fact that the probe is measuring a particular field, either magnetic or electric, which may be completely different than that observed in the far field. The probe will distort the field it is measuring, especially if located immediately adjacent to the hot spot.

One problem when using sniffer probes lies in not understanding what to look for during use. A "hot spot" detected by a probe may force one to spend considerable time in trying to fix a particular area of concern. Unfortunately, the radiated field detected may be from cables, harnesses, or digital components that are coupling to this particular point, especially on a PCB. This is because complex coupling paths may not be easily detected due to parasitic inductance or capacitance. Therefore, probes are best used for locating areas of intense RF fields instead of trying to measure an absolute value and compare this reading with regulatory requirements.

5.9.2 Commercial Probes

Several vendors market probes for use in EMC debugging and diagnosing. Typical commercial probe are shown in Figure 5.26. Most vendors provide both *E- and H*-field probes operating in the frequency range of 10 kHz–1000 MHz. Some probes have moderately small tips useful for diagnosing PCB noise measurement on pins of digital circuits and traces.

The *E*-field probes are used to locate the vicinity, or alleged source of radiated emissions, usually within a PCB. Once the location of the possible source of EMI is detected with a larger size probe, a smaller tip probe can be used to pinpoint the exact circuit that might be causing the problem. Using two probes in combination allows one to quickly find and suppress undesired EMI at the source before resorting to other means of suppression, such as shielding the overall system.

Loop probes allow detection of magnetic fields emanating from chassis seams, ribbon cables, and connector pins. When used on PCBs, loop probes identify where radiating loop currents are present. Locating a radiating loop area is probably more important (by experience) than using *E*-field probes. This is because measurements are taken in the near field. Magnetic flux is greatest in the near field. The electric field component may be so small in magnitude that it is impossible to measure. By knowing the loop perimeter, one can narrow in on circuits that are located in the center of the perimeter of the loop if the field intensity is of a sufficient magnitude. Magnetic flux propagates in a radial manner from a source driver. From here, an *E*-field probe can be used to isolate the exact pin or trace radiating after locating the problem area with an *H*-field probe.

5.10 DIFFERENTIAL-MODE PROBES

Differential-mode probes provide value when attempting to measure differential-mode voltage and/or current present between two points of interest. These two measurement points can be separate wire or cable assemblies or between voltage and ground pins on a component. Common-mode rejection is based on the design of the combiner module. In addition, use of a spectrum analyzer or oscilloscope is required. A signal combiner connects two probes together and provides as an output signal either the sum or difference of the two currents flowing through the two probes, depending on the orientation of the probes and type of combiner. The most popular combiner is one that uses an 180° subtracting mode of operation.

Voltage-type differential-mode probes allow one to measure differences present between signal and ground. Common-mode rejection of ground lead voltage and rejection of interference from adjacent fields are a primary advantage of differential-

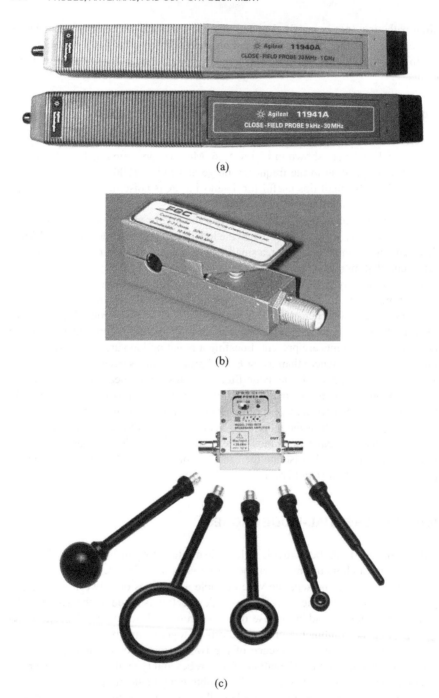

(a)

(b)

(c)

Figure 5.26 Commercial near field probes. (Photographs courtesy of (a) Agilent, (b) Fischer Custom Communications, Inc., and (c) EMC Test, Inc.)

mode voltage probes. Shields of differential-mode probes are grounded only at the measuring instrument. Any voltage difference along the cable shield and instrument ground is rejected as a common-mode signal.

Current-based differential-mode probes are useful to measure relative phase and amplitude between two common-mode currents providing differential-mode information. If the magnitude of current in two transmission lines is known, it becomes an easier task to find the source of the EMI problem. This can only happen if the probes used have a response that is matched to within a few percent of each other.

When two current probes are connected to a signal combiner, the length of the cable interconnects must be identical along with the addition of 50 Ω feed-through termination at the output of each current probe. Identical cables preserve both relative phases between signals. The terminator ensures the combiner sees 50 Ω on both ports. When using a signal combiner, remember to account for a typical 6 dB attenuation loss from the output of each current probe resulting from the 50-Ω terminator that was provided to match the coax impedance to the combiner.

Before using any differential-mode probe, it is important to perform a null test [1]. A null test allows one to learn if a problem exists. To perform a null test using a voltage probe, the two tips are connected together. There should be no signal present on the analyzer or scope. If a signal is present, a damaged cable may exist or the combiner has a problem. For current probes, there are two methods of testing. One method takes a single wire and probe. This single piece of wire is run both in and then out of the probe at the same time. If RF currents are on the wire, they will cancel and produce no output voltage from the clamp. The second test is for a single wire and two probes. Both probes are clamped on the wire physically placed adjacent to each other in opposite orientation. The combiner output should be reduced by 20–30 dB when both probes are used versus one probe only. This output level should be consistent across the frequency range of operation. Examples of how to perform null experiments are provided in Figure 5.27. Note that a combiner is required when two probes are used at the same time.

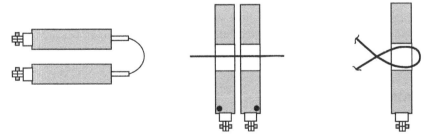

Voltage probe null procedure   Dual-reversed probe null procedure   Folded-wire null procedure
Both probe tips are shorted together   Note polarity of probe orientation Current in the cable cancels out

Figure 5.27   Performing differential-mode probe null tests.

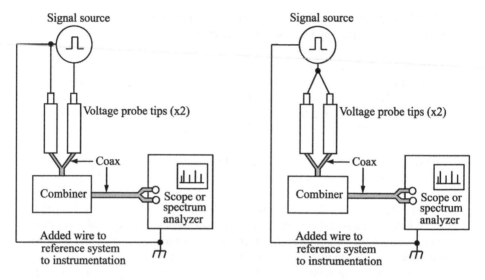

Figure 5.28 Using a differential-mode voltage probe.

When using a balanced differential voltage probe, be aware that this device does not work efficiently in high-impedance circuits. The reason is due to the impedance of the probe tip and combiner, which are designed to match the probe to the analyzer. This probe is optimal for noise measurements on low-impedance signal sources. With most high-frequency circuits being high impedance, this probe finds limited use in certain applications.

Examples on using a balanced coaxial differential voltage probe are shown in Figure 5.28 [1]. The probe can be used to check common-mode overloading or investigate the common-mode rejection ratio (CMRR) of a circuit or system after performing a null test, all with an optional ground lead connected.

A commercial differential voltage probe is shown in Figure 5.29. This probe can measure transients having rise times as short as 200 ps and pulse widths up to 70 ns. It is ideal for use in characterizing a wide variety of transients, including electrostatic discharge, coupled lightning, and EMP simulation. Differential current probes are the standard probe used for measurement of current, except two probes are required at the same time.

5.11 HOME-MADE PROBES

Those on a tight budget or who need a quick diagnostic tool may want to make their own sniffer probe. The primary difference between commercial and home-made probes lies in application and cost. Commercial sniffer probes provide calibrated information, which for the most part is useless when trying to locate a problem area on a PCB or a leak point in a chassis housing. Home-made probes can be made ex-

Figure 5.29 Commercial differential voltage probe. (Photograph courtesy of Fischer Custom Communications, Inc.)

tremely small, which makes for easier location of exact problem areas at the expense of sensitivity.

There are two types of probe application: frequency and time domain. Frequency-domain probes are used to detect radiated emissions while time-domain probes are used with oscilloscopes for signal integrity. If one requires a home-made time-domain probe, it should be a field effect transistor (FET) probe, which is not easy to build. A FET probe provides a high-impedance transducer used to measure signal characteristic within a transmission line without adding a capacitive load or affecting performance of the propagating wave.

In Appendix A, details are provided on how to build a quality probe useful for most troubleshooting purposes for minimal cost. These home-made probes are excellent for both magnetic and electric field measurements.

5.12 ALTERNATE TROUBLESHOOTING DEVICES

AM Pocket Radios. These *sophisticated* diagnostic tools are readily available everywhere, generally in every office cubical or under the bed of one's children. Use of an expensive diagnostic probe is not required. A pocket radio (AM mode) helps to isolate general problem areas when sophisticated test and measurement equipment is not available, such as commercial test laboratories, start-up companies, or customer sites. A low-cost, very old radio works best. This is due to the fact that circuitry is provided on a paper-phenolic PCB and transistors are used instead of sophisticated integrated circuits. The more primitive the radio, the better the diagnostic tool.

It is nearly impossible to locate the exact source of RF energy (component pin or trace) due to the broadband nature of the pocket radio. Radios are a quick method of locating leaks in seams of chassis-and-cable assemblies. If lucky enough, the exact problem area can be identified and remedial measures taken, such as patching up an enclosure seam with copper tape, placing a ferrite core over a cable, or rerouting cable harness assemblies.

The advantage of using an AM/FM radio lies in its simplicity and availability. It is difficult to transport test instrumentation onto a manufacturing floor or a customer site (when they report an EMC event). Anyone is capable of locating problem areas and applying a quick fix. The magnitude of emissions minimized by a quick fix cannot be ascertained; however, if the audio volume of the interference is lower, chances are that one has isolated the leak point. From here, an experienced EMC engineer can proceed to the exact area that drives or stimulates the propagation path.

A disadvantage of using an AM/FM radio is that the output of a pocket radio only provides an indication of the *modulation* present on the waveform on speaker. For EMC analysis, we are generally interested in the frequency and magnitude of the RF signal. When we tune in to the harmonic of a digital clock, detection may only occur by the way the radio *squelches* background noise, or reduces the amplitude of legitimate radio transmissions.

Broadband Field Sensors. Broadband field sensors are useful for on-site diagnostic purposes to determine if a problem exists with compatibility between systems or units. Several different types of sensors are available. The most common use of these probes is in anechoic chambers. The specialized broadband probes in chambers ascertain if the field intensity is uniform across the EUT. These sensors are not useful for diagnosing or troubleshooting EMI, as they are designed to detect very strong levels of intentionally generated RF energy. What are of value to engineers or technicians are portable hand-held field sensors (Figure 5.30).

Most broadband field sensors indicate only the total field strength of a signal without providing frequency information. The antenna is usually a tiny dipole or magnetic loop in three orthogonal axes, thus providing the combination of all polarizations of the fields being measured. The antenna feeds a wideband detector and an amplifier to indicate field strength.

Depending on the vendor, broadband field sensors measure electric, magnetic, or electromagnetic fields. Each probe operates over a specific range of frequencies. Calibrated data may be recorded to ascertain if a solution (rework) is effective.

A useful application for a field sensor is in an industrial or commercial environment to locate sources of strong magnetic fields from AC mains transformers (that provide power to the building), which may affect video monitors located nearby. In hospitals and other controlled environments, a magnetic resonance imaging (MRI) machine, for example, can cause serious malfunction of life support monitoring

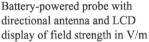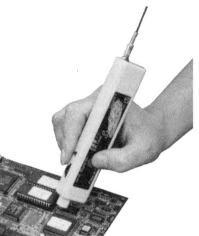

Battery-powered probe with
directional antenna and LCD
display of field strength in V/m

Battery-powered close-field
probe with LED indication

Figure 5.30   Broadband field sensors. (Photographs courtesy of Credence Technologies.)

equipment. Portable field sensors are valuable for determining the compatibility
level of electrical equipment in sensitive environments.

Other broadband field sensors are designed for determining whether electromagnetic fields present a hazard to human health. These probes may not be sensitive
enough to measure electric fields below 1 V/m or 120 dBμV/m. Therefore, this particular probe becomes useless when trying to measure low levels of field strength,
typical of most electrical products. This high intensity probe is useful for measuring
different types of fields at a customer's site to quickly determine whether fields present may be causing a product to malfunction.

Broadband field sensors can be used for quality assurance purposes to quickly
determine if known weak points related to radiated emissions have developed during manufacturing. In addition, one can validate products prior to shipment and
compare results to systems that have been accurately tested. Checking areas such as
cable terminations and connectors can reveal incorrect assembly (such as the use of
a pigtail instead of a 360° clamp for a cable screen, or incorrect bonding). Knowing
the pass or fail limits on the detector's display requires experience with how the detector responds to the kinds of problems that may exist and those that could lead to
a failure.

Broadband field sensors typically measure and present the *average* field level,
not the peak, so they cannot give any effective knowledge on weak points. High-
level, very low duty cycle pulsed RF signals may cause system failure, whereas the
broadband field sensor may barely respond to the (average) field level. The results
may be misleading and confusing.

5.13   FAR-FIELD ANTENNAS

In order to achieve EMC, measurements must be taken for both radiated and conducted fields to ascertain the magnitude of undesired RF energy that may cause harm to other electrical equipment. The probes described earlier in this chapter deal with either direct physical connection to a transmission line or that located in close proximity (near field). Magnetic or electric fields are determined based on the type and nature of probe used.

Full-compliance testing for radiated emissions requires antennas be designed for use in the far field. Electromagnetic compatibility standards specify emission limits only for the far field, with some exceptions. The far field is generally defined as greater than one-sixth of a wavelength, though the near field–far field transition boundary is a factor of ±2 of this value. For 30 MHz, this distance is 1.7 m.

For measurements in the far field, various types of antennas are available. An antenna can be used for transmitting or receiving electromagnetic fields and is designed to maximize coupling between an electromagnetic field source and a receiver. One does not need to understand theory on how antennas operate to get the job done. Many references (see the Bibliography at the end of the book) detail the design, analysis, and use for all types of antennas. The descriptions below provide insight into common types of antennas used for EMC testing.

Between 30 MHz and 1 GHz dipoles are recommended. Dipoles have a limited frequency range that makes testing very time consuming. Since dipoles have a known calculable response, they are used as the standard transducer for radiated emission site calibration.

In the near field one component, either the E-field or the H-field, dominates as a function of source characteristics and distance from the source. In the far field, the two are related by the wave impedance of free space. This was described in Chapter 2 and is again described by Eq. (5.5). In the near field, these fields cannot be directly related as their relationship is dependent on the source characteristics:

$$Z_0 = \frac{E}{H} = \frac{\mu_0}{\varepsilon_0} = \sqrt{\frac{4\pi \times 10^{-7} \text{ H/m}}{(1/36\pi) \times 10^{-9} \text{ F/m}}} = 120\pi \text{ or } 377 \text{ }\Omega \qquad (5.5)$$

Before using antennas for troubleshooting EMI or for conformity tests, one must select an appropriate transducer. There are other types of antennas available for very specialized applications, not detailed here. The antennas commonly used for measuring electromagnetic fields in the frequency range 20 Hz–40 GHz are summarized in Table 5.1. The common attributes of these antennas include small size, light weight, accurate calibration, and relatively low gain compared to communication antennas. An illustration of antennas used for the majority of EMC tests is provided in Figure 5.31.

Most test standards allow use of a broadband antenna which speeds up test time. The two most common antennas are the biconical and log periodic. Some antennas have an extended range of operation. It is often common to combine these two an-

Table 5.1 Summary of Commonly Used Antennas

Antenna Type	Frequency Range
Loop	20 Hz–30 MHz
Rod (vertical monopole)[a]	10 kHz–30 MHz
Tunable dipole (resonant and half wavelength)	30 MHz–1 GHz
Broadband dipole (biconical and cylindrical)	30–300 MHz
Log periodic (planar and conical spiral)	200 MHz–1 GHz
Bilog	30 MHz–1 GHz
Waveguide horn	1–40 GHz
Discone	1–10 GHz
Double-ridged waveguide[a]	1–18 GHz

[a]Not detailed in this chapter.

tennas into one that has a 50-Ω calibrated output, which saves test facilities the need to change antennas on a frequent basis.

A tuned dipole is specified as an alternative antenna with the advantage that its performance can be accurately predicted. Unfortunately, it can only be used at spot frequencies and is best suited for calibration of broadband antenna, site surveys, site attenuation, or other specialized needs.

An antenna must couple an electromagnetic field to a receiver. Electric field strength limits are specified in terms of volts (or microvolts) per meter at a given distance from the EUT. Receivers are calibrated in volts (or microvolts) with a

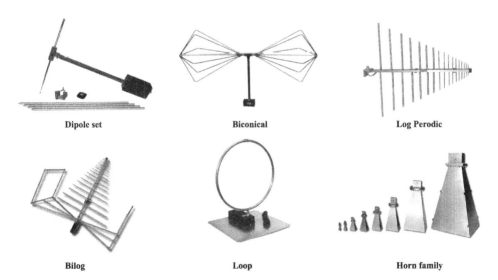

Dipole set   Biconical   Log Perodic

Bilog   Loop   Horn family

Figure 5.31 Typical antennas types (for various frequency ranges and use). (Photograph courtesy of A. H. Systems, Inc.)

50-Ω input. This means that the antenna must be corrected for a 50-Ω output at a given field strength for each frequency. Antenna factors are provided by the antenna vendor, which allows one to make necessary corrections. In addition, system sensitivity and polarization are important factors when using any antenna.

Antenna Factors. This is one of the most important parameters when using any antenna. Each antenna is provided by the manufacturer with a table of antenna factors, usually in dB/m versus frequency. To convert the measured voltage at the receiver's input into actual field strength, one must add the antenna factor to the measured reading per Eq. (5.6). The test engineer must perform this calculation using a calculator or software. In addition to the antenna factor being a function of frequency, the interconnecting coax is also frequency sensitive. A problem with antenna factors is that the antenna is usually calibrated at only a particular distance from a source transmitter. When using the antenna at a different test distance, the calibration chart from the vendor may be invalid for this particular test.

$$E_{fs} = E_{ant} + AF + CL \qquad (5.6)$$

where E_{fs} = actual field strength of signal
E_{ant} = measured field strength directly from antenna
AF = antenna factor
CL = cable loss (amount of attenuation within coax)

System Sensitivity. Serious problems can occur with spectrum analyzers during radiated emissions testing. This lies in the measurement distance used and the sensitivity of the receiver. Typically, the test distance for radiated emissions is specified at 10 m. Sometimes, it is only possible to take data at a closer distance. The minimal level that will be determined by the spectrum analyzer will be the noise floor of the equipment. For 120-kHz bandwidth, the noise floor is typically +13 dBμV. To this measured value, antenna factor and cable attenuation must be added to derive the overall measurement system sensitivity.

For test environments where the noise floor is greater than the specification limit, an alternative method of recording radiated energy must be performed. There are three options to overcome this sensitivity problem:

1. Reduce the distance to 3 m. Although the specification limit is increased by +10.5 dB, significant error may be present because the antenna may now be in the near field. This closer measurement distance still may not ascertain if an RF signal exists within the noise floor or is hidden in the ambient noise.
2. Use a low-noise preamplifier or preselector. These two items lower the effective system noise floor by a factor equal to the preamplifier gain less its noise figure. Typical gain is approximately 20–26 dB.
3. Use a test receiver. A receiver (not spectrum analyzer) has enhanced sensitivity with an extremely low noise floor.

Polarization. In the far field, electric and magnetic fields are orthogonal. Either field may be vertically or horizontally polarized, or somewhere in between. Actual polarization depends on the nature of the source emitter and reflections from metal objects located within the test environment, including the facility ground plane. A maximum response will be observed when the plane of polarization aligns with that of the incident field. A minimum value is observed when the planes are at right angles to the propagated signal. CISPR, EN, and FCC/DOC emission measurements must be made with antennas that can have their polarity changed by manual or automatic rotation.

Test Data Sheet. Table 5.2 shows a typical spreadsheet used to calculate actual field strengths from measurements recorded on a spectrum analyser or EMI receiver. This table includes calibration factors for both the antenna and the coaxial cable.

Almost all software programs perform the calculations automatically. If a signal is near the specification limit, a flag or alarm may notify the operator requesting that the signal be remeasured using a quasi-peak or average detector. This feature assists in identifying potential problem areas, thus saving considerable time during the testing process.

5.13.1 Common Antennas Used for EMC Testing

Tunable Dipole Antenna. The tunable dipole is used for electric field strength measurement, generally in the frequency range 25–1000 MHz. There are two types of tunable dipoles—the resonant dipole and the half-wavelength dipole. The resonant dipole is adjusted to a physical length slightly less than one-half wavelength in order to reduce the reactive part of the antenna impedance to zero. The antenna impedance of a half-wave dipole has both resistive and reactive components. In order to accurately use a dipole, physical length measurement of both elements must be performed at every frequency that is to be measured. As seen in Figure 5.31, vari-

Table 5.2 Typical Radiated Emissions Spreadsheet

Frequency (MHz)	Measured Signal (dBµV/m)	Antenna Factor (dB/m)	Cable Factor (dB)	Corrected Amplitude (dBµV/m)	EN 55022 Class B QP Limit (dBµV/m)	Margin (dBµV/m)	Polarity[a]
66.336	22.9	11.5	0.2	34.60	30	**+4.6 fail**	V
80.560	20.8	11.5	0.2	32.50	30	**+2.5 fail**	H
112.78	16.6	9.1	0.6	26.3	30	–3.7	H
166.23	19.6	7.8	0.8	28.2	30	–1.8	H
212.46	21.3	5.2	1.0	27.5	30	–2.5	V
244.68	31.9	4.6	1.2	37.7	37	**+0.7 fail**	H
280.90	27.0	2.6	1.2	30.8	37	–6.2	V

[a]V = vertical, H = horizontal

ous elements (tuning rods) are provided for use with a balun. The center support rod contains a matching network.

Since the length of a dipole must be adjusted for each frequency, use is limited for specific measurement applications at a select number of discrete frequencies. In addition, the physical size of the element in the vertical polarity may be extremely large. The distance spacing between the tip of the element and the ground plane may become electrically short or the distance between the center point of the antenna and the floor is less than the physical dimension of the element required for tuned operation. This limitation may preclude their use in some applications, such as vertically polarized fields and indoor facilities, including shielded rooms and anechoic chambers.

The impedance of tunable dipoles is affected by mutual coupling to the ground plane or reflecting surfaces, especially when the distance is less than a wavelength. The impedance of vertical polarized dipoles is also affected by coupling to its transmission line. Consequently, the antenna factor will differ between free space and actual use. Because of uncertainty due to mutual coupling, it is generally easier to use calibrated broadband dipoles such as biconicals. Biconicals exhibit negligible mutual coupling effects due to their small electrical length at lower frequencies. Long dipoles used at lower frequencies (e.g., 30 MHz) may exhibit measurement errors if the size of the field lobe from the EUT is much smaller than the length of the antenna. This is called *illumination error.*

Biconical Antenna. A biconical antenna is a broadband dipole that consists of two conical conductors having a common axis and vertex and that are excited or connected to a receiver at the vertex. When the vertex angle of one of the cones is at 180°, the antenna is called a discone. These antennas generally operate in the frequency range 30–300 MHz. When using a biconical antenna, polarization is important—vertical or horizontal. The antenna emulates a very broadband dipole, which makes it convenient for most EMC tests compared to the use of a dipole. In a dispute between a manufacturer and a regulatory body (agency), a tuned dipole antenna takes precedent, although it may present several decibels of illumination error.

Illumination error works in favor of compliance in most situations. Because a long antenna will not necessarily be fully immersed in the field, the output reading will be lower.

Some biconical antennas are cylindrical in form. Regardless of configuration, both antenna formats are compatible with automated instrumentation and permit swept-frequency measurements over their operating range, resulting in a significant reduction in measurement time compared with tunable dipoles that require mechanical adjustment at each frequency. Biconical antennas are suited for vertical polarization measurements due to their smaller size. In addition, biconicals exhibit less mutual coupling with nearby metal structures compared to tunable dipoles.

Log-Periodic Antenna. The log-periodic antenna has a structural geometry such that its impedance and radiation characteristic repeat periodically as the logarithm of frequency. These antennas commonly operate in the frequency range of

200–1000 MHz and are moderately directional when the source of the propagating wave is known. When the direction of the source is unknown, the antenna must be scanned in azimuth and elevation to locate the source, aligning the peak of the major lobe with the incoming wave. The boresite (the direction of the tip) must be rotated on axis in order to match the polarization of the incident wave. If multiple spatially discrete sources must be measured, this would become a very time consuming task.

Bilog Antenna. A bilog antenna is a single antenna that combines the features and electromagnetic characteristics of both biconical and log periodic into one assembly. The advantage of using a bilog antenna is time saved in changing antennas. Unfortunately, this antenna is physically larger than either a biconical or log periodic as a stand-alone assembly. Many variations exist in its design and use.

Loop Antenna. A loop antenna is sensitive to magnetic fields and is shielded against electric fields. This antenna is in the shape of a coil. A magnetic field component perpendicular to the plane of the loop induces a voltage across the coil that is proportional to frequency.

Electrically small loops are used to measure electromagnetic fields in the frequency range of approximately 20 Hz–30 MHz. A loop is considered electrically small if the length of the perimeter (or circumference for circular loops) is much less than the wavelength of interest. The most common loop antenna shapes are circular and rectangular.

While loop antennas respond to magnetic field components, the electric field element can also be estimated if the wave impedance Z_w is known ($E = Z_wH$). For plane waves in free space, the impedance value is 377 Ω (120π). Loop antennas are usually shielded against electric fields, which maintains symmetry and balance. The shield is made of nonmagnetic materials such as aluminum or copper with a small air gap provided.

Horn Antenna. Horn antennas are used to measure electromagnetic field strength in the frequency range above 1 GHz. These highly directional antennas have a narrow beamwidth. This means that the open end of the antenna must be pointed directly toward a source emitter. Any signal that is outside the area of the open end of the antenna will not be measured. Depending on the distance from the EUT, beam angles are typically 30°–45°. Any signal that falls outside of this range will not be detected by the antenna.

Since horns are extremely directional, they are used primarily for reception where the location of the radiation source is known or can be resolved. When the location is unknown, horn antennas must be scanned carefully in azimuth and elevation in order to center the major lobe to the direction of the incoming wave. Additionally, the polarization of the antenna must be varied to match the incident wave. The narrow beamwidth of horn antennas requires that angular scanning be performed at a rate slow enough to ensure that the receiver has time to respond. This process is time consuming, especially if many frequencies are to be measured.

One advantage of using horn antennas lies in their ability to have a high gain compared to omnidirectional antennas. The "gain" of a passive antenna really means "less loss" compared to a standard reference of an isotropic source. This high gain is sometimes required when low-level fields have to be detected and measured. Also, certain receiver noise figures have limitations in their front end regarding detecting low-level signal. Gains of horn antennas vary from approximately 10 to 30 dB over the frequency range of 1–40 GHz.

REFERENCES

1. Smith, D. 1993. *High Frequency Measurements and Noise in Electronic Circuits.* New York: Van Nostrand Reinhold.
2. Williams, T., and K. Armstrong. 2001. EMC Testing Part 2—Conducted Emissions. *EMC Compliance Journal.* April. Available: www.compliance-club.com.
3. Williams, T., and K. Armstrong. 2001. EMC Testing Part 1—Radiated Emissions. *EMC Compliance Journal,* February. Available: www.compliance-club.com.
4. Kodali, P. 2001. *Engineering Electromagnetic Compatibility,* 2nd ed. New York: IEEE.
5. Smith, A. 1998. *Radio Frequency Principles and Applications.* New York: IEEE/Chapman & Hall.
6. Williams, T., and K. Armstrong. 2000. *EMC for Systems and Installations.* Oxford: Newnes.
7. Williams, T., and K. Armstrong. 2002. *EMC Compliance Information.* Nutwood, UK.
8. Williams, T., and K. Armstrong. 2001. EMC Testing Part 5—Conducted Immunity. *EMC Compliance Journal.* December. Available: www.compliance-club.com.

CHAPTER 6

CONDUCTED TESTING

There are several modes of signal propagation. One mode of propagation is conduction through a physical interconnect (wire, cable, transmission line, or PCB trace). A second mode of propagation is radiation (through free space or a dielectric, detailed in Chapter 7). A third mode is the coupling of energy by an electric or magnetic field (also within Chapter 7).

This chapter provides information that relates to interference currents conducted within a transmission line or interconnects. Conduction is not limited to noise on a power cord but occurs on any type of transmission line carrying either alternating or direct current and may be observed in the frequency range of DC to 300 GHz.

Regulatory standards, be it commercial, military, or private, contain specification limits to ensure that the magnitude of undesired RF energy is not sufficiently large to cause harmful interference when the transmission line is carrying a signal of interest: steady state or time variant. Due to the nature of signal propagation, is this current common mode or differential mode? What is the magnitude of this current?

Understanding different modes of current propagation will assist in locating a problem area and help with implementing a design solution quickly. Knowing different measurement techniques and using proper probes and instrumentation can make the task of identifying signal propagation easier within interconnects.

6.1 OVERVIEW OF CONDUCTED CURRENTS

Interference sources, both external and internal to electrical equipment, and their respective power supplies can have RF energy couple both into and out of cables and interconnects. Coupling occurs by either inductive or capacitive means. Conducted emissions permit undesired RF energy to propagate from a source to a load through a transmission line, either the AC mains power cable or signal (data) lines between functional units.

Conducted current limits are aimed at limiting unwanted emissions present within any transmission line, such as data, control, and power cables. Radiated verification of emissions below 30 MHz is difficult to perform due to the size of the receiving antenna required, especially dipoles in vertical polarity. To perform tests at lower frequencies, conducted measurements are performed in lieu of radiated ones. Since cables are often the prime source of radiated emissions at longer wavelengths/low frequency, their effectiveness as radiators is a function of their length. At low frequencies, power cords can be likely culprits. Conducted noise voltages will be the source of radiated EMI. Owing to limitations in testing for low-frequency noise, specification limits below 30 MHz have been translated to conducted voltage levels.

Conducted emission limits are generally easier to meet than radiated ones. Measurement results are repetitive and commercial filters are available for in-line insertion to remove common- and differential-mode noise. The ports of a system are usually not well defined. Noise within the frequency spectrum is difficult to calculate with software. For differential-mode currents, which are well defined and controlled, the emission level can be calculated. Common-mode current is much more difficult to estimate due to parasitics. Therefore, the computation/estimation of conducted emissions is easily carried out only for differential-mode signals. This calculation involves the Fourier transform of signals and the comparison of spectra. Any out-of-limit signals require use of filtering.

Until recently, attention has focused on power cords as the primary source of conducted emissions. This is because CISPR requires that tests be performed on just this one cable. However, signal, data, and control lines can and do act as coupling paths, which may produce harmful interference. Extended lengths of parallel routed wires that are tightly bundled present not only a crosstalk concern but also EMI (emissions and immunity). More recent versions of test standards require that conducted emissions be performed on all external cable assemblies, even Ethernet if the cable is unshielded.

Resulting interference on cables may appear as either common mode or differential mode. To help distinguish between these two modes, consider an AC mains power cable. The differential mode is the RF noise between wires (e.g., high and neutral or phase 1 to phase 2). The common mode is RF noise between line/neutral and chassis ground. For signal, data, and control lines, common-mode interference noise is the primary area of concern and is measured at both ends of the cable using special interconnect devices. Power lines are treated differently, with the noise measured at the beginning of the line or where the mains cable connects to the facility's breakout or receptacle. Differential-mode emissions are generally associated with low-frequency switching noise from the power supply. Common-mode energy is the result of higher frequency switching components (digital logic signals), intercable coupling (crosstalk), or internal circuit sources (switching transistors and control components).

Before discussing conducted field propagation, it is necessary to understand the difference between common-mode and differential-mode noise.

6.1.1 Common- and Differential-Mode Currents on Wires and Cables

Cables are the primary contributor of unintentional emissions, both conducted and radiated. Common-mode propagation represents EMI current that flows in the same direction on all conductors in the cable at the same time. Differential mode refers to EMI currents flowing in opposite directions, as illustrated in Figure 6.1.

Note that some common-mode current returns through a path other than the intended transmission line ($2I_{cm}$). This return path may be another cable located adjacent to the source signal, a ground plane or other return path, even an unexpected, undefined path. Since common-mode current flows along the length of the cable, it is also called the "longitudinal" mode. Differential-mode currents, on the other hand, flow in opposite directions at the same time, hopefully with the same magnitude in order to cancel out undesired flux and fields. Differential mode is the "normal" mode of operation for power and signal current flow.

These two modes are again illustrated in Figure 6.2 [1]. Both modes can be present at the same time, regardless of the nature of the signal–power or data. In fact, power cable assemblies have two common mode cases—a general case where noise currents flow in the common mode on the power and neutral and return via the AC mains safety ground, and currents that flow on all lines in the same direction.

The distinction between modes is important for identifying undesired sources of EMI during troubleshooting and conformity tests. When one can determine the mode of operation, locating the problem area becomes a simple task. For example, line-to-line filters (X-capacitors) will reduce differential-mode noise with no effect on common-mode levels since lines are already propagating in the opposing direction. For the common-mode situation, line-to-ground capacitors (Y-capacitors) are required to filter the common mode, with little effect on the differential mode. Another example is a ferrite clamp over a cable, which will suppress common-mode

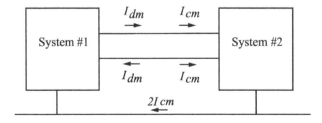

Figure 6.1 Comparison of common- and differential-mode current.

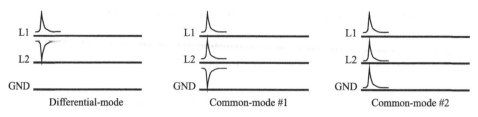

Figure 6.2 Common- and differential-mode transmission.

currents but have little effect on differential-mode currents. These modes provide clues as to the origins of EMI currents. For example, differential-mode currents suggest that conductors are connected to a source circuit, while common-mode currents imply that EMI energy has been coupled to the cable through radiation, crosstalk, common ground impedances, or imbalances.

Coupling modes can be determined experimentally with a current probe (Figure 6.3). A current probe is a loosely coupled transformer that has the primary windings of a transformer's coil connected to a coaxial connector. The probe is generally circular in configuration and may be solid or split core. A wire (can be a single wire or a bundled assembly) is passed through the core. This wire becomes the secondary of the transformer. The time-variant magnetic field present on the wire (Faraday's law) is coupled to the primary winding (Lentz's law). The output voltage is proportional to the input current—thus the name *current probe*. Current probes are available in a wide variety of configurations, frequency ranges, and power levels. For EMI analysis, a probe should be chosen that exceeds the frequency range of interest. At the top and bottom frequency range of the probe, the operation (calibration data) becomes skewed and nonlinear. Figure 6.3 illustrates a power cord; however, the concepts are the same for signal assemblies [1]. One wire is the source transmission; the other wire is the return.

To use a clamp-on current probe, a wire is inserted and the probe secured shut. More than one wire can be tested at a time. If current flow is measured, it can be either common or differential mode. Additional wires are then inserted. If current is still present, it is common-mode current. However, if there is no current flowing, differential-mode propagation exists since we now have "equal and opposite" currents flowing through the aperture of the current probe.

Using the current probe in this manner may be one of the more useful diagnostic techniques for locating undesired sources of RF energy. Many times, a radiated emissions problem can be solved with a current probe. Signal integrity concerns in high-speed transmission lines can also be solved using a current probe, especially when one does not have access to the control electronics.

Predicting common-mode currents from these types of probes is extremely difficult. Common-mode currents are not illustrated in schematics and assembly drawings. They exist by virtue of parasitic effects within a product. These parasitic effects include common ground impedance or unexpected capacitive or inductive

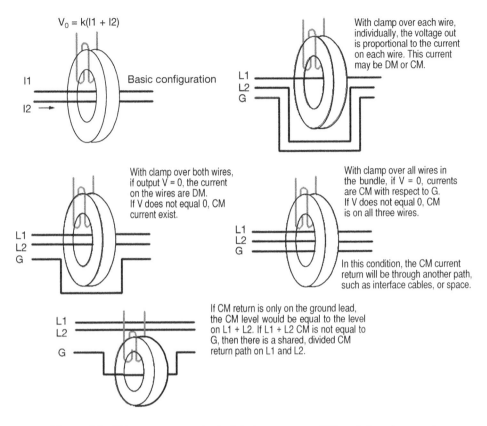

Figure 6.3   Using a current probe to detect common- or differential-mode currents.

coupling. Since intended currents are designed to be differential mode, conversion between a desired signal of interest and undesired EMI is complex. Details on this conversion are provided in Chapter 2. If we are able to determine conversion between differential and common modes, diagnosing and debugging electrical problems becomes easier. One must understand the problem before attempting to solve the situation, especially when one does not know what is happening, where it is happening, and why.

6.1.2   Coupling Paths for Conducted Emissions

Common-mode currents are the primary cause of high-frequency problems: emissions and immunity. Most failures are due to common-mode currents on cables and wire assemblies. Immunity failures occur due to externally induced EMI entering the product. The system disrupted by this externally received noise cannot distinguish between common and differential modes; it only reacts to RF energy at a particular frequency and amplitude that has caused the disruption. Even low-frequency EMI can be due to common-mode coupling, perhaps caused by improper earth grounding of AC mains.

However, differential-mode currents can cause EMI. For each signal line, a return path must be provided. If an imbalance exists between two signals paths (loss of differential-mode balance), common-mode noise will be created. If there are many single-ended transmission lines within a cable assembly, all with current flowing from a source to a load, the summation of current all traveling in the same direction can cause crosstalk. When dealing with EMI and cable assemblies, assume both modes are present until proven otherwise.

When dealing with conducted emissions, various coupling paths exist. These coupling paths are illustrated in Figure 6.4. Differential-mode current I_{dm} generated internal to a power supply, usually a switching type, will be measured as an interference voltage across the load impedance of each line with respect to earth connection. If digital switching noises (V_{dcn}), generally from digital components, have coupled into the power supply through C_c, this energy can also be observed on the power cord. The method of coupling includes capacitive coupling through the isolation transformer. The energy will appear between L_1/L_2 and E on the mains cable. Parasitic capacitance C_s is now present between units (power supply and system) referenced to earth ground. Radio-frequency (RF) noise is usually referenced to ground with C_s providing this reference.

In regard to signal cables, common-mode currents developed internal to the system will now have earth ground as one reference point. Free space, which is a vacuum with a media impedance, provides the mechanism for transfer of undesired RF energy. If one visually examines the configuration of the signal cable (voltage potential) and earth ground (0 V potential), it is observed that a monopole antenna is present.

In a real product, all coupling paths exist simultaneously. For most environments, one has no idea where the coupling occurred, as parasitics will always be present. The value of C_s is widely distributed and difficult to predict and is heavily dependent on the proximity of other metallic objects, especially cables that do not

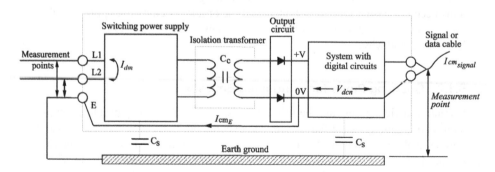

Figure 6.4   Coupling paths for conducted emissions.

have a shield. For those cables with a poor shield, additional coupling can be developed due to the higher capacitance that is present within the cable.

6.1.3   Conducted Emissions Test Requirements

The purpose of conducting tests on power, data, control, and signals lines is to ensure that RF currents do not enter the AC mains of the facility where the device is operated and, conversely, to measure the magnitude of noise that may be present on the mains from causing harm. If two or more systems are plugged into the same AC outlet (i.e., a computer and television), switching noise from the computer may cause unacceptable interference to either the audio or video section of the television. For this example, both emissions and immunity become a concern at the same time.

Various product standards exist to define the test setup and procedure to ensure conformity. Most standards are identical, or harmonized, which provides for convenience to manufacturers when designing and manufacturing products on a global basis.

Chapter 4 and the appendices provide details on setup and operation. In the next section, we will summarize key points and what needs to be considered when performing this test. Conducted emission procedures are straightforward but time consuming. Certain tests require expensive equipment; other tests do not.

Emissions testing should begin with conducted emissions. Signals detected during this process may be accompanied by harmonics that will fall into the radiated frequency range, usually above 30 MHz. When radiated emission tests are later performed, the power cord may act as a propagating antenna, radiating undesired RF energy onto other structures. Testing and corrective action for conducted limits can prevent problems that may arise later during the radiated portion of the tests.

6.2   PERFORMING CONDUCTED CURRENT TESTS

6.2.1   Engineering Investigation in Laboratory or Engineer's Office

Trying to test for the presence of conducted fields is different from testing for radiated fields. When ascertaining if a problem exists for radiated fields, simple instrumentation is available, such as an AM radio or sniffer probe carried within one's pocket. With conducted currents, nonportable instrumentation is required. Measurement data using simple tools is generally very accurate, at least within the boundaries of measurement uncertainty.

The cost to perform precompliance conducted current measurements can be expensive, especially when analyzers/receivers and RF amplifiers are required. Unfortunately, it is difficult to perform preliminary investigation in an office absent the LISN/AMN, receivers, and a metal reference ground plane. The use of an engineering laboratory or test facility is required. The best thing to do is to perform the formal test during the design cycle and take corrective action if necessary.

6.2.2 Test Environment

Conducted interference is the transference of undesired electromagnetic energy along a conductor. When the emissions are along AC mains, this type of interference is commonly called line-conducted interference (LCI) or the mains terminal voltage test.

For LCI testing, mains disturbance voltage between line and phase to ground from the AC power supply and back into the facilities is usually measured in the frequency range of 150 kHz–30 MHz. Military specifications and product specific standards may require an extended range of frequency.

For data, control, and signal lines, special test instrumentation is required. Depending on the complexity of the test setup, this type of measurement may be difficult to perform and is costly. It is thus prudent to incorporate appropriate filters and RF shields on all cables prior to sending the EUT out for testing.

If the test environment has deficiencies, test results may be invalid and may not provide benefit to the design engineer. For conducted currents, it is possible to automate the test run, unlike radiated fields that require a subjective analysis on the signal being measured. When using a shielded facility, automation software is simple to use; however, if tests are performed on an OATS or a bench within a laboratory, manual operation of the equipment is required, as externally inducted fields can affect results requiring unnecessary engineering analysis to be performed.

It is imperative that the worst-case emission level be determined. This is achieved by testing the EUT in both minimal and full-load conditions. If the EUT is a personal computer with eight adapter slots, the test must be performed with the minimal amount of adapter boards required for operation, such as a video board, as well as with all adapter slots fully populated. Power supplies perform differently under load conditions and many times are much worse in minimal than in maximum configuration.

A fully operating system, for example, a typical personal computer, should include the following during compliance testing:

- Full-speed DMA transfers on all floppy drives, hard disks, and CD/CDW/ CDR units simultaneously
- Sending continuous data or performing I/O on all parallel and serial ports and network connections
- All optional PCBs and features possible while executing fully their functional software

6.3 CONDUCTED EMISSIONS TESTING (AC POWER MAINS)

There are several ways to perform conducted emission testing. Details on each method are provided in Appendix B. Here we provide a summary of basic test setups, configurations, and problems.

The most common method of performing conducted emission testing (LCI) for products that consume less than 16 A AC (120 V AC range) or 10 A (230 V AC range) on input power cords is through use of a LISN or AMN. For data, control, and signal cables (low-voltage analog or DC levels), use of CDNs is needed. Specifications on LISNs, AMNs, and CDNs are presented in Chapter 5.

Peripherals and other assemblies can be plugged into a coupling network at the same time or powered separately. If a peripheral device is plugged into the LISN along with the EUT, all components will be tested simultaneously. If there is an EMI failure within one subassembly, it may be difficult to isolate without the use of a second LISN.

The LISN is used as a reference source that

1. provides a defined test impedance to the EUT,
2. prevents conducted interference from the EUT from corrupting the AC mains network, and
3. provides isolation of the EUT from ambient RF noise already on the AC lines within the facility.

A spectrum analyzer or test receiver compliant to CISPR-16 requirements is connected to a coaxial connector provided on the LISN or CDN. The EUT must be running in the worst-case operating condition based upon prior testing to ascertain this particular mode. All peripherals must also be operating at the same time if physically connected to the EUT.

There are two basic configurations for systems containing AC mains input: single phase and multiphase. The test procedure and setup are identical for all configurations. The time it takes to perform the test is dependent on the number of lines that carry AC mains voltages. Figure 6.5 illustrates the difference between these two

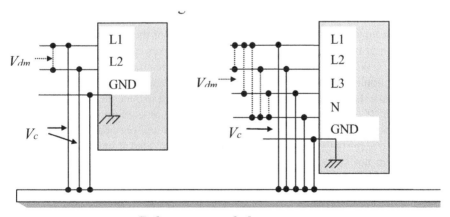

Reference ground plane

Figure 6.5 Single- and three-phase configurations for AC mains testing.

configurations. When performing the test, each line must be individually tested and recorded in the test report.

6.3.1 Potential Problems during Conducted Emission Testing

When performing LCI tests, several protection features must be implemented prior to energizing the test setup. This deals with risk of electric shock and damage to test instrumentation.

Electric Shock. When dealing with AC mains voltage, there is a chance of electrocution. Transducers, probes, and instruments may be connected directly to the AC mains. Equipment can be damaged or defective from prior use without knowledge of the test engineer. For example, when one clamps a voltage probe directly onto a 400-A mains terminal block and the blocking capacitor inside the probe is damaged, the full energy potential at the coaxial connector output can result in a shocking experience and/or instantaneous death. Every device must be connected to a safety ground to prevent this problem. All LISNs (as well as the EUT and associated peripherals) must be bonded securely to the ground plane and properly bonded to the facility's earth ground. Exceptions exist if protective means have been designed into the equipment to ensure that hazard to electric shock will never happen. *Consult a product safety professional or engineer for details on shock hazards.*

Instrumentation Damage. To prevent damage to instrumentation from DC voltages (very lethal, as inputs of analyzers and receivers are designed to have only an AC signal input), use of a DC block is recommended. A DC block prevents transient spikes or DC voltage on the coax, or transducer, from blowing out the front end of the instrument. The DC block does not change measurement results.

A DC block with an optional or separate 3- or 10-dB attenuator is recommended at all times. The purpose of the attenuator is to prevent extremely large signal amplitudes from damaging the input port of the analyzer or receiver. Many instruments provide for input RF attenuation to prevent overload of the intermediate-frequency (IF) section of the unit, however, strong transient spikes can still damage sensitive components. When using an input attenuator, the test instrument may have a correction factor applied via software to compensate for the attenuator. If the instrument does not have this software feature, the test report must include the value of external attenuation. If an attenuator is used for compliance or conformity testing, it becomes important whether the attenuator is calibrated and provided with a calibration chart to identify the real amount of attenuation provided at specific frequencies.

One item not known by many is the static charge that may be present on the coax. If the coax has been coiled and allowed to sit for any period of time, static charge may develop between the center conductor and the shield. When the coax is connected to the analyzer/receiver, this large static potential may jump onto the center conductor of the analyzer's input connector and cause permanent damage to the instrument. The coax cable assembly should always be discharged before installed onto the test instrument.

Analyzer Overload. When performing conducted emissions testing, care must be taken to ensure that the front end of the spectrum analyzer, if one is used, does not become overloaded. This is possible when an analyzer is used without a front-end RF preselector. This overload is caused by the 50/60-Hz component of the AC line voltage, resulting in wrong data as well as intermodulation products. To mitigate this problem, inserting a high-pass filter with a cutoff frequency of 1 kHz will reduce the 50/60-Hz signal from damaging the front end of the analyzer.

6.3.2 In Situ Testing of Systems and Installations

For certain products, testing must be performed in situ in a manufacturing facility or user location. Many systems have three phases with input current ratings greater than 10 A at 250 V AC (16 A at 115 V AC). Line impedance stabilization networks have a limited current rating. Once a system exceeds the current ratings of a LISN, alternate means of performing LCI testing must be considered.

Regulatory standards permit the use of alternate test procedures and coupling devices when regular means of testing is not possible. Figure 6.6 illustrates use of a voltage probe on an industrial product where the per-phase rating exceeded 100 A per phase. Three phases are clearly shown, with the probe affixed to the wire on the far right.

In lieu of a voltage probe, a current clamp can be used over each line being tested. When using this measurement technique, ensure that the current clamp is de-

Figure 6.6 Use of a voltage probe to measure mains interference in situ.

signed for high levels of currents and that saturation of the probe does not occur. If one hears a buzzing sound from the current probe or feels vibration at the same rate as the frequency of the circuit being measured, the probe is saturated and may cease to work properly or, worse, be damaged.

A word of caution: **Improper use of any probe on high-voltage AC mains circuits or carelessness of application can end in fatal results. Ensure that a non–electrically conducted floor or mat is provided for the person affixing the probe. This person must be standing on the mat. Use of heavy-duty electrician's gloves is also recommended. Never perform this test by yourself. Ensure that another person is nearby to provide emergency assistance if a problem develops.**

6.4 IMMUNITY/SUSCEPTIBILITY TESTS

Conducted immunity tests have the advantage of not requiring use of an anechoic chamber or expensive test sites; however, use of these facilities is recommended, especially if there is a concern about potential interference to other nearby equipment. In addition, noise currents induced into various power cords and signal cables may reradiate into the environment or cause harmful interference to communication systems.

There are disadvantages in having to perform these series of tests. Results may be questionable and may not accurately represent real situations for the environment of intended use. When the entire system is irradiated, all cables are subjected to RF currents. International regulatory standards, for the most part, require each cable to be individually tested with certain exceptions. Each cable represents a common-mode load on the EUT. This load must be artificially created by interface networks, detailed in the appropriate section herein. Networks for direct coupling to cables with many signal lines are difficult, if not expensive, to build. For this reason, multiple cables can be tested at the same time. In addition, these networks may adversely affect signal line characteristics.

There are numerous tests that can be performed for conducted immunity. Not all tests are required. For certain products, it may be impossible or not required to perform legally mandated conducted immunity tests, especially when these units do not have any external cables, such as battery-powered devices with high-speed digital circuits.

The primary immunity tests for the majority of products are in the categories of Information Technology Equipment (ITE); Industrial, Scientific, and Medical (ISM) Equipment; and consumer products. There are many other tests not detailed in this book. These additional tests have special requirements and are for systems and products that are used in specialty areas such as the military, aeronautical, or space-based satellites and systems. For almost all electrical products, the following identify the majority of immunity problems that may be observed in the field. The primary difference between ITE/ISM equipment and other categories of systems lies in specification limits and test requirements.

1. Electrical fast transients and bursts
2. Surge or high-energy line transients
3. Conducted currents
4. Voltage variation, sag, and dropout

6.4.1 Electrical Fast Transient and Burst Testing

Of all immunity tests, electrical fast transient and burst (EFT/B) testing is the easiest to perform and provides the most information related to potential emissions and immunity problems. Fast transient burst testing provides a broadband signal from the low-kilohertz range up to 60 MHz and beyond. Most digital logic components are sensitive to transient changes in logic signals. Transient pulses that couple onto transmission lines can be interpreted as a valid signal transition, causing functional problems. For analog devices, the envelope of the burst can be detected by sensitive amplifiers as if the signal was a pulse-modulated CW. Use of only an EFT/B generator is preferred by many engineers over other instrumentation when evaluating products to immunity standards. This is due to the broadband spectrum of energy that can be broadcast to the EUT along with variable voltage levels.

The fast transient/burst test aims to simulate AC mains disturbances by creating a showering arc of energy at the contacts of an ordinary AC mains switch or relay during operation. The inductance of the mains cable plus any inductance in the load causes a reflected voltage to be developed at the instant the current is interrupted. This reflected voltage rises until it is sufficient to break down the air gap at the contacts of the switch, creating an arc. When the arc stops, the reflected voltage level will reoccur, this time with the contacts of the switch slightly further apart due to the switch action in motion. During the opening of a switch or relay, transients generated in the switch can start with a low-amplitude signal and a high-frequency content, ending with a rising amplitude level signal and decreasing frequency content.

Bursts of fast transient pulses are capacitively injected into cables of all types. This waveform and a simplified drawing of the test generator are shown in Figure 6.7 Details of the electrical characteristics of the transient burst generator are found in IEC/EN 61000-4-4 [2]. The object of this standard is to establish a common reproducible basis for evaluating equipment when subjected to fast transients (bursts) on supply, signal, or control lines. Items that produce switching transients are those that contain inductive or relay contact bounce. The net result of this test is disruption to operating circuitry. This is intended to be a nondestructive test. Test severity levels vary based on whether the cables are mains voltage or data. Several levels of performance are defined in the standard, detailed in Figure 6.7 [3, 4].

The EFT/B-type disturbances are generated when inductive–capacitive circuits are interrupted. Inductive loads such as relay coils, motors, contactors, and timers are connected or disconnected from the line. When this happens, a spark is developed between the mechanical contacts of the switch. This arc is unstable in nature. This intermittent arc continues only as long as the voltage at the contacts exceeds the threshold level of electrical breakdown of the spark gap. Showering arcs result

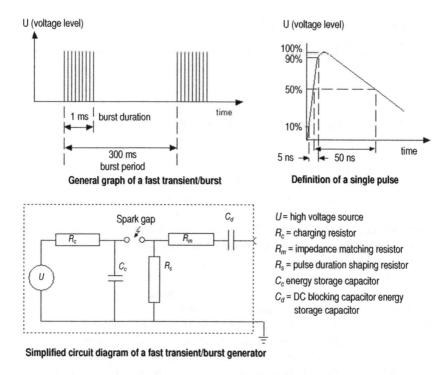

Figure 6.7 Electrical fast transient waveform, test generator, and test specifications.

from the operation of switches, contactors, and relays in industrial process operations. Similar transients are also observed with gas discharge lamps.

When more than one electrical device is connected to the same AC mains receptacle, which provides close coupling, removing power from one system while maintaining power to the other will often subject the device still powered to an EFT/B.

The EFT pulse train can interfere with the operation of equipment even when the intensity of a single pulse is not strong enough to disturb performance. Each pulse tends to charge the input capacitance (including stray capacitance) of logic devices. This charge effect may be of such magnitude that the stray capacitance will not have enough time to fully discharge between successive pulses. Under this condition, the voltage can slowly build up to a threshold level that can cause circuit malfunction.

Transient pulses affect a logic component through cable interconnects, regardless of the type of signal on the transmission line. Rarely does a radiated EFT pulse cause functional disruption; however, coupling of externally induced fields can occur at cable assemblies and be transferred directly into the system by conducted means. For this reason, EFT testing is performed on both signal and power lines.

Electrical fast transient and burst testing is usually performed in a controlled electromagnetic environment (i.e., a shielded enclosure) or a laboratory to ensure that the environment itself does not influence test results, especially when in the presence of strong radiated fields from broadcasting towers or power distribution subsystems. Climatic conditions do not substantially influence test results.

The test waveform involves calibration with a 50-Ω load. This is because the output of the generator includes a 50-Ω attenuator. The test *itself* is performed without any such limitation on the use of termination. Thus, the load into which the test wave is fed and the actual test wave shape are unspecified.

Test Networks

(a) *Coupling/Decoupling Network.* A CDN provides the EFT/B signal to the lines under test. An important requirement is that the interference signal needs to actually reach the EUT. On signal lines, the CDN prohibits noise from being injected back into the auxiliary equipment (AE). The same noise must not be injected back into the AC mains input of the test generator. The decoupler ensures that the burst signal is attenuated before it reaches other equipment both upstream and downstream of the EUT and auxiliary equipment which is not under evaluation. The CDN couples the signal from the EFT/B generator to each phase of the AC line though a capacitive coupling clamp.

(b) *Capacitive Coupling Clamp.* The capacitive coupling clamp couples the EFT/B waveform to cables without any galvanic or metallic connection. The coupling capacitance of the clamp is dependent on the size and material of the cables involved. Typical values of coupling capacitance between clamp and cables are 50–200 pF while the diameter of the cable ranges between 4 and 40 mm. Although a capacitive coupling clamp is the preferred method for injecting EFT/B waveforms, the clamp may be replaced by conducting tape or metallic foil that wraps the lines under test. This alternate means of coupling is acceptable as long as the capacitance of this arrangement is approximately equal to the capacitance of the standard coupling clamp.

(c) *Ground Reference Plane.* The ground reference plane is an important component of the test setup. Because the EFT generally has 5-ns rise time, widely varying test data are possible from poorly configured setups and from a misunderstanding regarding required generator performance. The reference plane contains tens to hundreds of picofarads of nearly inductance-free capacitance to free space, resulting in very low impedance for the rise time pulse. Permissible variation in the dimensions of the ground reference plane is indicated in different test plans and standards.

The main element of a burst generator is a high-voltage DC power supply that charges an energy storage capacitor through a charging resistor. When the energy storage capacitor voltage reaches a particular level, the spark gap in the generator will break down. The capacitor will discharge and the pulse waveform will be shaped by the output filter. The resultant waveform is depicted in Figure 6.7.

Operation of the charging circuit, or spark gap, is controlled to produce 15-ms bursts of transients at 300-ms intervals. The repetition rate of the transients is also controlled and is defined for the peak output voltage values. Both positive and negative polarities are required during testing. The test generator must be calibrated, which is detailed in the test standard.

Typical Test Procedure.   The specific manner for the most common type of EFT testing to be performed is detailed in individual product test standards. These procedures are provided in Appendix B for ITE and ISM equipment only. The test condition must mimic normal installation requirements for both table-top and floor-standing units. Earth grounding of the system must be in accordance with electrical codes.

The test voltage is applied using an appropriate coupling device. When the coupling clamp is used, the minimum distance between the coupling plates and all other "conductive structures" should be 0.5 m, except for the reference plane beneath the clamp. The length of the cable between the coupling device and EUT should be 1 m or less.

If the line current is greater than the specified rating of the coupling network, the test should be repeated as a field test. For this application, the output of the EFT/B generator is connected to each of the power supply phases and the reference plane. This reference plane should be approximately 1 m × 1 m, mounted as near to the EUT as possible and connected to protected earth. The generator is positioned on this plane. The length of the cable between the generator and EUT should not exceed 1 m.

For some applications, a 33-nF blocking capacitor is provided between the generator and appropriate supply mains line being tested. When the blocking capacitor is used, it is because an appropriate decoupling network is not available. Care must be exercised to ensure that there is no disruption of other equipment located within the vicinity of the EUT.

For field testing on signal lines, an appropriate clamp should be used. If use of the clamp is not possible for physical reasons, aluminum foil may be wrapped around the cable assembly with a capacitance value equal to that of the standard coupling clamp. In other application, direct coupling of the generator to signal lines can occur with 100 pF capacitors.

A test plan needs to be prepared that specifies the following:

- Test voltage level
- Duration of the test (i.e., >1 min per phase or line)
- Polarity of the test voltage (positive and negative)
- Number of applications of the test voltage

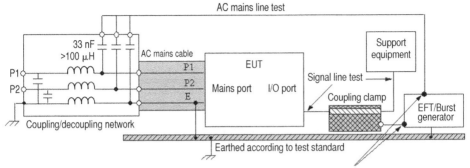

Figure 6.8   Test configuration setup for EFT/B.

- Circuits, lines and other cables to be tested, along with the coupling device required
- Representative operating condition of the EUT, including simulated signal sources
- Sequence of application of different test voltage levels

An example of a typical test setup for EFT/B testing is shown in Figure 6.8.

Standard IEC/EN 61000-4-1* provides for an agreement between manufacturer and user, including the voltage test level. In addition, the standard also states, in Section B "The selection of appropriate test levels is the task of the relevant product committee or is subject to an agreement between manufacturers and users; in all cases the technical-economical optimum is to be considered."

Proposed Diagnosis and Fixes. Test results must be recorded and analyzed carefully. Items of concern are operational failures. Details on each suggestion are provided in numerous reference books [3], as the magnitude of the fix may be extensive and is beyond the scope of this discussion.

The primary causes of failure include the following:

- Ungrounded metallic connector housings
- Incorrect or no bonding of metallic connector housing to the chassis assembly
- Signal wire traveling too far inside the system before filtering (i.e., pigtail)
- Missing or improper use of cable shield connection (braid, foil, drain wires)
- Incorrect placement, mounting, or selection of a power line filter
- Lack of proper common-mode filter capacitors
- Incorrect application and mounting of filter components
- Poor PCB layout

*IEC/EN 61000-4-1. Electromagnetic Compatability (EMC), Part 4: Testing and measuring techniques. Section 1: Overview of immunity tests. Basic EMC Publication.

Suggested ways to prevent failures are as follows:

- Address the list of potential problem areas described above.
- Apply common-mode ferrite clamps very close to cable entries. These ferrite clamps are generally the split-core version that allows for ease of use.
- Replace plastic connector housings with metal housings. Ensure the metal housing is bonded to the chassis and cable shield and not to the reference (ground) plane of the PCB.
- Install in-line filtered connector assemblies on critical signal lines.
- Improve PCB layout.

The key item to remember is that a capacitively coupled burst of energy is injected onto the signal or power lines. This means that it is critical to filter out this unwanted noise at the interconnect location before it causes disruption to operational circuitry. The energy is nondestructive, physically, but can cause total system failure electrically.

6.4.2 Surges

Surges (or high-energy line transients) test products to overvoltage levels caused by switching transients and lightning strikes. Various test levels are specified in regulatory standards depending on the environment and installation. It becomes the task of the engineer to determine the reaction of the EUT under specified operational conditions caused by surge voltages. The most common surge immunity standard for commercial products is IEC 61000-4-5 [5], although other standards exist for different types of surge events.

Surge testing according to IEC 61000-4-5, similar to the European Basic Immunity Standard EN 61000-4-5, is usually at levels of ±1 kV for line-to-line surges and ±2 kV for line-to-ground surges. However, AC mains power provided to typical urban buildings will suffer from line-to-ground surges of 6 kV at least once per year. This surge level is caused by normal (local) thunderstorm activity, not direct strikes. This high level of surge energy generally applies to buildings that do not have lightning protection systems. Rural buildings whose mains supply is provided by overhead conductors are usually exposed to tens or hundreds of 6-kV surges every year, depending on the length of their overhead line. The current surge waveform 8/20 called out in IEC-61000-4-5 is NOT representative of the surges generated by lightning strikes, but is typical of the output of the primary lightning protection components (10/350 waveform), representing the residual transient that may penetrate the equipment after the primary protection devices have suppressed the initial lightning strike current.

Sources of transients carried by electrical power supply lines are classified in two categories—switching (EFT/B in nature) and [2] lightning (surge) [4]:

1. Overvoltages usually caused by terrestrial phenomena such as lightning
2. Burst of high-frequency transient noise due to rapid switching of reactive loads (see Section 6.4.1 for a more in-depth discussion of this issue)

3. A sudden increase or decrease in the mains voltage caused by the switching of low-impedance loads

Transients identified in 1 are generally coupled to power supply input lines by means of a radiated electromagnetic field. Transients in 2 and 3 are usually coupled directly to the power supply mains.

Switching transients, can be caused by the following:

1. Major power system switching disturbances, such as capacitor bank switching
2. Minor switching activity near the instrumentation or load changes in the power distribution
3. Resonating circuits associated with switching devices such as thyristors
4. Various system faults, such as short circuits and arcing faults, to the earth grounding system of the installation

With regard to *lightning transients,* a typical lightning strike, or stroke, can last for over 1 s and consist of many discharges, sometimes over 10 discharges with each stroke. Along with this large number of discharges, each stroke can have an arc channel current between 2 and 200 kA (1% of the strokes can exceed 200 kA). Lightning can affect systems in the following manner [3]:

1. *Earth Potential Lift.* Soil has significant resistance. A lightning strike can cause large potential differences between areas that are normally at the same earth potential.
2. *Magnetic Induction.* A very high voltage surge can be induced into conductors by magnetic coupling from a lightning strike, even at a significant distance from the strike location.
3. *Current Injection.* Direct strikes to external equipment or cables often result in damage to equipment and can damage unrelated equipment due to side flashes in shared cable routes or terminal cabinets.
4. *Electrical Induction.* Electric fields of approximately 500 kV/m are possible before a lightning strike. This electrical induction may be observed up to 100 m from the eventual strike point. This induction can inject damaging current into conductors and devices.
5. *Lightning Electromagnetic Pulse (LEMP).* This far-field effect can be caused by intracloud lightning as well as by distant cloud-to-ground strokes. A side effect is thermal and mechanical concerns that can be associated with intense energies affiliated with a lightning strike.

The following mechanisms are responsible for generation of surge voltages by *terrestrial transients* [3]:

1. *Lightning strike near objects located on earth ground or within the cloud layers:* This strike produces high-intensity electromagnetic fields that can induce a voltage on the conductors of both primary and secondary circuits.

2. *Ground current flow from lightning resulting in cloud-to-ground coupling:* Current flow enters the common ground impedance path of grounding networks, causing voltage differences across both its length and breadth.

3. *A rapid change in voltage levels when a primary gap-type arrester operates to limit the primary voltage coupled through the capacitance of transformers:* This voltage surge is in addition to that coupled into secondary circuits by normal transformer behavior.

4. *A direct lightning strike to high-voltage primary circuits injects high current levels into primary circuits, producing voltages flowing either through the ground resistance, causing a change in ground potential, or through the surge impedance of primary conductors:* A portion of this voltage will couple from the primary side of the transformer to the secondary side by capacitance, transformer behavior, or both. The results appear in low-voltage AC mains circuits.

5. *A direct lightning strike to secondary circuits:* Under this condition, very high levels of current and voltages may exceed the withstand capability of equipment and conventional surge protection devices rated for secondary circuit use.

Surges that exist within power transmission lines, even unidirectional surges, can excite the natural resonance frequency of a system. Therefore, surges are typically oscillatory in nature and may have different amplitudes and waveforms at various locations along the power distribution line. These surges propagate as a voltage transient between phases or between phase (line) and neutral or ground. Transients can result in arc-over if transmission lines are physically located close to each other or to a grounded conductor within a system.

Power distribution systems are expected to deliver a constant voltage and frequency. If the load impedances remain constant and the line is not exposed to externally induced electromagnetic fields, everything is balanced, which is often not the case. Residential and industrial facilities present different impedances that vary in magnitude and phase. Every time an item is switched on or off, transients are injected into the power lines in addition to natural phenomena such as lightning. It is these surges and phase shifting that cause systemwide failures.

Telecommunication Equipment. From July 1, 2001, almost all newly supplied commercial and light industrial systems, including telecommunication equipment provided to Europe, must be declared to meet EMC standards that include surge testing. Telecommunication equipment uses a standardized lightning waveform, with typical peak levels of 2 kV from each line to ground and 1 kV from each line to each other line for AC power. Where cables can be longer than 10 m, they may also require surge tests with respect to ground, typically 1 kV peak [6].

Unfortunately, there is no coordination between the surge test levels required by the EMC immunity directive for Europe regarding overvoltage protection and exposures predicted by lightning standards. Equipment meeting the EMC directive's surge immunity standards cannot be used for all installations throughout Europe or the world, along with guarantees for a reliable system and network. Telecommunication equipment will still need to be evaluated according to its zone or risk category within a given structure and either moved to a zone with lower exposure or fitted with additional system protection devices or other protective measures as required. The reason for this "lack of coordination" is due to the fact that surge immunity requirements deal with the residual surge after the primary protection device, "converting" the 10/350 lightning waveform, which more closely resembles a real lightning strike, to the 8/20 waveform specified in test standards.

Instrumentation. To create a test environment for surges, a combination wave generator is required. A simplified generator circuit is shown in Figure 6.9. The values for the components are specified to deliver a 1.2/50-μs voltage surge (under open-circuit conditions) and a 8/20-μs current surge into a short circuit. The generator's output has an effective output impedance of 2 Ω. This type of generator contains both an open-circuit voltage and short-circuit current waveform and is referred to as a *combination wave generator* or *hybrid generator*. A unique waveform with 10/700 μs characteristics is specified for telecommunication equipment.

For the waveform specification there are two numbers. The first number (i.e., 1.2) refers to the rise time of the pulse up to a maximum value in microseconds. The second number (i.e., 50) is the period where the signal decays to an operational level, which is specified as 50% of the peak value. An example of this waveform characteristic is shown in Figure 6.10. When testing mains inputs, the surges are applied (as a positive or negative voltage) at all zero crossings and at the 90° and 270° phase angle of the AC mains waveform. Time is allowed between each impulse to avoid overheating surge protection devices.

The frequency spectrum of the surge test is much lower than that in the EFT or ESD tests. Thus, the test setup does not require a reference plane but does require an earth connection. Be aware that the surge current can reach kiloamperes; therefore the wiring between the generator and the EUT must be robust.

Important safety note: The instantaneous power and total energy in a surge test can be extensive. Should a failure occur, there is enough energy in the circuit to cause electronic devices to explode with potential harm to the operator. For this reason, surge testing should only be carried out in a controlled environment where only the operator is permitted to be present. Those who wish to witness the test must do so at a distance, preferably protected by a fiberglass barrier. Fire extinguishing equipment should be readily available and there should be quick access to AC mains breakers [4].

Performing Tests. Once we have a generator that can provide a specific waveform, this energy must be injected into a cable assembly. This is accomplished with a CDN. This network shall not significantly influence the parameters of the test environment, which include any aspect of the waveform characteristics or its tolerances.

For power supply circuits, capacitive coupling allows the test voltage to be applied from line to line or one line to earth at a time. All possible combinations of

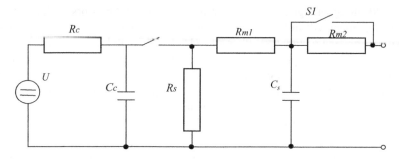

U   High-voltage source
Rc Charging resistor
Cc Energy storage capacitor (20 μF)
Rs Pulse duration shaping resistor (50 Ω)
Rm Impedance matching resistors (Rm1 = 15 Ω; Rm2 = 25 Ω)
Cs Rise time shaping capacitor (0.2 μF)
S1 Switch closed when using external matching resistors

Figure 6.9   Surge generator simplified circuit diagram.

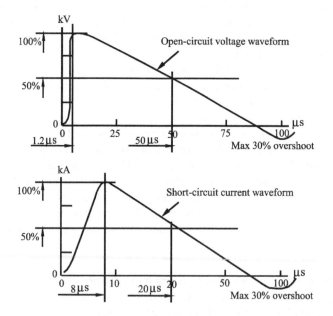

Figure 6.10   Surge waveform characteristics.

coupling are to be tested. For a single-phase system, the test will be L1–L2, L1–GND, L2–GND, and L1–L2–GND, where the "L" represents line voltage. For a three-phase unit, the test will be extended to include L1–L2, L1–L3, L2–L3, L1–GND, L2–GND, L3–GND, and L1–L2–L3–GND.

For a network of interconnection lines, coupling occurs via capacitive means using either a capacitive coupler or an arrester. Details on these particular coupling units are found in test standards, as very specific requirements exist.

Coupling to signal lines can be a problem, as it must be invasive. There are no commercial clamp-type networks available for this test. However, for signal lines affected by a 0.5-μF capacitor connected to them, it is permissible to use a gas discharge tube (surge arrester) coupling instead. This method is shown in the test standard IEC/EN 61000-4-5 for shielded lines. The surge is effectively applied longitudinally along the shield by coupling it directly to the EUT at one end of a noninductively bundled 20-m-long cable with the farthest end grounded. This test is carried out with no series resistor. The surge current in the cable shield will be limited to several hundred amperes.

Due to the number of possible configurations possible for test setup, refer to IEC/EN 61000-4-5 [5] for details. These configurations include capacitive coupling for AC/DC power (all lines to lines, lines to earth), three-phase alternating current, unshielded interconnect lines (generator output earthed, coupling capacitors, and arresters), and unshielded symmetrically operated lines (telecommunication lines, via arresters). Surge testing can be performed in any location. Use of a shielded enclosure is not mandatory or discouraged.

Appendix B provides greater detail on actual test procedures and equipment setup.

Testing at Both High and Low Levels. Depending on system design, an upset can occur at any level of stress testing. Testing solely at maximum levels as specified by the test standard may provide inclusive results. A fraction of the maximum test level needs to be applied, increasing in value until the maximum value is reached. Typically, surge tests start at 500 V and increase in 500-V steps up to the maximum voltage level. One can apply a higher surge level to ascertain the breaking point of the system if desired.

When surge protective devices (SPDs) are provided, it is recommended that tests be performed at stress levels both below and above the rated values of the SPD. This is because actual surges with levels less than maximum are more likely to be effective than at higher surge levels. It has been found that systems that meet the maximum surge test levels can suffer failure with less powerful surges. For example, a gas discharge tube (GDT) will only fire when the voltage threshold is reached. When the maximum surge test level is applied, the GDT will allow an overvoltage for a few microseconds before triggering, protecting the circuit. However, at a lower level, the surge voltage may be insufficient to activate the GDT. This condition will allow the surge to be applied to the circuit for its full duration, causing possible damage.

When surge testing data lines, the actual operating signal source should be provided or simulated. The auxiliary host equipment must contain back-filter protection from harmful damage to the lines.

In Situ Testing. In situ testing of surge requirements is not difficult on small equipment and in fact is not specified in the test standard. This test is not common outside the laboratory. In situ is relatively simple to set up if portable test gear is available. No reference plane or shielded room is required. The test standards describe "system level testing" recommended to demonstrate reliability in an installation rather than compliance with any specific regulations. However, it may be possible to get an EMC Competent Body to agree to accept on-site testing when following the Technical Construction File (TCF) route to EMC compliance, especially for custom equipment intended for a specific site.

The surge waveform has a much lower bandwidth than either ESD or EFT pulses; therefore radiated coupling is rarely a problem. The major concern is that the surge is a high-energy destructive transient. It becomes important to implement decoupling networks (back filter) to ensure that energy is only applied to the EUT. Trying to apply a surge without this network inserted into a power supply cable shared by other systems would be a potential disaster.

If the connection to the EUT can be broken to insert a suitable CDN, ancillary or adjacent apparatus should be properly isolated from the surge. With this network, surge testing in situ becomes feasible.

Diagnosis and Fixes. Should a system failure occur due to a transient surge, a fix or redesign is required. Surge isolation and protection devices can be used to prevent damage.

Surge protection devices cannot protect a system from the direct effects of lightning. The magnitude of the current is too massive and simple semiconductor or passive components are unable to withstand this level of surge. It becomes mandatory to correctly design and route cables as well as provide for the installation of high-current-rating components and filters. To protect against lightning, additional expense for special components sometimes can be significant. For example, a typical SPD that is affixed to a routed data cable in a building will not protect against a side flash from a down conductor and may not even be sturdy enough to protect against magnetic induction from a nearby down conductor or external communication line such as used for terrestrial telecommunication purposes. Surge protection devices should not be fitted where there is a risk of fire or explosion, unless special precautions are taken to prevent this hazard. Refer to vendor data sheets and application notes on installation of SPDs and requirements for protection against fire or electrical shock.

Isolation involves separation of circuits and systems away from each other, properly grounded to earth, each with individual protection. The difference between SPDs and isolation is minimal, with more reliance on diverting the energy charge away from devices before it enters the system, whereas isolation becomes the next level of protection.

6.4.3 Conducted RF Current Immunity

Sources of harmful disturbance, which is basically an electromagnetic field from both intentional and unintentional radiators (RF transmitters), can affect operation of electrical products, especially if the product has interconnect cables (e.g., AC mains, data, control, and interconnect). The physical dimensions of equipment that is being disrupted, which is generally a subassembly of a larger unit, are assumed small compared with the wavelengths of propagating fields. Cables behave as passive receiving antennas and are usually several wavelengths in length.

When network cables are provided, susceptible equipment becomes exposed to currents that flow through the network. Cables that connect to an active driver/receiver may be in resonant mode ($\lambda/2$, $\lambda/4$, open or folded dipoles) and can be represented by coupling and decoupling devices having a common-mode impedance of 150 Ω with respect to a reference plane.

Conducted immunity tests involve the injection of RF currents into a cable or wire. This current simulates fields from intentional RF transmitters. The disturbing fields (*E* and *H*) result from voltage and current levels generated by test equipment.

The use of CDNs to apply a disturbing signal to a single cable (or bundle for certain applications) is performed while not exciting other cable assemblies near the injection point. The test is supposed to represent a real-life situation that may exist with a range of different amplitudes and phases.

Figure 6.11 illustrates a comprehensive test environment for conducted immunity testing using the two primary methods of testing: current injection and voltage injection.

6.4.3.1 Coupling Methods. Three primary methods of injecting RF current into cables are detailed in IEC/EN 61000-4-6 [7]. The preferred method is direct voltage injection via a CDN. The CDN (detailed in Chapter 4) has extremely low insertion loss and therefore requires little power. Another alternative is the clamp-

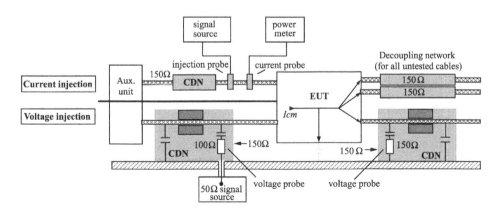

Figure 6.11 Conducted immunity test environment.

on current probe, which is easier to use but has a much higher insertion loss, requiring that a calibration factor be applied to the test results, mandating use of a higher power amplifier if the losses are too large. These probes require more drive power to energize the cable. Between these two devices is the EM clamp, which contains ferrite cores that are clamped over the cable of interest. The EM clamp provides both *E*- and *H*-field coupling to the cable and is more efficient than either the voltage or current probe.

For any method used to inject RF current into a cable, the common-mode impedance at the end of the cable remote from the EUT must be defined. For each configuration, or type of cable, a common-mode decoupling network or ISN is provided at the far end. These networks ensure the impedance of the transmission line and isolate any ancillary equipment from affecting the RF current on the cable. This type of configuration is similar to a LISN (AC mains emissions), except that the technical specifications for a LISN do not match that of those mandated for the ISN. When using direct voltage injection, the ISN couples RF voltage onto the cable. This is a more complicated manner of testing from that of using bulk current injection. Bulk current injection requires only use of a simple probe. Test houses generally stock a wide variety of CDNs for almost every cable configuration. If an ISN or CDN is not available, one will have to be manufactured and calibrated, which is not an easy task to perform.

The current probe has advantages over voltage injection. This is a nonintrusive test. The probe clamps over the cable upon which common-mode current is injected. This is an attractive method for systems with many conductors and cable assemblies. One disadvantage is that stray capacitance between the probe and cable is generally undefined. Not knowing the value of stray capacitance may cause the cables to be overstressed with higher levels of RF power. In addition, stray capacitance limits the probe to a specific usable frequency range, since at higher frequencies both the inductive coupling path and cable common-mode impedance can be affected. The EMC clamp does not have these problems, at least to the magnitude that the current probe has.

Another uncertainty regarding test levels to ensure conformity for either voltage or current injection deals with the amount of power delivered by the amplifier to the injection probe. The amount of power is dependent on the common-mode impedance at the EUT port. If a low-impedance connection is present, the power output from the amplifier must be increased, which means that the system may be overtested (if a voltage source is used) or undertested for a current source (high impedance).

To alleviate this problem, precalibration of probes is required. Details of calibration procedures are provided in the test standard and summarized in Appendix B. A power meter is connected to one end of the calibration unit and a 50-Ω terminator to the other end (Figure 6.12). A known power level from an RF amplifier is sent to the probe. The magnitude of power is adjusted to provide the desired test level on the power meter. Since this is a frequency sweep test, the power level of the amplifier is left untouched. The RF output amplitude from the signal generator that is fed to the RF amplifier is adjusted at every frequency to ensure a constant power level is provided to the meter. It is easier to read a digital readout on a signal generator

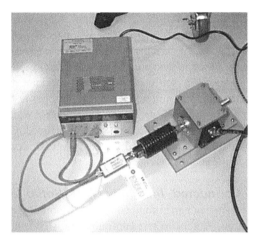

Figure 6.12   Calibration of bulk injection probe with power meter.

for accuracy instead of rotating a dial blindly on an amplifier. A calibration chart must be maintained and used during the test. Upon completion of calibration, one only needs to monitor the frequency and amplitude level on the signal generator.

The current (bulk) injection probe is removed from the calibration jig and secured over the wires to be tested (Figure 6.13). The wires/cables will have fixed impedance. With an impedance mismatch existing between probe and cables, the am-

Figure 6.13   Bulk current injection probe and regular current probe (monitoring purposes).

plifier must provide a smaller or a greater amount of RF power to the probe to ensure that the proper power level is available at all times. To determine the power level on the cable(s), a standard current probe is located adjacent to the bulk injection probe. The current probe is then connected to a power meter or spectrum analyzer. If the signal generator is set to its calibration set point, the value of power measured on the meter becomes a nonconcern if the proper test level is being applied. If the measured power level is greater than the level at which the product is to be tested, the signal generator's amplitude output is reduced. Once the resonance point of the cable assembly returns to normal, the amplitude level must be reset to its original calibration point.

6.4.3.2 Typical Conducted Immunity Test Setup and Equipment.

There are two concerns when performing conducted immunity tests: test equipment and setup configuration. A generic test configuration is shown in Figure 6.14 for both table-top and floor-standing systems. The placement of the ground reference plane is either on the floor or on a table 0.8 m above the reference (ground) plane.

Equipment and Networks. The specification for test equipment and coupling/decoupling networks is provided in IEC/EN 61000-4-6 [7]. There are many types of configurations and networks provided for this immunity test. It is to be noted that there are two concerns when performing this test [3]:

1. The coupler for injecting a common-mode current must be noninvasive, that is, preferably a current clamp. To prevent loading of a cable under test by the mirror impedance of the RF generator, the clamp must be a weak-coupling current transformer. For example, with a 4 : 1 turn ratio, the generator impedance observed on the tested cable is in series with $50 \, \Omega (\frac{1}{4})^2 \cong 3 \, \Omega$, which is an

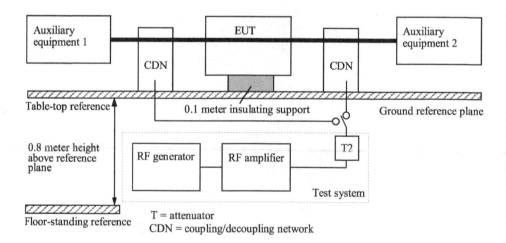

Figure 6.14 Generic conducted RF test configuration.

acceptable value. Because of this mismatch, a significant loss of power in the coupler happens. A typical value for probe loss is −10 dB. This means it takes 10 W of power to inject 1 W into a typical EUT cable. This corresponds to simulating 10 V/m RMS plus 20% AM-modulated incident field on a cable located approximately 0.50 m above the reference plane.

2. When EUT cable lengths approach a quarter-wavelength resonance, these cables in themselves can become a radiating antenna dispersing the current. Much of the injected power is wasted in developing a propagating field instead of entering the EUT. This means that the original problem of preventing radiated harm to the environment will be reestablished, which is exactly what we do not want. For this reason, use of a shielded enclosure is recommended.

With the two concerns noted above, there is a significant benefit of using BCI:

1. With the exception of quarter-wave resonances, the injected field enters the cable assembly without being propagated into space as a strong radiating field.

2. If using a proper test setup and environment, with cable routing of 3–5 cm above the reference ground plane, a reduction in the cable-to-near-field coupling will occur below 600 MHz. The BCI method is not recommended for use above 230 MHz. By coupling cables to the ground plane, a reduction in radiated energy by approximately 6 dB or more is present.

3. The effects of a human body near the test setup or EUT will not affect the RMS value of power being injected into the cable. This makes it easier to perform the test when a shielded enclosure is not available.

The discussion below is a summary of the equipment and specifications described in the test standard. The standard allows three types of transducers: the CDN, the EM clamp, and the current injection probe. The standard also allows direct injection that is not commonly performed at test laboratories [3].

(a) *Test Equipment.* The test system consists of several components that supply the input port of each coupling device with a disturbing RF signal at the required level specified in the test plan. This system can be a fully integrated unit or separate components connected together (Figure 6.14).

(i) *RF Signal Generator.* This generator must be capable of covering the frequency band of interest, generally 9 kHz–230 MHz. If a generator is designed for operation from 9 kHz to 1 GHz, this same unit can be used for radiated immunity tests. Most international test specifications require AM modulation of a 1-kHz sine wave with a modulation depth of 80%. Details on this waveform are provided in Chapter 7. The signal generator should have an automated sweep capability of 1.5×10^{-3} decade/second and/or manual control. Should an RF synthesizer be used instead of a signal generator, this unit shall be programmed with frequency-dependent step sizes and dwell times. The tracking generator of a spectrum analyzer may be adequate and should accept amplitude modulation, be programmable to sweep by increments, and include halt-on-step functions when needed.

(ii) *Wideband Power Amplifier(s)*. A power amplifier is required to amplify the input signal from the signal generator/synthesizer to appropriate voltage levels if the generator/synthesizer is unable to provide sufficient drive power. Due to the design of the amplifier and limited range of operation, more than one amplifier may be required. The amplifier power depends on (a) the efficiency of the coupling network and (b) the E-field level required for stimulation purposes. Based on a typical coupler plus a mismatch loss of 10 dB, Table 6.1 details the amount of power required, including modulation margins to actual susceptibility levels with a 6-dB margin of compliance when the BCI clamp is used.

(iii) *Attenuator* T_1. This attenuator is typically in the range of 0–40 dB with adequate frequency and power ratings to not only control the disturbing test source output (can be internal to the signal generator/synthesizer) but also prevent a reflected wave from a transmission line mismatch causing damage to the front end of the RF signal generator.

(iv) *Attenuator* T_2. This attenuator is a fixed-value device ($\geq$6 dB, $Z_0 = 50$ Ω) with sufficient power ratings to prevent or reduce the mismatch from the power amplifier to the network. This attenuator shall be located as close as possible to the coupling device. This attenuator can be provided internal to a coupling network, with the ability to be switched out of the circuit should the broadband power amplifier remain within specification under any load condition.

(v) *RF Switch*. The RF switch permits the disturbing test signal to be switched on and off when measuring the immunity level of the EUT. This switch can be incorporated internal to the generator/synthesizer.

(vi) *Low-Pass Filter (LPF) and/or High-Pass Filter (HPF)*. These filters may be required to avoid interference with certain types of equipment, that is, RF receivers caused by (sub)harmonics. When required, these filters should be inserted between the broadband amplifier and attenuator T_2.

Table 6.1   Power Requirements for BCI Clamp

E-Field Requirements (V/m)	AM Modulation (%)	Open Voltage to Coupler (V)	Amplifier Power for 150 Ω Common-Mode Impedance (W)
1	50–80	1	0.16–1
3	50–80	3	5–10
10	50–80	10	60–100

(b) *Coupling and Decoupling Networks.* Coupling and decoupling devices are required for appropriate coupling of a disturbing signal over the entire frequency range of interest, with a defined common-mode impedance at the EUT port or various cables of the EUT. These CDNs can be combined into one assembly. The primary specification for the output impedance of the CDN is defined for two frequency ranges:

150 kHz–26 MHz $150 \, \Omega \pm 20 \, \Omega$
26–80 MHz $150 \, \Omega + 60 \, \Omega - 45 \, \Omega$

(c) *Direct-Injection Clamp.* For applications where a CDN cannot be inserted, direct injection may be used. The disturbing signal from the test network is injected onto the shield of cables and coax with a 100-Ω resistor.

    When using this test method, a decoupling circuit shall be inserted as close as possible to the injection point. For simple shielded-cable configurations, the decoupling circuit and 100-Ω resistors may be provided into one assembly. This combination network may be installed into a CDN designed specifically for a particular shielded-cable configuration.

(d) *Clamp Injection.* When a clamp injection device is provided, the coupling and decoupling functions are separated. Coupling is achieved by a clamp-on unit while the common-mode impedance and decoupling functions are established at the auxiliary equipment (AE).

    When an EM clamp or current clamp is used, the procedures in the test standard must be carefully followed as certain restrictions on calibration and operation exist. When using clamp injection, the resulting injected current on the cables must be monitored and corrected using a second current probe. The opening of the clamp is recommended to be between 2.5 and 4 cm in diameter, or large enough for the biggest cable to be tested.

(e) *EM Clamp.* The EM clamp is another version of clamp injection. This unit is located over the cable to be tested. Both electric and magnetic fields are injected into the cable under test. This is a different test from the current or clamp injection, where only one type of electromagnetic field is provided. The test setup for the EM clamp is provided in Figure 6.15.

(f) *Bulk Current Injection Probe.* This probe injects currents equally into the cables on both sides of the device. This may permit for more radiation from longer cable runs toward the "ancillary equipment." A problem with BCI is that the injected power has no directionality; it both stresses AE as much as it tests the EUT. If the ancillary equipment is susceptible, it can be protected by fixing a clip-on ferrite core 200 mm or more in length to the AE side of the BCI transducer. Doing BCI testing inside a shielded room and placing the AE outside the shield room, or running interconnections through filtered bulkhead connectors in the wall of the room is one way to protect AE.

(g) *Current Probe.* A current probe establishes magnetic field coupling to the cable being tested, similar to clamp injection. When using a current probe, care should be taken to ensure that the higher harmonics generated by the power amplifier do not appear at levels higher than the fundamental signal levels at the EUT port of the coupling device.

It is recommended that the cable be positioned directly in the center of the current clamp to ensure uniformity of field penetration. These probes are not designed for high-power applications and can be damaged if subjected to large RF fields. The main use for a current probe is to monitor the power level from the injection clamp/probe.

(h) *Calibration Jig.* This device is used to calibrate the test equipment and probes/clamps prior to performing the test.

(i) *Directional Coupler.* This is an optional device, highly recommended for monitoring the power delivered to the probe in the forward direction. Reverse power is rejected by the coupler.

(j) *Spectrum Analyzer or EMI Receiver.* This instrument monitors the power injected into the cable assembly. The current probe detects the amount of power present. In addition, the analyzer allows one to ascertain if the current level is appropriate or if too much power is being injected into the cable.

(k) *Decoupling Networks.* These networks are optional, as they have less accuracy due to mismatched lines that are to be tested. Decoupling networks provide a constant common-mode impedance of 150 Ω above 100 kHz, which helps reduce mismatch oscillation and high VSWR. It is recommended that decoupling networks be provided with a connector that matches the type of I/O cables used by the EUT (e.g., D-sub, RJ-11, RJ-45).

Clamp Injection Application. When using clamp injection, the AE shall present the common-mode impedance as close as possible to the required specification. Each AE used with this test procedure shall represent a functional installation. To ensure proper common-mode impedance, the following are required:

- Each AE shall be placed on an insulating support 0.1 m above the ground plane.
- All cables connected to each AE, other than those to the EUT, must be provided with decoupling networks located no further than 0.3 m from the AE. These cables shall not be bundled or wrapped and shall be kept between 30 and 50 mm above the reference plane.
- The cable length between the AE and clamp injection device shall be as short as possible (0.3 m) to improve reproducibility at higher frequencies (≥30 MHz).
- At each AE, the decoupling network for the specific cable being tested shall be replaced by a CDN terminated at its input port with 50 Ω, if possible.

Important Note Regarding Calibration. The test voltage given in the specification for the EM clamp is the open-circuit voltage to be applied. Paragraph 6.4.1 and Annex A, paragraphs A1 and A3, in IEC61000-4-6 [7] indicate how the test levels

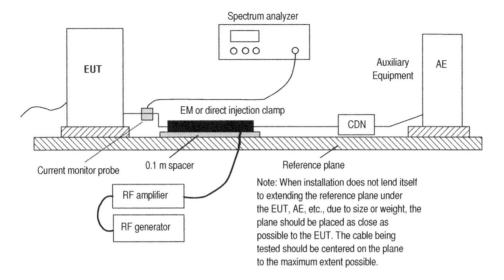

Figure 6.15 Conducted RF disturbance using an EM or direct-injection clamp.

are set up in both 150/50- and 50-Ω systems to compensate for the loading effect. Paragraph 7.3 of IEC 61000-4-6 indicates how the current injected into the line can be monitored to ensure that the equipment is not overtested.

If the current probe injection method is used, the calibration data (input power to the probe vs. test voltage) should be included in the test report. The number of calibration points should be selected to ensure that the voltage level between test frequencies measured at the port of the current probe calibration jig varies by no more than 1 dB with a constant input to the power amplifier. A minimum of five points per decade shall be provided. For variations in excess of 1 dB, the input value to the amplifier shall be adjusted to restore the test voltage to its original value. The frequency and the voltage measured on the forward power port of the coupler shall be noted.

When performing the test at each calibration point of the signal generator, the output of the generator shall be adjusted to ensure that the forward power into the RF coupler matches the calibration value. A typical calibration setup is shown in Figures 6.16 and 6.17 when using only one test receiver or spectrum analyzer.

6.4.3.3 Performing Typical Conducted Immunity Test. Details of how to perform conducted immunity tests are provided in Appendix B, where systematic procedures are given. There are two important items related to RF conducted immunity testing: monitoring the EUT and in situ testing.

Monitoring the EUT. The functional performance degradation allowed during and after conducted RF immunity tests may be specified by European product-family standards (e.g., EN 55024) [8]. When applying European generic standards EN

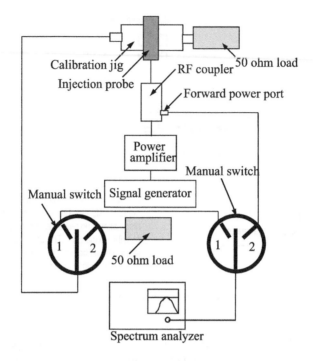

Figure 6.16 Typical test setup for calibration of injection current probe.

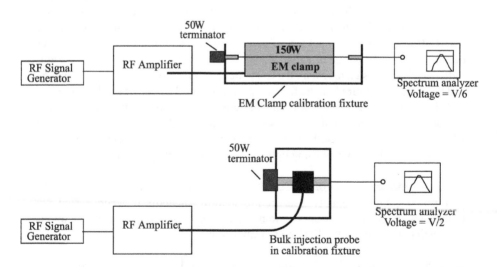

Figure 6.17 Typical test setup for calibration of EM clamp and BCI probe.

61000-6-1 or EN 61000-6-2, performance of the unit must be worse than the specification in the manufacturer's data sheet. These specifications should represent what its users would find acceptable given the marketing claims for the product. Thought should be given on the functional performance of the product to be tested, especially if it is to be tested in a shielded room with no observers inside.

In Situ Testing. The product specification standards describe the methods that have been developed for in situ testing for conducted immunity. The differences between injection methods are minor. The BCI clamp is often easier to employ at a customer site.

A problem with in situ testing is that it may be impractical to arrange for a reference plane as specified in the standards. Consequently, capacitive parasitics can have a significant effect at higher test frequencies. Although this is often acceptable for a Technical Construction File for an individual apparatus, it is less satisfactory if used to predict the conducted immunity level of the same or similar type of apparatus when installed at a different site.

Another serious concern with radiated fields across the frequency spectrum is potential interference with communication systems, machinery or process control systems, and implanted electronic devices such as pacemakers. Disruption can have lethal consequences. Appropriate precautions *must* be taken to make sure no harm is caused by the radiated field in both software and hardware. Always take precautions when there is a possibility of significant financial loss being caused by interference from in situ testing.

If the radiated emission from an in situ test for conducted immunity is significant, that is, when the cable being tested is longer than $\lambda/10$ of the highest frequency injected, it may become necessary to shield the system with a shielded enclosure. If there are no safety or financial loss implications when performing radiated immunity tests outside a shielded enclosure, it may be necessary (for legal reasons) to first obtain a special license from the government agency responsible for preventing interference to communication systems, for example, FCC/DOC (United States and Canada).

6.4.3.4 Diagnosis and Fixes. When a failure is observed, the following is a recommended guideline. Test results must be recorded and analyzed carefully. The following items of concern are operational failures [3]:

- Ungrounded metallic connector housings
- Incorrect or no bonding of metallic connector housing to the chassis assembly
- Signal wire traveling too far inside the system before filtering (i.e., pigtail)
- Missing or improper use of cable shield connection (braid, foil, drain wires)
- Incorrect placement, mounting or selection of power line filters
- Lack of proper common-mode filter capacitors
- Incorrect application and mounting of filter components
- Poor PCB layout

Closed-Field Probes for Localizing Problem Area. Suggested fixes include the following, which are similar to those for EFT/B testing:

- Apply common-mode ferrite very close to cable entries. These ferrites are generally the split-core version that allows for ease of implementation.
- Replace plastic connector housings with metal housings. Ensure the metal housing and cable shield is bonded to the chassis and not to the reference (ground) plane of the PCB.
- Install in-line filtered connector assemblies on critical signal lines.
- Remove all pig-tail connections from connector housings.
- Change the cable from unshielded to shielded. Properly bond the shield at both ends of the cable to chassis ground, not logic ground.
- Improve PCB layout.

6.4.4 AC Mains Supply Dips, Dropouts, and Interruptions

The effects of poor AC mains voltage quality may cause downtime of electrical equipment worldwide, along with a significant financial cost to users [9]. Dips, sags, brownouts, swells, voltage variations, dropouts, and interruptions are the primary cause of poor supply quality. The issue of AC mains power quality is based on the international/European standard IEC/EN 61000-4-11 [10]. This basic test standard deals with dips, sags, brownouts, swells, voltage variations, dropouts, and short interruptions in the AC mains supply for a specific range of products. For conformity to the EMC directive of Europe, compliance is based on generic or product-family standards that define actual test levels. Other standards in Europe exist for power supplies used in land, sea, and air vehicles, although these standards are not legally harmonized or mandated under the EMC directive.

Dips are short-term reductions in supply voltage caused by load switching and fault clearance in the AC supply network. Dips can also be caused by switching action between AC mains and localized sources used in uninterruptible power supplies or emergency power backup systems. Examples of dips are 30% for 10 ms and 60% for 100 ms. Figure 6.18 illustrates a 40% dip for 20 ms (one mains cycle). A dip of 40% is equivalent to a reduction in supply voltage to 60% of its nominal value.

Rapid fluctuations at low levels in the supply voltage can occur from once a minute to 30 per second. These fluctuations are identified as "flicker." Flicker can cause visual effects in electric lighting systems. Flicker is generally caused by fluctuating industrial loads such as arc furnaces, motor switching, welding equipment, or highly inductive products. Printers are included in the list of devices that can cause flicker. A detailed discussion on flicker is provided later in this chapter.

Dips may not be observed as isolated events. Sometimes a fast sequence of dips is present. Most immunity tests apply dips abruptly, starting and finishing at a zero crossing. In reality, dips may be observed at any point in the mains cycle, from 0° to 360°. Short interruptions are defined as an 80–100% reduction in supply voltage

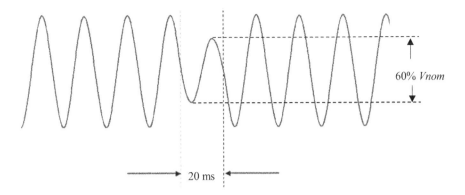

Figure 6.18 Typical AC mains voltage dip (40% of maximum value).

(dips of 80% or more) lasting for up to 1 min. The term *dropout* is commonly understood to mean the same as *dip.* Interruptions in the mains supply of more than 1 min are simply called *interruptions.*

Poor power quality is random in nature and is characterized in terms of deviation from the rated voltage and duration of the cycle. Voltage dips and short-term interruptions are not always abrupt, as the reaction time of rotating machinery and protection elements provided within a distribution network minimizes these harmful effects. Many data-processing systems have built-in power fail detectors to minimize disruption or sufficient input capacitance in the input circuit to provide adequate power to switching circuits during a temporary loss of input power; it is necessary to protect and save processor operations and memory content. Without this protection, many products will continuously be nonfunctional. Use of an uninterruptible power supply (UPS) is another way to prevent problems with poor power quality.

Figure 6.19 illustrates a short interruption that is 60-ms long (three mains cycles at 50 Hz). Like dips, short interruptions are caused by load-switching events. Short interruptions are also caused by switching action between AC mains and UPSs.

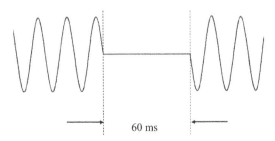

Figure 6.19 Typical AC mains interruption cycle, 60-ms period.

Test Equipment. IEC 61000-4-11 [10] requires a test simulator to be capable of providing up to 16 A RMS per phase at rated voltage, provide 23 A at 70% of rated voltage, and 40 A at 40% rated voltage for a duration of up to 5 s. The limit for peak inrush surge current is up to 500 A for 220–240 V AC mains, and up to 250 A for 100–120 V AC. Exempt from this test requirement are electrical and electronic equipment for connection to DC networks or 400 Hz AC power supplies.

The voltage rise (and fall) time during abrupt changes with a 100-Ω resistive load must be between 1 and 5 μs (Figure 6.20). The output impedance of the test voltage generator must be low to achieve this condition, even during voltage transition dips.

For dips and interruption tests, it is necessary to verify voltage transition level times and inrush current capability. Most modern oscilloscopes are capable of observing the voltage levels and transition times. For verifying inrush current, a bridge rectifier and a suitably rated 1700-μF capacitor connected through a switch and an appropriate current transformer are required. The parallel discharge resistance should be chosen to allow several *RC* time constants between tests. The test for inrush current is performed by switching the generator from 0 to 100% at both 90° and 270° to ensure sufficient peak inrush current drive capability for both polarities. In addition, the rise (and fall) time as well as overshoot and undershoot should be verified for switching at both 90° and 270° from 0 to 100%, 100 to 70%, 100 to 40%, and 100 to 0%. Phase angle accuracy shall be verified for switching from 0 to 100% and 100 to 0% at nine phase angles from 0° to 260° in 45° increments. It is also required to validate the waveform from 100 to 70% and 70 to 100%, as well as from 100 to 40% and 40 to 100%, at 90° and 180° phases.

EUT Performance Criteria and Selection of Test Levels. For dips and short interruption tests, the performance criterion varies depending on the duration or severity of the interrupt. Under certain conditions, loss of function is allowed pro-

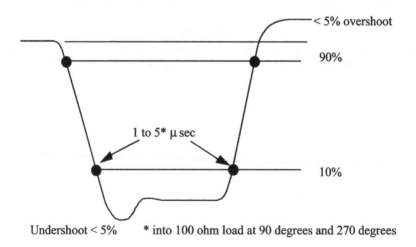

Figure 6.20 Rise and fall times of requirements for dips and interrupt tests.

vided the system is self-recoverable or can be restored by control operations. Under less severe conditions, degradation of performance is permitted; however, the unit must continue to operate as intended after the test. The product must not become unsafe under any conditions of operation. The performance criterion is different for various products. Refer to the product standard specification to determine the maximum amount of performance degradation permitted.

Table 6.2 [10] details the preferred test levels and duration for voltage dips and short interruptions, where U_t refers to the test specification level. Table 6.3 [10] provides the timing of short-term supply voltage variations.

6.4.4.1 AC Mains Supply Sags/Brownouts.

Sags (brownouts) are slow varying changes in voltage, sometimes over periods of hours. These disturbances are usually identified as a *voltage variation.* Voltage variation is an optional test requirement for conformity to the EMC directive, but is highly desired by many manufacturers to ensure the quality of their products.

A brownout is another name for sag and is the term most used in North America. Brownouts can cause the input mains voltage to drop down to a low level for an extended period of time. An illustration of a brownout event is provided in Figure 6.21, where the voltage-versus-time profile for a typical brownout is 50% of nominal voltage [8].

Table 6.2 Voltage Dips and Short Interruption Test Levels

Test Level (% U_t)	Voltage Dip and Short Interruptions (% U_t)	Duration (in period)
0	100	0.5[a]
		5
		5
40	60	10
		25
		50
70	30	x

[a]For the 0.5 period, the test should be made in positive and negative polarity, i.e., starting at 0° and 180°, respectively.

Notes:
1. One or more of the test levels and durations may be chosen.
2. If the EUT is tested for voltage dips of 100%, it is generally unnecessary to test for other levels for the same duration. However, for some cases (safeguard systems or electromechanical devices) it is not true. The product specification or product committee will give an indication of the applicability of this note.
3. "x" is an open duration. This duration can be given in the product specification. Utilities in Europe have measured dips and short interruptions of duration between ½ period and 3000 periods, but durations less than 50 periods are most common.
4. Any duration may apply to any test level.

Table 6.3 Timing of Short-Term Supply Voltage Variations

Voltage Test Level	Time for Decreasing Voltage	Time at Reduced Voltage	Time for Increasing Voltage
40% U_t	2 s ± 20%	1 s ± 20%	2 s ± 20%
0% U_t	2 s ± 20%	1 s ± 20%	2 s ± 20%
	x	x	x

Note: x represents an open set of durations and can be given in the product specification.

A potential problem with low voltage levels lies with AC motors and certain types of DC motors. These motors can stop spinning, overheat, or have their insulation damaged. A damaged motor can result in an increased risk of fire and electric shock. In addition, damage to electrical equipment can happen. For this reason, most motors are sometimes protected by components that shut the motor off when the voltage level exceeds a set value. When the safety feature triggers, the system has downtime that can result in significant lost revenue to the owner.

Typical Test Setup and Procedures. The setup for sag/voltage variation is detailed in Figure 6.22. The test conditions are the same as for dips and interruptions described above. This test can be performed manually with a variable transformer (variac) or using a programmable power source. A timer is required to determine the length of time for the sag. Extreme accuracy is not required; a wristwatch will

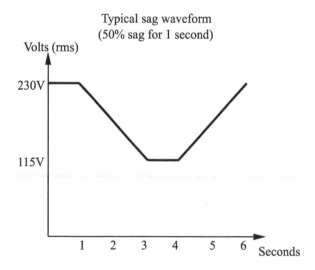

Figure 6.21 Typical sag (brownout) waveform.

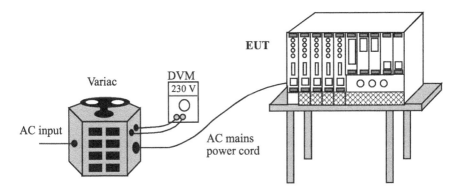

Figure 6.22   Test setup for sag/voltage variation (EN 61000-4-11 [9]).

do. The mains voltage during the test can be varied smoothly or varied stepwise providing the steps are no greater than 10% of the nominal mains voltage. If using a programmable power source, ensure that the sag occurs at the zero crossing of the mains waveform.

For each sag test (voltage level, time duration, ramp-down and ramp-up conditions), three separate tests are performed with a 10-second delay between tests.

The test setup and other requirements do not require a special test site. It is possible to perform full compliance in situ or where the unit is physically installed. If there are out-of-specification issues with the mains supply or other parameters, it is often possible to perform precompliance tests with reasonable accuracy and repeatability.

The EUT must be monitored to determine if any degradation is observed both during and after each test. At the completion of the test, a full functional check should be performed. The test standard (relevant generic or product family) will specify the performance degradation criteria to be applied in each case. Dimming of displayed images or illumination may be permitted during the test.

Where the EUT's rated supply voltage range does not exceed 20% of its lowest voltage, any voltage within that range can be used. If the EUT has a wider voltage range, two sets of tests are required—one at the lowest rated voltage range and one at the highest.

It is recommended that two sag tests be performed, one at 40% and one at 0% of the supply voltage, each lasting for 1 second, with the supply voltage ramped down and then up again over a period of 2 seconds in each test. At time of writing, most generic and product-family harmonized EMC standards do not specify or mandate this test, but mention it as an optional test that is *recommended.*

In an effort to improve product reliability, testing on power quality should be performed. During actual use, when installed in a hostile RF environment, sags can last for minutes or hours at a time. If certain components are susceptible to a brownout condition, such as AC or DC motors, tests lasting greater than 1 second are recommended to ensure reliable operation.

6.4.4.2   *Swell Testing.* A swell is a slow increase in voltage that can last up to a few seconds, as opposed to surges and transients which are fast increases in voltage usually lasting under 1 ms. Electrical power engineers refer to surges as swells. The word *surge* typically refers to a high-energy level applied to a system, such as lightning. To prevent confusion, the word *swell* is used to indicate an increase in voltage. Most power supplies operate with a ± 6–10% voltage tolerance. It is common to find a wide voltage range daily throughout the world using various distribution networks [9].

Not only does a low voltage level present a problem, a high voltage level (swell) is also a serious concern. Insulation breakdown and fire hazard may result. Mains surge protection components such as avalanche clamping devices may be designed into the motor to start conducting as close to the maximum expected mains voltage as possible in an effort to protect the device. Consequently, the leakage current at voltages above mains nominal can cause overheating and damage if the swell lasts for more than a few seconds.

Testing for swells may help improve product reliability. Regulatory standards do not investigate voltage variations that are above the nominal mains voltage; thus any tests performed are done as part of a quality assurance program.

If this test is performed, the same variac, as for sag, should be used provided the device can operate at a voltage level that is higher than the mains voltage. A step-up transformer may be required. Most test laboratories use a programmable synthesizing power source that meets the specification requirements detailed in the test standard (as a recommended test). The test setup and conditions are identical to those for sag.

6.4.4.3   *Three-Phase Equipment—Compliance Testing.* If a three-phase test generator or three-phase variac is unavailable, three-phase equipment can be tested for dips, interruptions, and voltage variations using three sets of single-phase test equipment [9]. Phase-by-phase testing is preferred, although some equipment may require simultaneous testing. This requires a common or synchronized control of the three single-phase test systems. The test standard (generic or product family) will specify whether phase-by-phase or simultaneous control is required.

When using simultaneous control for dips and interruption testing, the ±10% specification for the zero-crossing performance can only be met for one phase. The test generators must be of the type that can switch at any phase angle.

6.4.4.4   *Diagnosis and Fixes.* For those who design products incorporating a purchased power supply, there is very little that can be done to prevent effects of voltage dips, dropout, interruption, sag, and swells except replace the existing power source by one built by another company. If power conditioning is required for the system, a capacitive bank of regulated voltage and current sources can be incorporated at a significant cost or an external UPS will be needed.

Those who are affected the most are manufacturers of power supplies. Power factor correction circuitry may alleviate some of the problems. For others, a redesign in the logic circuitry and magnetics must be performed. The complexity of the design changes is well known to designers of power systems and is beyond the scope of this introductory material.

6.4.5 Power Line Harmonics

Loads connected to electrical supply systems may be broadly categorized as either linear or nonlinear. Linear loads include induction motors and incandescent lamps and may exhibit a high or low power factor. In either case, input current consumption is present only at the power line's fundamental frequency. In contrast, nonlinear loads such as rectifiers or switch mode converters draw significant current at the harmonics of the fundamental frequency. Harmonic currents flowing through electric power supply systems can cause voltage distortion.

IEC 61000-3-2 [11] applies only to equipment rated up to 16 A, which at a power level for 230 V AC is 3680 W. Most products may never reach this level of power consumption. An AC power source for high-power applications and highly distorting loads is expensive, not to mention physical size and weight of the test instrumentation. The harmonic frequency range extends only up to 2 kHz (the 40th harmonic of 50 Hz). Therefore the test does not need to use RF measurement techniques; however, some aspects of the measurement are not entirely obvious.

Voltage distortion caused by a growing penetration of nonlinear loads has in many cases been unnoticeable in power distribution systems without serious consequences. In other cases, mitigation steps have been required to avoid compromised power quality. In North America, the issue of power line harmonics is not as serious of a concern as in Europe due to the phasing of the power distribution network. However, in Europe power line harmonics is a significant threat to public utility companies.

A large number of nonlinear loads connected to a utility power system can cause problems in distribution networks. The severity of the problem depends upon local and regional supply characteristics. These characteristics include the size of the loads, the quantity of these loads, and how the loads interact with each other (e.g., harmonic cancellation, attenuation, and dilution).

Utilities are clearly concerned about future problems from increased concentrations of nonlinear loads that will result from the growing proliferation of electronic equipment. Experience shows that higher concentrations of nonlinear loads tend to increase the number and severity of problems. Problems from high concentration of nonlinear loads in industrial systems are well documented.

Industrial environments can have many three-phase systems and nonlinear loads drawing high levels of load current. The effect on transformer operation when multiple loads are connected is that each load generates triplen harmonic currents within the neutral conductor. Harmonic currents travel into the transformer's secondary windings and are reflected back into the primary section of the transformer. These reflected currents circulate within the windings of the transformer, causing overheating, shortened service life, and catastrophic failure.

Control of lower order power line harmonic emissions from nonlinear loads is a serious concern that requires cooperation between utilities, equipment manufacturers, premise owners, and end users. Utility providers desire to prevent problems by restricting emissions from systems connected to the public supply network. End users want low-cost, high-performance, and trouble-free operation without wasting valuable power from the utility company.

The EMC standard EN 61000-3-2 (Ed. 2:2000) [11] includes requirements for measuring harmonic emissions for electrical and electronic equipment with an input current up to 16 A per phase and supersedes the earlier EN 60555-2 standard (which had a more limited scope). For flicker, EN 61000-3-3 [12] is specified. It has been amended many times, each with an overlapping period of conformity. Confusion exists on the types of products to be tested, the waveform, whether limits are relaxed, classification of specific products, and the like. This test standard changes on a yearly basis, making it difficult for one to perform legally mandatory conformity tests on a moving target—the standard.

Although the harmonic frequency range extends only up to 2 kHz (the 40th harmonic of 50 Hz), the test does not require RF measurement techniques. There are, however, certain aspects of the measurement procedure that are not entirely obvious to the test engineer.

6.4.5.1 How Harmonics Are Created and Related Concerns. Due to finite impedance (a complex number) of mains distribution systems, nonlinear loads can generate harmonic currents. These currents create harmonic voltage distortion and can overload neutral conductors and distribution transformers. The effect of this overload can cause nuisance tripping of protective devices, creating functional problems and causing a potential fire due to overheating of the wires. Harmonics can also cause a variety of reliability problems with other electrical equipment. When the AC mains voltage suffers from harmonic distortion, linear loads may become affected, which in turn can permit functionality issues to exist.

Examples of reliability problems due to power line harmonics include faulty operation for the following:

- Fluorescent lamps
- Three-phase power converters
- Microwave ovens and oven systems with magnetrons and klystrons or other RF-generating components
- Transformers and induction motors
- Arc welders, furnaces, and similar equipment

Additional problems that are unique to harmonics include the following:

- Elevated RMS current
- Circuit breaker trips
- Nuisance fuse operation
- Reduced equipment life

- Transformers, conductors, circuit breakers, and power switches
- Equipment malfunction
- High-frequency currents
- Reduced effective power factor utilization

The primary culprits of harmonic distortion are switch-mode or DC/DC power supplies. The source of noise is the rectifier (AC/DC) circuit, which can be in either of the above power conversion techniques. These supplies are used in almost every product found in residential, commercial, and industrial environments. As more units are connected to the mains distribution network, an unbalanced phasing of the voltage lines causes harmonic currents to be injected into the neutral conductor. These currents will add up into one very large amount of current, causing problems for the utility company. In the past, when there were few systems connected to the distribution network, the subject of harmonics was not a concern.

Harmonics are produced by the diode–capacitor input section of power supplies. The diode–capacitor combination rectifies the AC input power into the DC voltage used by circuits. For example, a personal computer uses DC voltages generated internally to power various circuits and printed wiring boards. This power supply draws input current from the AC mains during the peak or crest of the voltage waveform, thereby charging a capacitor internal to the power supply to the peak of the line voltage. The DC equipment requirements are fed from this capacitor and, as a result, the current waveform becomes distorted since the power supply does not utilize the full waveform on the input AC sine wave.

The harmonics in the electric power distribution system combine with the fundamental (i.e., 50/60 Hz) to create distortion. The level of distortion is directly related to the frequencies and amplitudes of the harmonic current. All harmonic currents combine with the fundamental current to form total harmonic distortion. (THD). The THD value is expressed as a percentage of the fundamental current, and any THD value over 10% is significant enough for concern.

Wherever there are large numbers of nonlinear loads, harmonics in the distribution system may be present. It is not uncommon for THD levels in industrial plants to reach 25%. Normally, THD levels in office settings will be lower than in industrial plants; however, office equipment is much more susceptible to variations in power quality.

Odd-number harmonics (third, fifth, seventh, etc.) are of greatest concern in the electrical distribution system. Even-number harmonics are usually mitigated because the harmonics swing equally in both the positive and negative directions.

On balanced three-phase systems with no harmonic content, line currents are mutually shifted 120° between phases. For this configuration, the phases mutually cancel each other, resulting in very little neutral current, assuming that currents are identical in each individual phase and the phase angles or power factor relationships are approaching unity with reference to neutral (earth ground reference). When there is distortion in any of the phase currents, harmonic currents will increase and the cancellation effect becomes compromised. The usual result is that neutral current THD is significantly higher than desired. Triplen harmonics (odd multiples of 3) are additive in the neutral line and can quickly cause dangerous overheating of motors and transformers.

In theory, the maximum current that the neutral will carry is 1.73 ($\sqrt{3}$)times the phase current. If the wire gauge is not sized correctly, overheating will occur. Higher than normal neutral current will also cause a voltage drop between the neutral line and AC mains ground, which can be well above normal values for optimal operation.

Triplen Harmonics. Triplen harmonics are a byproduct of current being injected back into the distribution system due to an unbalanced network. Triplens are observed at multiples of 3 (3, 6, 9, 12, etc.). This means that for a 50-Hz distribution network, harmonics at 150, 300, 450, and 600 Hz are possible. Odd harmonics present harm to electrical systems, whereas even harmonics do not. Motors and select electrical products do not operate well at frequencies other than their designed rating and may in fact be damaged if high-frequency triplen currents are provided to the input terminals.

Triplens add in the neutral conductor and can produce up to 1.73 times the phase current that should be present within any installation. Single-phase distribution networks generate a greater amount of triplen harmonics than three-phase systems. This is due to the balance of the phase load on each input line. Depending on the phasing of the current, a cancellation effect between harmonics is possible. This situation does not occur frequently.

An illustration of the first few harmonics is provided in Figure 6.23. Note that the fifth harmonic, in reality, is the sixth harmonic (fundamental plus the fifth har-

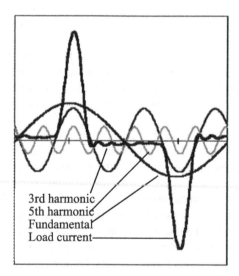

Figure 6.23   Triplen harmonic waveforms.

monic). As observed, the magnitude of the 3rd harmonic is slightly below the magnitude of the fundamental frequency with the applied load. The fifth harmonic is below the third harmonic.

Voltage Distortion. Voltage distortion is expressed as THD, which compares measured harmonic content to the magnitude of the fundamental frequency. Voltage distortion is a function of harmonic current interacting with an impedance ($V_h = I_h \times Z_h$). Voltage distortion can be expected to be high when either harmonic current or system impedance is high. This will appear as a flattening of the voltage peaks.

Voltage and current distortion both tend to be highest where the equipment is connected to the electrical system (i.e., at the mains receptacle). Typical voltage distortion levels range from 1–2% at the input of a commercial building to as much as 15% or higher for systems connected to the primary distribution network. Voltage distortion may be acceptable at the input to the building yet be totally unacceptable at the point where a computer or disk drive is connected.

Load Profiles and Associated Harmonic Current Spectra. A bar chart is another way of presenting harmonic profiles or spectrums. Figure 6.24 shows the difference between single-phase nonlinear loads (phase–phase) and three-phase, six-

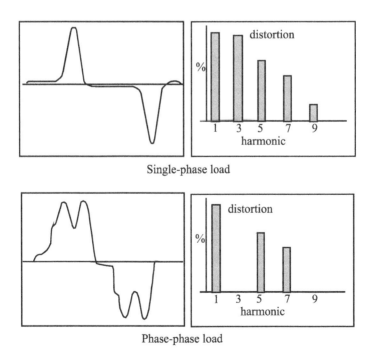

Single-phase load

Phase-phase load

Figure 6.24 Load profiles and associated harmonic current spectrums.

pulse nonlinear loads. The important characteristics of a single-phase nonlinear load are as follows:

- They have a broad range of harmonic frequencies (3, 5, 7, 9, 11, . . .).
- The third harmonic will be in phase with the other phases. The harmonics will add in the neutral conductor.
- The fifth- and seventh-harmonic currents will be high enough in most instances to require special filtering.
- Triplen harmonics (3, 9, 15, 21, . . .) and other zero-sequence currents will result in a neutral-to-ground voltage. This voltage will couple into the primary delta winding of a delta–wye transformer, causing voltage distortion and excessive losses (heat).

Noteworthy characteristics of a phase–phase or three-phase nonlinear load (six-pulses) are as follows:

- Both are rich in fifth- and seventh-harmonic currents.
- There is a noticeable absence of third and other triplen harmonics.
- These harmonics, being negative and positive sequence in nature, will flow through transformer windings.
- Twelve-pulse nonlinear loads draw predominantly 11th- and 13th-harmonic currents as their first two characteristic harmonics.

Skin Effects. Harmonic frequencies are always higher than the fundamental frequency. At higher frequencies "skin effects" becomes a concern. Skin effect is a phenomenon in which the higher frequency currents cause electrons to flow toward the outer portion of a conductor. This reduces the ability of the conductor to carry current by effectively reducing the cross-sectional diameter of the conductor, thereby reducing the ampere capacity rating of the conductor. Skin effect increases as the frequency and amplitude increase. This is the reason higher harmonic frequencies cause a greater degree of heating in conductors.

Rotating Machinery. Voltage harmonics may cause problems in electrical distribution systems. Motors are typically linear in design. When the source voltage supply contains harmonics, the motor will draw this harmonic current. The result is typically a higher than normal operation temperature, thus shortening service life.

Motors that are connected directly to the AC mains will try to rotate at the speed of phase rotation of the various harmonics. These motors will try to run at three times the fundamental frequency for the third harmonic, six times for the sixth harmonic, and so on. Harmonic distortion of the supply waveform will cause a decrease in efficiency and a rise in winding temperature. Higher levels of distortion can cause serious overheating with consequent shutdown (assuming protection devices are in place), excessive acoustic noise, vibration, and damage to bearings that wear out long before their expected life cycle.

Different frequency harmonic currents can also cause additional rotating fields in the motor. Depending on the frequency, the motor will rotate in the opposite direction (countertorque). The fifth harmonic, which is very prevalent, is a negative-sequence harmonic causing the motor to have a backward rotation, shortening the service life.

Noise can be picked up in computer networks, communications equipment, and telephone systems when harmonics are at audio or radio frequencies. With the increase in speed of computer networks, the future will bring these systems into the frequency range where they will be more affected by harmonic-generated noise. This noise is inductively or capacitively coupled into the communications and data lines through various means.

The majority of generators and transformers base their operating characteristics on a nondisturbed 50/60-Hz waveform. When waveforms are rich in harmonics, shortened service life or complete failure is sure to result.

Standard for Power Line Harmonics. To ensure compatibility with the EMC Directive of Europe, a test standard has been issued. This standard, EN 61000-3-2:1995, *Electromagnetic Compatibility (EMC), Part 3-2, Limits for Harmonic Current Emissions (Equipment Input Current Up to and Including 16A per Phase)*, is the most confusing standard to be adopted by the European Commission related to electrical products. Several variations exist, although at the time of writing many of these editions are obsolete. A historical background on power line harmonics is presented.

Table 6.4 identifies which versions may be listed on the EU Declaration of Conformity and the latest date at which one must declare amendments or new issues. Note that a standard may be used to support an EMC Directive Declaration of Conformity as soon as it appears on this list.

Equipment Classification. Equipment to be tested to IEC/EN 61000-3-2 [11] specification are classified in four categories. The definitions provided below are from the 2000 issue and differ considerably from previous issues in many areas.

Table 6.4 Power Line Harmonic Standards

Reference of the Superseded Standard	Date of Cessation of Presumption of Conformity of Superseded Standard
EN 61000-3-2:1995+A1:1997+A2:1998	Required 01.01.2001, not valid beyond 01.01.2004
EN 61000-3-2:1995+A1:1997+A2:1998+A14:2000	May be applied at any time; mandatory 01.01.2004
EN 61000-3-2:2000	May be applied at any time; mandatory 01.01.2004

- Class A: Equipment not specified in one of the three other classes shall be considered as Class A equipment.

    Balanced three-phase equipment

    Household appliances excluding equipment identified as Class D

    Tools excluding portable tools

    Dimmers for incandescent lamps

    Audio equipment

Note: Equipment that can be shown to have a significant effect on the supply system may be reclassified in a future edition of the standard. Factors to be taken into amount include the following:

    Number in use

    Duration of use

    Simultaneity of use

    Power consumption

    Harmonic spectrum, including phase

- Class B: Portable tools
- Class C: Lighting equipment
- Class D: Equipment having a specified power less than or equal to 600 W of the following types:

    Personal computers and personal computer monitors

    Television receivers

Note: Class D limits are reserved for equipment that, by virtue of the factors listed in the Note under Class A above, can be shown to have a pronounced effect on the public electricity supply system.

A special wave shape on the AC mains input that mandates testing be performed to harmonic requirements is defined by an envelope detailed in Figure 6.25. This wave shape identifies rectifier–capacitor power supply input circuits that normally draw operational current for less than a third of the AC sine wave for one half-cycle. The harmonic limits quoted in the standard are absolute values for Class A products, regardless of input power. Equipment with an input rating greater than 600 W that falls within the Class A category will cause greater harm to the power distribution network in a proportionate manner. Most instrumentation cannot handle greater that 16 A of current without damage.

For Class D products, the input waveform to the power supply is measured to determine if the unit falls within the category of the special wave shape of Figure 6.25. If the wave shape exists, the amount of active input power must be applied to the limits. This process is difficult to achieve, especially if the input current or harmonic content is constantly changing. Class D products are considered to have the greatest impact on power distribution networks.

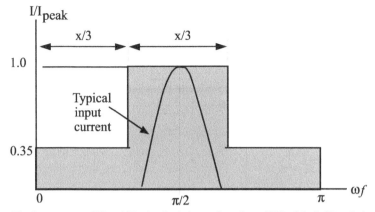

The input current falls within the shaded area for at least 95% of the half-period.
Center line $\pi/2$ corresponds to the peak value of the input current.
ωf does not represent the zero-voltage crossing point.

Figure 6.25 Special wave shape for IEC 61000-3-2, Class D.

Special Considerations for Class D Products. Class D limits are specified in milliamperes per watt. According to Amendment 14 (A14), the average emission limit is compared to specification limits based upon the maximum amount of measured power in each observation time window during the test. Harmonic currents and active input power are measured under the same test conditions but need not be measured simultaneously.

Test Instrumentation. IEC 61000-3-2 defines the method of measurement of harmonic current emissions for equipment with input current up to and including 16 A per phase. The test setup is simple, shown in Figure 6.26. The basic components are as follows:

- AC power source
- Current transducer
- Wave analyzer

AC Power Source. An AC power source with very low distortion is required to perform harmonic emission testing. This system must be calibrated for accuracy and contain very low distortion (a pure sine wave), high-voltage stability, programmability, and a low-impedance output. For most test environments, the power quality from the public mains distribution is not able to meet these stringent requirements. A power source that can provide the required AC input waveform might be required.

The test specification requires the voltage level to be stable within ±2% of the test value. The frequency must be accurate to 9.5% of nominal. Harmonic distortion

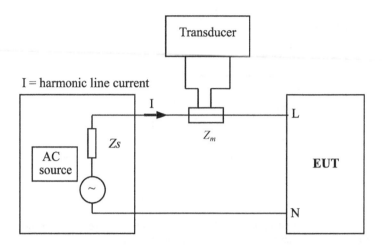

Figure 6.26 Mains harmonic emissions test setup.

must be <0.9% at the third harmonic, 0.4% at the fifth, 0.3% at the seventh, 0.2% at the ninth, and 0.1% at all other harmonics. Due to the finite source impedance of the AC mains network and the variability of loads receiving this AC mains power, the local mains source may not be capable of providing harmonic distortion mandated in the standard.

The impedance of the transducer, Z_m, should create a voltage drop of less than 0.15 V peak. The source impedance is not specified. The test environment and instrumentation do not allow errors at any harmonic frequency of more than 5% of the permissible limits.

Typical test equipment may include a power amplifier driven by a 50- or 60-Hz sine wave oscillator. The amplifier contains negative feedback to maintain low output impedance. The output of the amplifier may be fed through a power transformer for voltage step-up purposes. If a power transformer is used, the transformer reactance must not affect the output impedance at higher harmonic frequencies. The amplifier must be capable of operating under a full range of loads.

The maximum allowable transitory harmonics for Class B equipment can be approximately 40 A. While certain systems can substantially exceed this level, the power source should be able to provide current without distortion. The same power source is used for Class A, B, C, and D systems. If the measured harmonics exceed or are below the distortion level, voltage distortion becomes a minor concern. Distortion only becomes important when the product is at the specification limit. Figure 6.27 illustrates what distortion current looks like compared to AC mains input.

Current Transducer. A current transducer is used to couple harmonic current I_n to a measuring instrument. This transducer can be either a current shunt or a transformer. The transducer has an impedance Z_m that is added to the output impedance

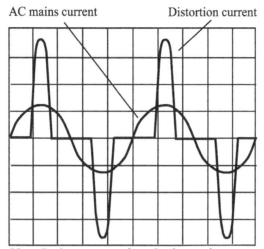

AC mains current   Distortion current

Note: Peak current can be > 3x the maximum
expected sinusoidal current for a given power rating

Figure 6.27   Distorted EUT current waveform.

of the power source. Together, the two elements must cause negligible variation in the load current harmonic structure. A resistive current shunt of less than 0.1 Ω impedance with a time constant less than 10 μs is acceptable. A safety consideration to be aware of is that a shunt, by itself, does not provide galvanic isolation from the measuring circuit.

Current transformers provide galvanic isolation. These transformers should ensure necessary electrical strength capabilities with an insulation material to prevent hazards of electric shock during a single fault condition from the EUT. Unlike resistive current shunts, current transformers must be calibrated at each harmonic frequency. Erroneous test results are possible when the transformer is saturated. Saturation may also include DC current present in the power cord.

The tolerance for error is required to be less than 5% of the permissible limit. Maintaining this accuracy from the lowest to the highest measurement level puts severe demands on the dynamic range of the current transducer and associated input signal processing. To ensure accuracy and to prevent saturation, one may use two transducers, one for high current and one for low current values.

Wave Analyzer. A wave analyzer measures the amplitude of each harmonic component from $n = 2$ to $n = 40$. This instrument can be a frequency-domain type using selective filters, a spectrum analyzer, or a time-domain method to perform digital computation to derive the Discrete Fourier Transform (DFT). The original version of EN 61000-3-2 [11] defines requirements for time-domain instrumentation rather differently from those for the frequency-domain type. However, Amendment A14 to this standard deleted this option and replaced it with the measurement techniques described in IEC 61000-4-7 [13]. This standard is used to define reference information related to instruments for harmonic measurements. Amendment A14 prohibits frequency-domain instruments, mandating only DFT analysis.

When harmonic components fluctuate while the measurement is being made, the response at the indicating output should be that of a first-order low-pass filter with a time constant of 1.5 seconds. IEC 61000-4-7 includes specific details of the time constant smoothing function.

Test Conditions. Special test conditions are required for the equipment listed in EN 61000-3-2. This equipment includes various household appliances, TV receivers, audio amplifiers, VCRs, and lighting equipment. Independent lamp dimmers and other phase control devices should be set for a firing angle of 90°. Information technology equipment must be tested with the equipment configured to its rated current.

A significant change to EN 61000-3-2 occurred with Amendment A14:2000. Earlier versions of the standard required equipment be operated to give maximum harmonic amplitude for each successive harmonic frequency by manual or programmable means. This procedure would require extensive test time. It is unrealistic to maximize current draw and operational characteristics for each harmonic up to the 40th harmonic. The harmonics most likely to create harm to power distribution systems are the lower order harmonics. It is THD that is important. Higher order harmonics generally have little effect on THD. Amendment A14 removed this test requirement.

Performing the Test. The EUT should be operated in a mode that produces maximum total harmonic current (THC) under normal operating conditions. The harmonic current limits apply only to line currents and not to currents in the neutral conductor. Preliminary operation of motor drives by the manufacturer may be needed before the tests are performed to ensure that results measured are consistent with normal use. Harmonic currents and active input power should be measured but need not be measured simultaneously.

Measurement is performed as follows:

For each harmonic order, measure the 1.5-second smoothed RMS harmonic current in each DFT window.

Calculate the arithmetic average of the measured valued from the DFT time windows over the entire observation period.

When equipment enters or stops operation, either manually or automatically, harmonic currents and power are not taken into account for the first 10 seconds following the switching event. The EUT should not be in stand-by mode for more than 10% of any observation period. The test standard provides additional details on test requirements and observation periods.

6.4.5.2 Diagnosis and Fixes. The following are steps to assist in isolating problems encountered when harmonics are present in an electric distribution system (Step 1 is easy, steps 2–6 are more complicated):

1. Inventory all equipment that may produce excessive harmonic currents.
2. List the nonlinear (reactive-phase angle) loads that are on each branch circuit.
3. Record true RMS current in each phase at the service entrance.
4. Record the neutral current of the transformer secondary.
5. Compare the measured neutral current to the anticipated current due to phase imbalance. If the phase currents are equal, the vector sum of the neutral currents will add to zero. If excessive amounts of triplen harmonics are present in the neutral, neutral current may exceed the phase current.
6. Measure each feeder for harmonic content. A voltage THD reading is also useful.

For installations and systems, the following provides guidance:

1. Use only equipment that has been previously tested to IEC/EN 61000-3-2.
2. Install primary transformers that are oversized for high-voltage to low-voltage installations.
3. Make the neutral conductor one wire size (gauge) larger than all other phases.
4. Equalize load sharing between phases. This does not minimize or prevent power line harmonics but reduces heat generation within the neutral line.
5. Distribute the AC mains power cable to different locations by using feeders or transformers.
6. Provide power factor correction circuits to the input of all power supplies.

6.4.6 Voltage Fluctuation and Flicker

Voltage fluctuation, or flicker, is of great concern around the world; however, only Europe has mandatory regulatory requirements for products connected to power distribution networks. The regulatory standard that mandates limits on flicker is IEC/EN 61000-3-3 [12]. Flicker is defined as the "impression of unsteadiness of visual sensation induced by a light stimulus whose luminance or spectral distribution fluctuates with time" [Ref. 12, Section 3.6.].

EN 61000-3-3 applies to electrical and electronic equipment having an input current up to and including 16 A per phase and intended to be connected to public low-voltage distribution systems between 220 and 250 V AC at 50 Hz line to neutral. For systems with nominal voltages less than 220 V AC, line to neutral, and/or a frequency of 60 Hz, regulatory limits have not been considered. This standard applies to almost the same type of equipment as harmonics standard EN 61000-3-2 by limiting the voltage variations generated across a reference load. This reference load is defined as follows:

- Relative change in voltage characteristics (d_t)
- Relative change in voltage levels (maximum, d_{max}, and steady state, d_c)
- Short-term flicker value, P_{st}
- Long-term flicker value, P_{lt}

Within power distribution networks, series impedance develops voltage fluctuations, which in turn causes varying current loads that are provided to electrical equipment. This equipment may all be connected to a common distribution point or AC receptacle in one room. If the amplitude of the voltage fluctuation is greater than the tolerance level of a product's power supply, flicker can be observed in luminaries and can cause harm to those sensitive to flashing lights.

Regulatory limits do not apply to emergency switching or interruptions. In addition, the flicker values limits P_{st} and P_{lt} do not apply to manual switching or voltage changes occurring less frequently than once per hour. The voltage change limits d_{max} and d_c apply to these occasional events. Because of this requirement, a limit on allowable inrush current for any apparatus exists, even where the power is turned on manually.

EN 61000-3-3 requires d_{max} be 6% for manual switching and 7% for attended equipment or equipment which will be switched on no more than twice a day. For manual switching equipment, conformity is assumed without testing if the maximum RMS input current, including inrush over each half period, does not exceed 20 A. If inrush current measurements are performed, 24 inrush events are recorded with the highest and lowest value discarded. The average value of inrush current is calculated for the remaining 22 measured events.

Equipment that typically produces flicker in normal operation includes any device that switches varying loads during its operating cycle. Many household appliances fall into this category such as printers, hotplates, and heaters whose temperatures are controlled by burst firing; that is, power is provided to the heater for a few cycles of the mains supply at a time (the on/off ratio of the bursts controls the temperature). If the heating load is substantial, this kind of equipment easily fails flicker limits.

6.4.6.1 Description of Short-Term Flicker.

Flicker is caused by short-term fluctuations in the voltage supply. This fluctuation causes the eye-to-brain connection to perceive a change in the luminaries that provide light. Those with seizure disorders are more sensitive to flicker and can incur a medical situation if exposed to long-term flicker. For this reason, the change in line voltage must happen over a period of several minutes, taking into account the shape of the voltage change characteristic and cumulative irritating effect of the flicker. The amount of flicker can be easily measured using a flickermeter.

A flickermeter indicates short-term flicker, P_{st}, observed over a period of 10 minutes. During this period, the part of the operating cycle in which the EUT produces the greatest amount of voltage changes is not allowed to exceed a value of 1.

6.4.6.2 Instrumentation. The primary instrument used to measure flicker is the flickermeter. This meter must comply with specification detailed in IEC/TR 60868 [14] and is connected as shown in Figure 6.28. This figure gives the circuit for a three-phase supply. This meter has essentially the same block diagram and characteristics as the harmonic analyzer. Due to similarity, both harmonic and flicker analyzers are often packaged together.

The measured variable is the voltage across the point of supply rather than the current drawn. The source impedance of the supply generator is defined in the test standard. This specification ensures that load current changes in the EUT will produce a defined voltage change. The data from measurements is compared against published regulatory limits.

The accuracy of this test setup must be better than ±8% of the maximum value for relative voltage change. The measurement errors can be distributed between the reference impedance and the harmonic analyzer as long as the total amount of flicker remains within regulatory limits.

To evaluate effects of voltage changes in the distribution network, the RMS level is measured over successive half-periods of 10 ms each. This time frame allows the system to display a time-dependent view of the voltage change, normalized to

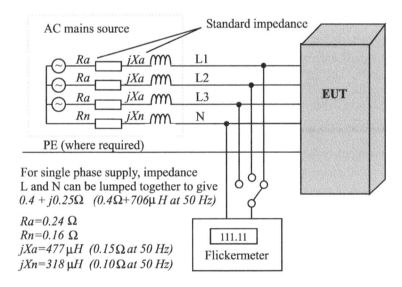

Figure 6.28 Reference network for three-phase supplies for flicker measurement.

give the relative change in voltage levels. From this measurement, two items of interest become evident:

- The relative steady-state voltage change d_c, which is the difference between two adjacent steady state voltages separated by at least one change (steady-state is defined as persisting for at least 1 second)
- The maximum relative voltage change d_{max}, which is the difference between maximum and minimum values of the voltage change characteristics

Test Limits. The limits apply to voltage fluctuations and flicker at the supply terminals of the EUT, measured or calculated. These limits are subject to change (refer to the test standard for newer requirements):

- The relative steady-state voltage change, d_c, shall not exceed 3%.
- The maximum relative voltage change, d_{max}, shall not exceed 4%.
- The value of $d(t)$ during a voltage change shall not exceed 3% for more than 200 ms.

The limits above are multiplied by 1.33 for manual switching or events occurring less often than once an hour. If the voltage changes are caused by manual switching or less than once per hour, the P_{st} and P_{lt} requirements do not apply.

Test Conditions. The test for flicker shall not be made on equipment that does not produce significant voltage fluctuations or flicker (maintains a steady-state input current draw). This applies to the majority of products produced. For those products that affect the voltage level every time they switch on and off, compliance to IEC 61000-3-3 is mandatory.

Equipment required is as follows:

- Test supply voltage source
- Reference impedance
- Equipment under test
- If necessary, a flickermeter

The relative change in voltage, $d(t)$, may be measured directly or derived from RMS current. To determine P_{st}, one of the methods described in the test standard is to be used. In case of doubt, the flickermeter is the referenced standard.

The test voltage shall be the rated voltage value of the EUT, which is generally 230 V AC single phase or 400 V AC three phase, maintained within ±2% of nominal value, at 50 Hz ± 5%.

Measurement accuracy must be ±1% or better, unless the active and reactive current phase angle is used, in which case accuracy shall not exceed ±2%. The relative change in voltage, d, must be better than ±8% with reference to the maximum value of d_{max}. Total impedance of the circuit shall be equal to the reference impedance.

The stability and tolerance of the total impedance of the network shall ensure overall accuracy of ±8%.

The observation period to assess flicker is defined below. This definition includes all parts of the operation cycle in which the EUT produces the most unfavorable sequence of voltage changes:

Short-term flicker: $P_{st}T_p = 10$ min
Long-term flicker: $P_{lt}T_p = 2$ h

For P_{st}, the cycle of operation shall be repeated continuously. The minimum time to restart the EUT must be included in the observation period when the system stops automatically at the end of an operational cycle that lasts less than the observation period.

For P_{lt}, the cycle of operation shall not be repeated when the cycle of operation is less than 2 h, or when the EUT is not normally used on a continuous basis.

REFERENCES

1. Gerke, D., and W. Kimmel. 1995. *Electromagnetic Compatibility in Medical Equipment.* Piscataway, NJ: IEEE Press and Buffalo Grove, IL: Interpharm.

2. IEC/EN 61000-4-4. 1995. *Electromagnetic Compatibility (EMC), Part 4. Testing and Measurement Techniques. Section 4. Electrical Fast Transient/Burst Immunity.* Basic EMC publication.

3. Mardiguian, M. 2000. *EMI Troubleshooting Techniques.* New York: McGraw-Hill.

4. Williams, T., and K. Armstrong. 2001. EMC Testing Part 3—Fast Transient Burst, Surge, Electrostatic Discharge. *EMC Compliance Journal,* June. Available: www.compliance-club.com.

5. IEC/EN 61000-4-5. 1995. *Electromagnetic Compatibility (EMC), Part 4. Testing and Measurement Techniques. Section 5. Surge Immunity Test.* Basic EMC publication.

6. Williams, T. 1996. *EMC for Product Designers,* 2nd ed. Oxford: Newnes.

7. IEC/EN 61000-4-6. 1996. *Electromagnetic Compatibility (EMC), Part 4. Testing and Measurement Techniques. Section 6. Immunity to Conducted Disturbances, Induced by Radio Frequency Fields.* Basic EMC publication.

8. EN 55024: 1998. *Information Technology Equipment—Immunity Characteristics—Limits and Methods of Measurement.*

9. Williams, T., and K. Armstrong. 2000. EMC for Systems and Installations—Part 6, "Low-Frequency Magnetic Fields (Emissions And Immunity); Mains Dips, Dropouts, Interruptions, Sags, Brownouts and Swells. *EMC Compliance Journal,* August. Available: www.compliance-club.com.

10. IEC/EN 61000-4-11. 1994. *Electromagnetic Compatibility (EMC), Part 4. Testing and Measurement Techniques. Section 11. Voltage Dips, Short Interruptions and Voltage Variations Immunity Tests.* Basic EMC publication.

11. IEC/EN 61000-3-2. 1995+Amendment A14:2000. *Electromagnetic Compatibility (EMC), Part 3-2. Limits for Harmonic Current Emissions (Equipment Input Current Up to and Including 16A Per Phase).* Basic EMC publication.

12. IEC/EN 61000-3-3. 1995. *Electromagnetic Compatibility (EMC), Part 3-3. Limitation of Voltage Fluctuations and Flicker in Low-Voltage Supply Systems for Equipment with Rated Current ≤ 16A.* Basic EMC publication.

13. EN 61000-4-7: 2002. Electromagnetic Compatibility (EMC)—Part 4: *Testing and Measurement Techniques—Section 7: General Guide on Harmonics and Interharmonics Measurements and Instrumentation, for Power Supply Systems and Equipment Connected Thereto.*

14. IEC/TR2 60868: 1986. *Flickermeter—Functional and Design Specifications.*

15. ANSII C63.4. Multiple dates. *American National Standard for Methods of Measurement of Radio-Noise Emissions from Low-Voltage Electrical and Electronic Equipment in the Range of 9 kHz to 40 GHz.*

16. Williams, T., and K. Armstrong. 2000. *EMC for Systems and Installations.* Oxford: Newnes.

17. Williams, T., and K. Armstrong. 2001. EMC Testing Part 1—Radiated Emissions. *EMC Compliance Journal,* February. Available: www.compliance-club.com.

18. Williams, T., and K. Armstrong. 2001. EMC Testing Part 4—Radiated Immunity. *EMC Compliance Journal,* August/October. Available: www.compliance-club.com.

18. IEC/EN 55022:2000+A1. 1999. *Limits and Methods of Measurements of Radio Interference Characteristics of Information Technology Equipment.*

19. Williams, T., and K. Armstrong. 2000. EMC for systems and installations—Part 5, Lightning and Surge Protection. *EMC Compliance Journal,* August. Available: www.compliance-club.com.

20. Williams, T., and K. Armstrong. 2000. EMC for Systems and Installations—Part 7, Emissions of Mains Harmonic Currents, Voltage Fluctuations, Flicker and Inrush Currents; and Miscellaneous Other Tests. *EMC Compliance Journal,* August. Available: www.compliance-club.com.

CHAPTER 7

RADIATED TESTING

Similar to conducted currents and their manner of development and propagation, another area of concern for electrical engineers lies with radiated fields. The manner in which propagation of a radiated field occurs is through a dielectric that is capable of supporting field propagation. The dielectric that is most familiar to us is free space. Water is also a dielectric that supports field propagation as well as the material used to construct a printed circuit board.

Transmission lines provide a path for current to flow, steady state or time variant, propagating from a source to a load. Think of this in terms of a road or highway. The road is the path on which the automobile travels. It is the automobile that carries the item of concern (electrical information in the form of an electromagnetic field or a passenger in the form of a biological compound).

Transmission lines only provide either a voltage- or current-drive mechanism to drive an antenna, either dipole or loop configuration. Other forms of radiation are possible, although the majority of signal propagation is observed through these two most common antenna types. Electromagnetic radiation is a category of field propagation and is investigated within this chapter as well as electrostatic or ESD fields.

Radiated RF energy consists of both magnetic and electric fields. With each field there are two primary modes of signal transfer: common mode and differential mode. It is impossible to determine the type of current flow when the field is measured by an antenna. The common mode is generally the more prominent mode of interference propagation.

One field will be dominant, either electric or magnetic. Both fields exist simultaneously; however, one field will have a larger amplitude or field strength than the other except at some physical distance from the source of the energy. This distance

Testing for EMC Compliance. By Mark I. Montrose and Edward M. Nakauchi
ISBN 0-471-43308-X © 2004 Institute of Electrical and Electronics Engineers

is where both fields create a propagating plane wave that has the wave impedance value of approximately 377 Ω (impedance of free space). Details on the plane wave are found in Chapter 2. When testing a product for compliance purposes or troubleshooting a system, it is helpful to know which field is dominant. Antennas, probes, and instrumentation are designed to measure signals in a particular manner. If the wrong transducer is chosen, it may be impossible to ascertain if the emission detected is really a problem or not.

The purpose of this chapter is to provide information on radiated fields that propagate from a transmission line, interconnect, or digital component. Undesired radiated RF energy is generally observed within the frequency range of approximately 100 kHz–300 GHz, which is the frequency range most often used for communication purposes.

Regulatory standards, be it commercial, military, or private, contain specification limits to ensure that the magnitude of undesired RF energy is sufficiently low so as not to cause harmful interference when a propagating RF field is present, be it telecommunication, broadcast (radio and television), or when used as a sensor.

Knowing different measurement techniques and using the right antennas, probes and instrumentation can make the task of identifying radiated emissions easier and with greater accuracy.

7.1 PERFORMING RADIATED TESTS

The material in this section deals with radiated emissions and immunity events. Conducted emissions are discussed in Chapter 6. Before discussing the process of performing radiated testing, one must understand what needs to be accomplished in the most efficient manner. Test specifications were developed to guarantee system EMC in almost every anticipated location or environment. A high level of confidence must be present prior to production to ensure compatibility with other electrical equipment and communication services.

Noncompliance can happen at any time. The stages of testing and analysis during development are as follows, in order of preference. This is illustrated in Figure 7.1:

1. *Engineering Investigation in Laboratory or Engineer's Office.* This includes measurements with instrumentation or simulated analysis with computers. Various subassemblies are first tested for self-compatibility before integration into a final system. If simulated analysis is performed on PCBs, related to propagating fields, results may not reflect actual field propagation or far-field results with interconnect cables attached, especially if a metal enclosure is provided. This is due to parasitics that are generally unknown to the engineer; capacitive effects between a PCB and metal chassis.

   Time-domain simulation will optimize the signal integrity of digital circuits for functionality purposes; however, trying to simulate an operational PCB (with thousands of nets switching synchronously) for radiated emis-

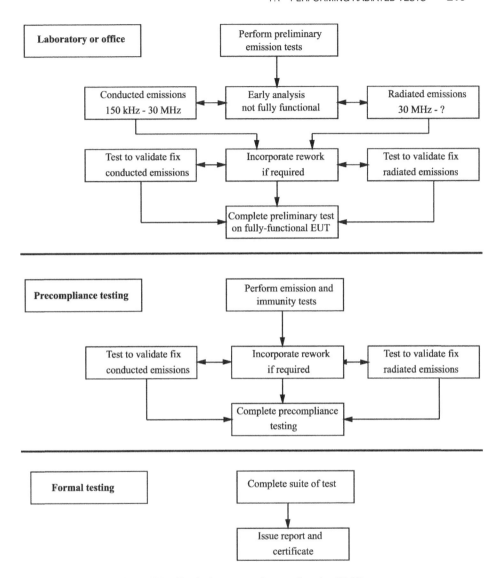

Figure 7.1   Typical process when performing EMC tests.

sions within an enclosure cannot be easily performed and is nearly impossible to achieve.

1. *Precompliance Testing.* This refers to in-house testing using a fully populated system and test setup similar to what the unit would be subjected to during formal qualification tests. These systems are generally preproduction in nature.

2. *Formal EMC Qualification Tests.* This is when the product goes for formal certification approval.

The sooner in the design process that analysis is performed, the easier it becomes to implement mitigation techniques or to redesign the product. By the time the system goes for formal qualification, it may be impossible to make any changes; therefore one is stuck with a noncompliant product resulting in no revenue for the manufacturer.

To optimize the sequence of analysis during development or precompliance, it is important to use resources efficiently. The following are suggested:

1. Perform emission tests first, then immunity. Immunity tests may be inadvertently destructive.

2. It is not mandatory to monitor malfunctions. In fact, invalid software operation may be useful as long as the system is powered on and operated in its worst-case noise mode of operation. Clocks and periodic signals have a higher priority of concern at this stage of development and test.

3. Only after making the system EMI appear to be compliant, related to clocks in the standby mode, can evaluation be performed with operational software.

4. Radiated emission tests are more complex, requiring extensive time to perform and the ability to detect signals from the EUT versus the ambient. Distinguishing signals is an art of engineering, requiring talent and experience.

5. Troubleshoot the system for noncompliant signals and then retest.

6. Incorporate rework into a new model of the product and repeat the steps above. Only if a high confidence level exists that the system will pass formal qualification tests should one attempt to take the product to a third-party testing laboratory. If doubts exist regarding the possibility of failure, diagnostic work using immunity test equipment can be performed.

7.1.1   Engineering Investigation in Laboratory or Engineer's Office

All engineers are encouraged to perform careful levels of analysis during the development cycle. Analysis includes functionality, manufacturability, and compliance to various regulatory requirements, including EMI and immunity. During the design cycle, when hardware does not exist, use of simulation tools is the only possible means of analysis.

This chapter details how to perform an investigation while in the development cycle when hardware is built. Any measurement technique is acceptable, as the magnitude of RF disturbance is irrelevant, after one gets the product to function as desired. The element of concern is whether any RF energy can be detected, which is the easiest investigation with which to begin. Use of probes and sniffers become valuable tools during this time. Expensive instrumentation is not mandatory, nor is use of calibrated current or voltage probes; however, their use provides much more information than a probe made out of a paperclip and shrink tubing.

7.1.2   Precompliance Testing

Precompliance analysis involves testing to relevant EMC standards using techniques that allow one to understand how the product will function prior to being sent to a formal laboratory for certification. These tests should use procedures and techniques as described in relevant standards. Most EMC standards describe test methods themselves or refer to other documents where test methods are provided in detail. For example, radiated emissions for compliance to Europe's EMC directive usually involves test methods described within EN 55022 or EN 55011 (CISPR-22 or CISPR-11 [1, 2]). These standards mandate testing at an OATS. Because ambient noise is becoming more common, especially with the onslaught of digital broadcasting, it is difficult to find a location for an OATS that does not suffer from high levels of ambient noise. For this situation, use of shielded test facilities and chambers is becoming popular; however, the test standard still mandates formal testing at an OATS regardless of the data taken in a chamber for radiated emissions.

Precompliance testing utilizes full-compliance methods without adherence to strict test procedures. Not testing strictly to formal test standards can save considerable money and test time. The important item about precompliance testing is to understand all test procedures and what errors can be introduced when not performing tests "by the book." The spectrum of precompliance tests runs from rough and quick methods based broadly on test standards, capable of generating results ranging from very inaccurate to excellent (depending on the capabilities of the test engineer and understanding what is being done and why), to those that give results indistinguishable from full-compliance testing.

Deviation from exact methods and use of low-cost equipment not compliant to CISPR-16 [3, 4] can result in measurement errors. These errors can lead to wasted time and effort, resulting in being late to market (overengineering, reengineering, higher cost to manufacture) or excessive financial risk (poor reliability, high rate of customer returns and warranty costs, possibly even suspension from the European market either voluntarily or by order of enforcement officers). Low-cost testing with measurement errors is not low cost at all, especially if buyers of the product decide to test it themselves and then report nonconformity to authorities under false pretence.

It is possible, without much effort, to perform precompliance testing in a manner that saves considerable time and money without commercial and/or financial risks. It is not just plugging in the equipment and taking data; one needs to know what the errors are during the measurement process.

There are two ways to determine measurement errors in a precompliance test. One is to follow the same procedure as for a full-compliance test. This includes measuring the normalized site attenuation (NSA), obtaining calibration data for all equipment, cables, and antennas, and working out the measurement uncertainty calculations prescribed by appropriate test procedures.

The second method is to compare test results with a known good product that has been fully evaluated. For this baseline system, there is no urgent requirement to know anything about the test site or measurement uncertainties. The compliant system is taken from the test laboratory after formal EMC testing, never opened or modified, and archived. This system is generally called a "golden" (reference) unit. The differences between official compliance and precompliance are generally correction factors. A correction factor may be required for all measurements made at the precompliance site.

Ground planes made of metal are designed to provide repeatable references and images from the plane. The purpose of having different height scans is to maximize both reflected and direct signal propagations. Some engineers use "precompliance" test equipment without a ground plane. In addition, the same engineers may not bother with changing the height of the antenna required for formal certification testing. To compensate for lack of adherence to published test standards, some will reduce the limit line by 6 dB, assuming that compensation for lack of a ground plane and antenna height scanning can be accommodated for. What is not taken into consideration is possible field cancellation from ground plane reflections. This can change the measured value by as much as 25 dB.

Some precompliance antennas are supplied with tripods or stands that can give a fixed or only two different antenna heights. Taking the worst-case reading at each frequency, from measurements taken at all possible scanning heights possible, cancellation effects are avoided. The worst-case emissions should always be recorded, even if full-height scanning is not conducted.

Precompliance testing inside a building, where a CISPR ellipse is not possible, may suffer from unknown reflections giving frequency-dependent errors perhaps between +6 and -25 dB for each reflection. This variation can vary from day to day as people move equipment and furniture around. This equipment and furniture do not have to be in close proximity to the test area. In this environment, verifying site attenuation and performance on a regular basis with a known good product or a comb generator will help discover problems, allowing one to achieve some degree of repeatability.

Another problem with performing precompliance measurements in buildings is ambient noise caused by other electrical equipment. Some facilities suffer from ambient noise that far exceeds the regulatory limit line over a large portion of the frequency spectrum. This makes precompliance measurements almost impossible. Moving outside to a parking lot, a sports field, or landscaped area may reduce ambient noise from inside the building but does nothing for the external ambient problems typically observed at an OATS. This can be virtually overcome by using a correlation analyzer that can provide true ambient cancellation in real time. Refer to Chapter 3 for more details on the capabilities of a correlation analyzer.

A CISPR-16 quasi-peak (QP) detector is generally available in higher end spectrum analyzers. Low-cost spectrum analyzers or receivers usually provide good results when measuring emissions from clock harmonics and similar sources of emissions. For measurement of broadband emissions from the brushes of DC motors or low-rate pulse signals, such as a strobe light running at one flash per second, an incorrect reading may be measured using a wrong detector. Testing a known good product is one manner of calibrating various detectors during the precompliance process.

Antennas occasionally become damaged, cables and coax crack, connectors break, and instrumentation goes out of calibration. Before performing any level of emissions testing, ensure that all instrumentation and support equipment is in excellent working order and calibrated on a frequent basis.

7.1.3   Performing Precompliance Analysis

When performing precompliance radiated emission tests, one generally assumes that use of antennas is required. There are several ways to perform precompliance testing.

For many diagnostic or even qualification tests, use of a current clamp and immunity test equipment assists in locating problem areas quickly. An example of using current probes/clamps for radiated emissions is shown in Figure 7.2.

In the scheme of testing, immunity tests should be done first. Why would one want to perform immunity tests before the traditional emissions sequence? The answer is simple. It is sometimes easier to cause a disruption to the EUT with an externally induced event than to ascertain the possibility that a radiated emission signal will exceed a specification limit at a certain frequency. Most of the time the same component, or cable assembly, will be the source of failure for both emissions and immunity.

Regardless of the order in which tests are conducted, eventually all immunity tests will have to be performed. Doing immunity analysis early in the design process will assist in shortening the design cycle. Emissions and immunity are exact opposites of each other, that is, RF energy either enters or leaves a product. Solving one problem may result in preventing another problem from developing.

Immunity tests can be easier to perform than emission tests; for the most part, however, they require special instrumentation and test environments. Many tests can be performed without use of a shielded room or anechoic chamber. In many

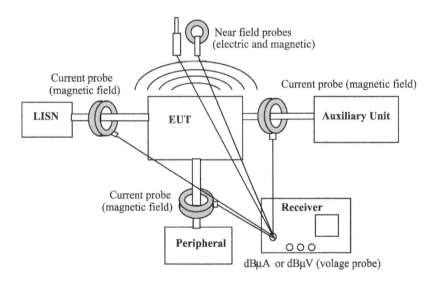

Figure 7.2   Alternate radiated emission testing using current and near-field probes.

cases, immunity tests using conducted methods, CW, or pulse stimuli are only feasible outside a formal EMC laboratory. Advantages of immunity testing include the following:

1. Isolation and hardening of victim components, circuits, cable assemblies, and transmission lines
2. Validation of effectiveness of line filters, gaskets, metallic enclosures, harness routing, and manufacturing processes related to possible coupling paths from either radiated or conducted fields
3. Verification of robustness of software to recognize errors, glitches, and watchdog circuitry, in addition to other functions

7.1.4 Formal EMC Qualification Tests

This level of investigation is for official certification and testing of products per regulatory requirements. Tests are performed in accordance with published standards and procedures. Examples on how to perform actual testing is provided in Appendix B. By the time one attempts formal EMC testing, if proper design techniques were implemented and verified as functional, the product should be compliant.

In compliance testing three aspects are to be considered (see Chapter 4 for concerns):

- Quality of test site
- Quality of test equipment
- Correctness of procedures

A problem with EMC measurement at an OATS is that some emissions from the EUT may be swamped by strong broadcast and other legitimate radio transmissions that coexist at the same frequency within the ambient environment. This is often not a problem for near-field and current probe measurements. Electromagnetic compatibility receivers and spectrum analyzers may have provisions for a demodulated output. The output of the modulator is sent to a speaker in the receiver or to headphones. It thus becomes useful to listen to the demodulation of a modulated signal to determine if the measured signal is an ambient or an actual signal from the EUT. Voices or music indicate a broadcast station or private mobile radio.

To determine if the measured signal is from the EUT, or ambient, a simple procedure exists—turn off the EUT. If the signal disappears, then this is probably a valid emission. Certain products may still emit significant levels when in standby mode (e.g., inverter drives for AC motors). It may be necessary to actually remove all AC mains or DC power from the EUT several times to discriminate between emissions and ambient signals.

Some spectrum analyzers and EMC receivers have a two- or three-channel display: A, B, and C. Saving a spectral plot in memory within one channel and using a second channel to investigate the spectrum again, this time with the EUT powered on or off, allows one to determine actual signals related to the EUT. If using the A–B function, ambient cancellation is performed. One disadvantage of attempting ambient cancellation with spectrum analyzers is that transient signals can pop up suddenly. Although these transients are not from the EUT, difficulty in determining actual emissions from transients can become a time-consuming task. Transients are signals that come and go almost at random; often, in the very high frequency (VHF) bands. They are usually caused by portable or mobile radio transmitters.

A better alternative is to use ambient cancellation techniques using a correlation analyzer as described in Chapters 3 and 9.

Underground parking lots or cellars can have quieter RF environments for performing open-area tests, although they may exhibit cavity impedance or reflection conditions. In actuality, the further away from a noisy RF environment for an OATS installation, the better the performance will be.

To resolve problems with ambient signals for radiated emissions, use of shielded enclosures are preferred. These enclosures are recommended only for precompliance work. Only when a product is compliant within a chamber or a shielded enclosure, when known frequencies are recorded can one go to an OATS and fight their way through the ambient environment. Another option is to use a correlation analyzer as mentioned previously. This type of analyzer performs "true" ambient cancellation in real time, unlike the A–B technique described earlier with its inherent problems.

There are alternate methods of performing radiated emissions tests. Most EMC standards require a measurement distance of 10 m. For precompliance purposes, it is common to measure at 3 m and increase the limit line by 10 dB. Measuring at a closer distance allows use of smaller test sites, sometimes within an empty portion of a corporate building. A smaller test distance improves the signal-to-noise ratio of measured signals. An improved signal-to-noise ratio becomes important when working with lower cost spectrum analyzers or EMC receivers. It is important to remember that a system may be compliant at 3 m and noncompliant at 10 m due to signal reflection from the ground plane and other nearby metallic structures or other factors. Correlating data do not always provide accuracy between the two distances.

The NSA requirements for a 10-m OATS, as specified in ANSI C63.4 and EN 55022 (CISPR-22 [2]), are shown in Figure 7.3.

Such OATSs are not difficult or expensive to set up and can be temporary structures with metal mesh ground planes that can be rolled up and stored away. Parking lots are acceptable as temporary site for use at night or on weekends.

7.1.5 Instrumentation Error

Errors can be caused by instrumentation problems affecting accuracy of measurement. During preliminary testing the system may appear to have minimal EMI, thus giving a false sense of security. The system is certified and placed into production. Perhaps unknown to the test engineer, another person (not necessarily an engineer)

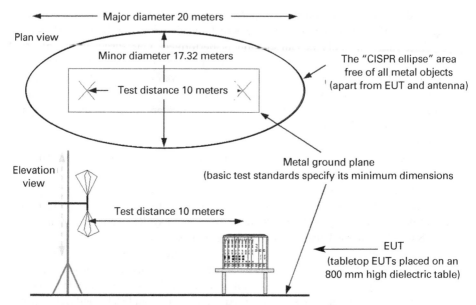

Antenna (various types) height-scanned over 1-4 m to find the maximum
emissions at each frequency (for both vertical and horizontal polarisations)

Figure 7.3   General OATS requirements defined in EN 55022 (CISPR 22). (Illustration
courtesy *EMC Compliance Journal* [5].)

may have used the equipment for a specialized use the day before and failed to restore the instrumentation setup to its original state.

The following is a list of potential problems from instrumentation one should be
aware of:

(a) Incorrect bandwidth selection on the receiver or spectrum analyzer.

(b) Improper detector (peak, quasi-peak, average)

(c) Inappropriate scan time for the range of frequencies to be examined—too
fast or too slow

(d) Video filtering applied that may affect the amplitude of the measured signal

(e) Overload of the front end of the receiver or spectrum analyzer by out-of-
band signals, including 50/60-Hz components

(f) Vertical linearity affected by incorrect switching steps or not set at the correct reference level

(g) EMI and ambient signals added together to create a false reading. Narrow-
ing the bandwidth will help separate these signals into their discrete components, which can then be measured individually.

(h) Insufficient sensitivity of the receiver relative to ambient background noise

(i) Broadband noise preventing detection of narrowband noise spikes

7.1.6 In Situ Testing of Systems and Installations

Not all systems can be easily tested at an OATS or in a shielded chamber/enclosure. Such systems that are physically large, require three-phase power, have input current greater than 100 A/phase, or need ancillary support (vacuum and water pumps, gas panels and interconnects, RF generators, extensive bulkhead interconnects, etc.) must be tested in-house. Testing may be performed in a manufacturing facility, with minimal room to walk around the systems. There is sometimes not much space to set up antennas to take radiated emission data without RF field corruption from nearby systems. In situ testing thus becomes an art on knowing how to perform tests with on-site limitations, measurement uncertainty, and substitution of test instrumentation (i.e., current probe instead of a voltage probe), resulting in very questionable accuracy.

In many situations, a single piece of equipment cannot be evaluated by itself. Under these conditions, a larger installation can be considered the EUT. Technically, the EMC Directive mandates no harm to communication equipment. An entire facility can be tested as a single entity for the purpose of meeting the EMC Directive. For this type of test, an antenna is placed 30 m away from the exterior of the building and data recorded at various antenna positions (360° around the site). This level of testing prevents one from having to test each individual component and assembly to meet the essential requirements of the directive. An example of this type of installation includes AC power distribution subsystems, railway stations, telecommunications farms, and nuclear power plants.

The main causes of EMI in most systems and installations are electronic modules (PCBs) and their interconnecting cables. For example, when testing radiated emissions above 30 MHz, a cable length of 4 m is considered electrically long but is acceptable for in situ testing. This length of cable will generally be representative of the final installation, where this cable may be hundreds of meters long.

If a system is tested in situ in a manner that represents its final installation and then installed correctly within the environment that the unit is expected to operate in, a good degree of correlation should exist between the laboratory or manufacturing facility and final installation. This level of installation makes on-site testing unnecessary.

Appendix B provides information regarding test procedures and setup for in situ testing. Generally, most products tested in situ must have a TCF written that explains why testing had to be performed in-house and why certain test requirements were waived. Performing in situ testing is more difficult than at a commercial test facility. Those who perform in situ testing must be specially trained and certified competent.

The problems of on-site testing are the same as those of precompliance or formal testing where the test site is not ideal. For example, Figure 7.4 illustrates an in situ test using a biconical antenna that is physically close to large metal assemblies that will affect radiated field propagation. Unfortunately the option of comparing one system to another for the purposes of verification analysis is not available.

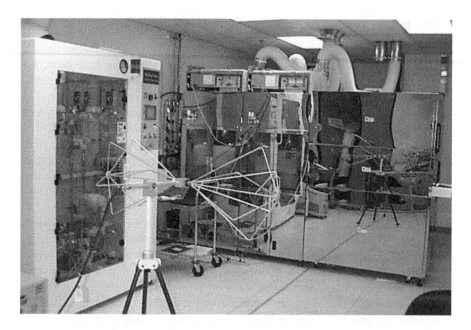

Figure 7.4   Sample in situ test of an industrial product.

There is some debate as to whether in situ testing can be used to represent the compliance status of systems that are not tested at an open-area site. CISPR 11 (EN 55011 for Europe) states [Ref. 1, Section 1.2]:

> Measurement results obtained for an equipment measured in its place of use and not on a test site shall relate to that installation only, and shall not be considered representative of any other installation and so shall not be used for the purpose of statistical assessment.

In contrast, CISPR 16-2 [4] suggests that where a given system has been tested at three or more representative locations, the results may be considered representative of all sites with similar systems for the purposes of determining compliance. The FCC rules have a similar condition. For compliance with the EMC Directive via a TCF route, a manufacturer may write a TCF around whatever degree of on-site testing is desired if the chosen Competent Body agrees with the documentation.

7.2   IMMUNITY/SUSCEPTIBILITY TESTS

Tests for electric (IEC 61000-4-3 [6]) and magnetic (IEC 61000-4-8 [7]) field disturbance are required to evaluate the susceptibility of equipment to radiated fields present in the operating environment.   Threats to products are generally from local transmitting sources, principally mobile security radios. Most mobile radios have an output power rating of not more than 5 W. Some threats may also exist from other sources such as citizen band (CB) and amateur radio, military and broadcast communications, and cellular telephones. These threats are confined to a limited band, generally in the frequency range of 80–1000 MHz. The RF intensity near electrical equipment will also be limited by the control of access and the operation of such transmitters. Some environments prohibit the use of radios at all times. Such locations include hospitals and semiconductor manufacturing facilities.

In addition, radiated fields may be present within a particular environment caused by arc welders, thyristor drivers, fluorescent lights, and switches operating inductive loads. These fields can couple into products directly through slots or openings within an enclosure and capacitively couple to cable assemblies or other parts of a system. The net result of subjecting a system to these fields may be disruption of operating circuitry. These effects are generally nondestructive, although it is possible to overstress input filters and sensitive components.

Power line magnetic field immunity is one test that is not performed on a regular basis, as only those products susceptible to magnetic field disturbance are affected. This includes video monitors and units containing sensitive Hall effect sensors and transformer-based control systems.

The concept for radiated immunity testing is to develop a constant uniform field over a specific volume in which the EUT is located. If a uniform field is placed across the device, it is assumed that the test will be repeatable when fully exercising all operational circuits of the EUT. Achieving an adequately controlled RF field and preventing coupling to a victim's circuit is the goal for EMC. In large systems, the field will in probability not be uniform in regard to uniformity illumination.

There are several common tests performed on equipment to ensure electromagnetic compatibility with radiated fields. The most common are detailed as follows:

1. Radiated immunity
2. Electrostatic discharge (belongs in both the conducted currents and radiated fields category)
3. Power frequency magnetic fields

7.2.1 Radiated Immunity

A typical test configuration for radiated immunity testing is shown in Figure 7.5 [8]. The basic requirements include an RF signal source, broadband power amplifier, transducer (antenna), and test chamber. This equipment allows one to generate a constant RF signal source along an entire plane of the EUT. Monitoring of the field strength ensures that excessive amount of energy is not impressed on the equipment. Test facilities normally integrate all instrumentation with computer control to automate the frequency sweep and leveling functions.

The effective radiated power present from a transmitting source is related to field and distance by Eq. (7.1). In this equation we also calculate the power density.

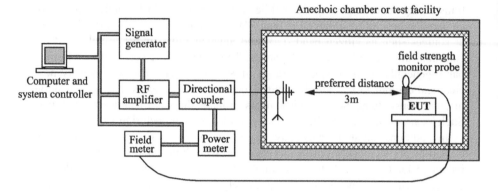

Figure 7.5 Typical radiated immunity test setup.

$$E = \frac{\sqrt{30\ \text{ERP}}}{r} = \frac{\sqrt{30PG}}{r} = k\frac{\sqrt{P}}{r}$$

$$\text{ERP} = \frac{E^2 r^2}{30}$$ (7.1)

$$S = \frac{\text{ERP}}{4\pi r^2} = \frac{PG}{4\pi r^2} = \frac{E^2}{120\pi}$$

where S = power density (W/m^2)
ERP = effective radiated power
E = field strength V/m
P = forward power
G = antenna gain (or constant)
R = distance from transmitting source (m)
$K = \sqrt{30G}$

Table 7.1 illustrates the approximate value of k in Eq. (7.1) depending on antenna configuration.

Table 7.1 Antenna Factors of Various Configurations

Type of Antenna	k
Isotropic	$\sqrt{30}$
$\lambda/2$ dipole antenna	7
$\lambda/2$ rod antenna on ground plane	10
Short-rod antenna with top cap	$3\sqrt{10}$
Folded half-wave dipole	7
Broadside dipole array	$2\sqrt{30}$
Colinear dipole array	$2\sqrt{15}$

For example, a 5-kW transmitter has a field strength intensity at 5 km from a victim system of 0.2 V/m. However, the directivity of the antenna, type of terrain, weather, and other environmental factors can affect this value. If the immunity level of a system is 3 V/m, this broadcast tower should not cause functional disruption as the magnitude of the propagated electromagnetic field is well below the threshold level of 3 V/m.

The possible level of radiated interference from adjacent equipment located in the same room or nearby is considerably less than that from intentional transmitters, since such equipment is limited in its RF emission profile to levels considerably below that which would be present to cause a realistic threat. A level of 36 dBμV/m at 30 m equates to approximately 66 dBμV/m at 1 m ignoring near-field effects. This is less than 0.5 V/m at this 1-m distance. Any induction effects that did exist would be limited by the enclosure material falling off rapidly (inversely with the cube of the separation distance).

Cables that penetrate the equipment enclosure can also act as antennas. Cable-to-cable coupling may be minimized by layout and separation. Upper frequency coupling (above 80 MHz) is well simulated by radiated electromagnetic field testing on the system, including enclosures and cables. Optimum coupling into machine cables with a typical length of 1–3 m will be from the mid to the very high frequency (VHF) range up to the ultra high frequency (UHF) range.

An inexpensive and qualitative method for testing coupling into cables is to make use of a signal generator and a 5–10 W amplifier to directly couple this noise into a 1–3 meter length wire secured (by tape or other means) to a cable(s). If the immunity problem is located in the frequency band allocated for amateur radio, an amateur radio transmitter with a dummy load can be used as the noise source. This radiated immunity issue can now be reduced to a benchtop test. This test can also be performed using an ESD simulator (as the noise source) discharging into the tape-secured test wire. This method produces a greater broadband source of noise rather than a specific or well-defined frequency.

7.2.1.1 Modulation.
When performing radiated immunity testing, there are two types of signals commonly used—CW and AM. An unmodulated signal is a pure CW tone. A modulated signal commonly specified by test standards is a 1-kHz sine wave with 80% modulation. An example of this waveform is shown in Figure 7.6. A modulated RF signal is often a more severe test than an unmodulated signal for certain types of circuits and components. This is most noticeable on display terminals. In addition, peak modulation voltage level is greater than a pure CW tone.

For European compliance purposes (CE mark), transmitters (walkie-talkies) may be used for in situ testing instead of performing the radiated immunity test within an anechoic chamber for certain industrial products or systems that cannot be moved to a shielded enclosure. Many industrial products cost millions of dollars and can weigh several thousand pounds. It is potentially not probable to perform a radiated sweep across the frequency spectrum in a commercial or industrial environment without causing harm to communication systems. In order to be permitted to per-

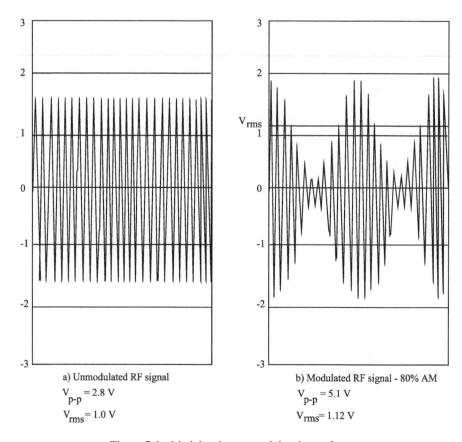

a) Unmodulated RF signal
$V_{p-p} = 2.8$ V
$V_{rms} = 1.0$ V

b) Modulated RF signal - 80% AM
$V_{p-p} = 5.1$ V
$V_{rms} = 1.12$ V

Figure 7.6   Modulated vs. unmodulated waveforms.

form this alternate type of test using portable transmitters, prior discussion with a European Competent Body must be held and a TCF written.

Hand-held transmitters simulate the types of RF fields generated by portable electronics commonly used throughout the world by security personnel, maintenance technicians, firefighters, and users of cellular telephones. These types of transmitters are most likely to be used near equipment that may be disrupted by externally induced electromagnetic fields.

When using portable radios for radiated emission tests, one must know the exact field strength present. Calculation of field strength can be determined from measuring the actual field from the radios [Eq. (7.2)] using appropriate antennas and receiver:

$$V/m = 10^{(dB\mu V/m-120)/20} \qquad (7.2)$$

where $dB\mu V/m$ is the measured field strength at a distance in meters between the antenna and transmitter.

7.2.1.2 Harmonic Issues.   Radiated immunity tests are now required for personal electronic devices using the 2.4-GHz and/or 5.8-GHz frequency range. This frequency range is considerably higher than the typical 1-GHz frequency limit referenced in numerous test standards. Higher test frequencies mean a higher frequency RF amplifier is required.

When using higher frequency amplifiers, which are more expensive than regular amplifiers, care must be taken when selecting the proper unit. It is easy to initially select a wideband amplifier that would cover the entire additional frequency range from 1 GHz to about 10 GHz. This would allow the anticipation of ever-increasing higher frequency usage and to save money since only one more amplifier would be needed. However, selection of a wide-bandwidth amplifier could add an additional problem during immunity testing that was not previously considered.

Harmonics will always be present during testing. The signal generator used as the source will contain harmonics in addition to its fundamental frequency. The RF amplifier will also not only amplify harmonics but also can generate or add to the harmonic noise level. The radiation characteristics of the immunity transmit antenna could permit these harmonics to be just as high in radiated amplitude to approximately 20–30 dB below the fundamental level. Antennas generally radiate high-frequency components more efficiently. In-band harmonics are those that fall within the amplifier's operating range. As an example, if the fundamental frequency is 300 MHz, the second harmonic is 600 MHz and the third harmonic is at 900 MHz. If the upper frequency limit of the amplifier is 1 GHz, in-band harmonics can pose a potential problem with this 300-MHz primary frequency. With a primary frequency of 600 MHz, the third harmonic is 1800 MHz, which is out of band and suppressed by the natural roll-off characteristic of the amplifier. With a wider band amplifier, harmonics will be observed throughout the operating spectrum.

Why the concern for the third harmonic? Most amplifiers have a push–pull configuration for the final output stage. A circuit configuration of this type tends to suppress the second harmonic, thus leaving the third harmonic. During troubleshooting, is it generally the fundamental frequency or a harmonic that causes immunity failure. One could spend a great deal of time attempting to solve a radiated immunity problem at the wrong frequency. In-band harmonics are sometimes a concern with specification limits. The roll-off characteristic of the RF amplifier causes the harmonic output power to be reduced in value. For higher immunity frequencies, a smaller bandwidth amplifier might be desirable. The third harmonic for 1 GHz is 3 GHz. This would not be a problem for a 2-GHz-bandwidth amplifier. The next amplifier might be 2–4 GHz wide, in which case the third harmonic of 2 GHz is 6 GHz, which is also out of band and, hence, it will have a reduced third-harmonic component.

Two solutions are available. One is to select smaller bandwidth amplifiers. This will cost additional money. However, the second alternative is to possibly perform the radiated immunity scan twice: once with the signal source starting at the low-frequency end and going upward in frequency and a second time starting at the upper frequency end and going downward in frequency. This increases time and cost by doubling the test time or by spending troubleshooting time at possibly the wrong frequency.

7.2.1.3 *Monitoring of Immunity Field Level.* This is a controversial area where opinions are varied. The method for monitoring the field intensity level is somewhat dependent upon the test method and/or standard chosen. For meeting EN requirements (e.g., IEC 61000-4-3/EN 61000-4-3), a "substitution method" is described. The substitution method allows the field level to be "calibrated" with the EUT to establish a given field. This includes measuring the forward power to the transmitting immunity antenna. After the calibration step the EUT is placed at the test location. The radiated scan is started with these calibrated levels. The rationale is that whatever distortions the EUT will cause to the field will be what occurs in a real-world environment.

A second method of monitoring is to measure the field strength intensity in real time during testing and adjust the field to always be at a specified level. This is sometimes referred to as "closed-loop monitoring." One concern with this approach is that the measured field level will be highly dependent upon the location of the field-monitoring sensor. Also, does one specify having one, two, three, . . . monitors depending upon the size of the EUT and its physical characteristics? The EUT will distort the field. The field levels read by each sensor will be different for each size room and each size EUT. The difficulty in using this method would be in defining where the sensors are to be positioned or how many field-monitoring sensors are necessary or required. Another issue will be increased test time since at each step the field must be level for each frequency. In leveling, the field could exceed the standard for some amount of time as well.

7.2.2 Electrostatic Discharge

Electrostatic discharge (ESD) is a high-amplitude event that causes permanent damage, latent failures, or disruption in functional operation. Performing an ESD test is a powerful method of determining potential failures over an extremely wide spectrum of frequencies. The 1-ns (or faster) rise time pulse of an ESD event translates to a 320-MHz spectra. When applied in a conducted manner, this pulse creates an extremely strong electromagnetic field. Compared to the electrical fast transient (EFT) test (Chapter 6), which essentially excites cables, ESD also evaluates box shielding and PCB layout deficiencies as well as manifests currents in cables.

The problem of static electricity accumulation and subsequent discharges becomes more relevant for uncontrolled environments. In addition, the widespread application of equipment and systems in various environments exist. This equipment may be subjected to electromagnetic fields whenever a discharge occurs from people standing nearby. In addition, discharges develop between metal objects such as tables and chairs in the proximity of the equipment.

The effects of operator discharge can cause a simple malfunction of equipment or damage of electrical components. The dominant effect is attributed to the following parameters: rise time, current waveform, and duration. Details on these parameters are found in Chapter 2.

The ESD test simulates effects of discharges directly or indirectly to a system. Most of the time, people are charged to a high voltage level by triboelectric charging, usually due to rubbing contacts between their shoes or clothing and dissimilar materials used for flooring, storage, and the like.

7.2.2.1 *General Information.*

Equipment, systems, subsystems, and peripherals may be involved in static electricity discharges due to environmental and installation conditions. These conditions include low relative humidity, use of low-conductivity (artificial-fiber) carpets, vinyl garments, and the like.

Generation of electrostatic charges is favored by the combination of various fabrics and a dry atmosphere. There are many variations in the charging process. A common situation is one where a person walks on carpet. With each step, the person loses or gains electrons from the body to the fabric. Friction between the person's clothing and a chair can also produce an exchange of charge, often in the thousands-of-volts range. A conducting carpet provides no protection unless the operator is adequately earthed to it.

Equipment may be directly subjected to discharges of voltage up to several kilovolts, depending on the type of synthetic fabric and the relative humidity of the environment.

Energy transfer is a function of the discharge current rather than the electrostatic voltage that exists prior to the discharge. Discharge current is typically less than proportional to the predischarge voltage in the higher level ranges.

7.2.2.2 *ESD Waveforms.*

Electrostatic discharge is typically broken down into two primary types of discharge: human (direct) and furniture (air), illustrated in Figure 7.7. Human discharge is characterized by a fast rise of current, approximately 1 ns, up to a peak of 10 A, followed by a decay back to zero. With this waveform, significant RF energy is present up to 300 MHz. Furniture discharge has a slower rise of current to a peak of 40 A, followed by damped oscillations. For this waveform RF energy is observed up to 30 MHz.

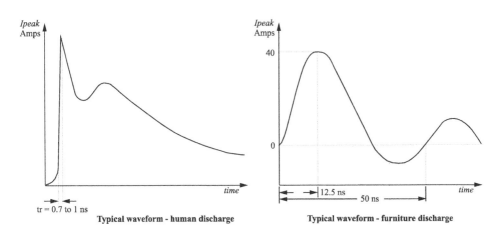

Figure 7.7 Typical ESD waveforms.

7.2.2.3 Triboelectric Series. Electrostatic discharge is a natural phenomenon that affects materials of different potential. This difference in potential is a result of accumulated electric charges. Static electricity is generated when two materials rub against each other, each with a different dielectric constant. When an excessive number of electrons accumulate, they discharge to another material with a lower concentration of electrons to balance out the charge. The effect of the discharge, resulting in electromagnetic disruption, can vary from noise and disturbance in audio or measurement equipment to complete destruction of sensitive components, including electric shock to a person who has been energized.

Electrostatic discharge introduced through a conductive transfer mode must be considered in terms of current flow, not voltage. It is like a burst dam—it is the water flow that does the damage, not the pressure that was behind the dam before it burst. Voltage is merely a convenient metric of the electrostatic "pressure" before the ESD event occurs [9]. Electrostatic discharge introduced through direct or indirect means must be viewed with respect to both electric and magnetic fields derived from voltage and impulse current components.

A buildup of energy must develop before a discharge results. When two materials are rubbed together, with at least one being a dielectric, an accumulation of a positive charge builds up on one material while the other material receives a negative charge. The farther apart on the triboelectric scale, the more readily the charge will accumulate. The triboelectric scale (Table 7.2) represents, at the atomic level, the charge density distribution related to polarity. Table 7.2 is to be used only as a

Table 7.2   Triboelectric Series

Positive charge	
1. Air	19. Sealing wax
2. Human skin	20. Hard rubber
3. Asbestos	21. Mylar
4. Rabbit fur	22. Epoxy glass
5. Glass	23. Nickel, copper
6. Human hair	24. Brass, silver
7. Mica	25. Gold, platinum
8. Nylon	26. Polystyrene foam
9. Wool	27. Acrylic rayon
10. Fur	28. Orlon
11. Lead	29. Polyester
12. Silk	30. Celluloid
13. Aluminum	31. Polyurethane foam
14. Paper	32. Polyethylene
15. Cotton	33. Polypropylene
16. Wood	34. PVC (vinyl)
17. Steel	35. Silicon
18. Amber	36. Teflon
	Negative charge

guideline and is accurate to a degree, depending on material composition. The material at the top of the table, item 1, easily gives up electrons and therefore acquires a positive charge. The materials at the bottom of the table, item 36, absorb electrons, thus accumulating a negative charge.

7.2.2.4 Typical Test Setup.

The equipment to be tested can be evaluated anywhere without the need for a shielded enclosure. When testing, it is advised to not touch the EUT. The static discharge may jump to the test engineer and cause a shock. For air discharge events, direct contact with a system may not cause physical harm to test personnel. It is still recommended that physical contact with the EUT not happen during test cycles.

A typical test setup contains the following components. Precautions should be taken to ensure that the radiated magnetic field does not interfere with instrumentation or other sensitive equipment near the test setup:

- Ground reference plane (GRP)
- Equipment under test
- ESD simulator and appropriate test electrode (tip)

There are two basic setup configurations for performing ESD testing: table top (Figure 7.8) and floor standing in the laboratory (Figure 7.9). The test procedure is identical, only the manner in which the test is performed differs. According to the international test specification for ITE and ISM equipment (IEC 61000-4-2 [10]), tests must be performed in compliance with the manufacturer's test plan and shall specify the following:

- How the test is carried out
- Verification of the laboratory reference conditions
- Preliminary verification of the correct operation of the EUT
- An evaluation of the test results

There are different types of discharge mechanisms (Figure 7.10): *direct* versus *indirect* and *contact* versus *air.*

The "direct" discharge is when one makes physical contact with the EUT and draws an arc that may or may not be observed. The "indirect" discharge involves applying the ESD pulse to a horizontal coupling plane (HCP) and/or vertical coupling plane (VCP). These planes will reradiated a magnetic field in proximity to the EUT. This radiated field can cause nondestructive disruption in operation, whereas a direct charge into a component, via the chassis enclosure or other means, can destroy a silicon wafer or other discrete components.

Contact discharge is performed by directly touching a device or system. On the other hand, air discharge is observed by the EUT from a high-energy propagating field. Direct damage to components does not generally happen with an air discharge event.

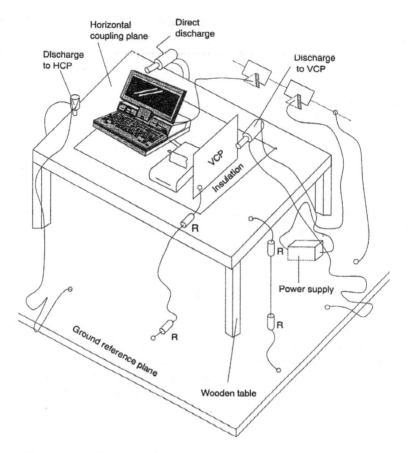

Figure 7.8 Test setup for ESD testing (table-top equipment). (Illustration courtesy of IEEE Press [11].)

Some aspects of the test plan are only valid during an evaluation period. Examples include locations affecting choice of points for discharge:

- Metallic sections of cabinets both electrically isolated and directly connected to ground
- User interfaces, such as control/keyboard areas, screen displays, indicators, slots, grills, switches, knobs, buttons, or other operator-accessible areas
- All I/O housings where the hand touches the connector during cable installations (Discharges to pins within the connector are not required and in many applications not desired!)

Direct discharges should only be applied to those points and surfaces accessible to operators during normal usage, including maintenance personnel. Areas that

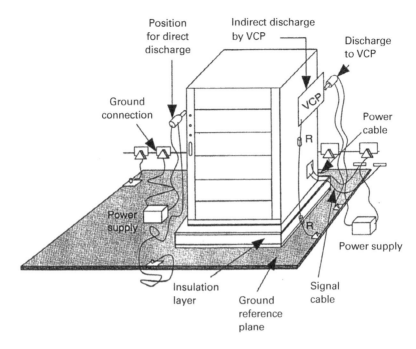

Figure 7.9 Test setup for ESD testing (floor-standing equipment, laboratory setup). (Illustration courtesy of IEEE Press [11].)

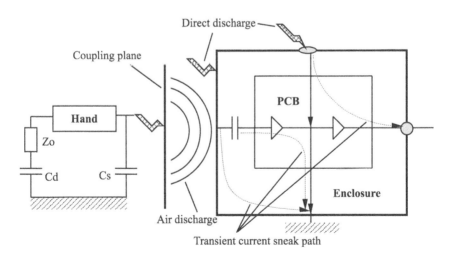

Figure 7.10 Direct and air discharge coupling.

are accessible to maintenance personnel internal to the system need not be tested, as maintenance personnel should have already discharged themselves when approaching this section of the EUT. The ESD simulator should be gradually increased from a minimal to a maximum voltage level to ascertain the threshold of failure. Fifty single discharges shall be applied in the most sensitive polarity on the unit, allowing at least 1 second between discharges. The discharge points may be selected during the exploration phase using a repetition rate of 20 discharges per second.

For contact discharges, the tip on the test simulator needs to have a sharp point (Figure 7.11). A pointed tip is required to pierce through paint and plating, thus ensuring optimal contact with the EUT. The tip should be in contact prior to the discharge switch being closed. When painted metal surfaces exist, the sharp tip must puncture the surface. Where such a source is intended to be insulating, such as a plastic bezel, a direct-contact discharge cannot occur. For this type of test, the air discharge method is required. The air discharge tip has a rounded end which simulates a typical human finger that could touch the EUT or other metallic reflecting surface (Figure 7.11).

For the air test, the discharge switch is closed and the round tip of the simulator approaches the EUT as fast as possible, consistent with not actually causing mechanical damage when the tip touches the coupling plane or EUT. This test most closely approximates the real-world environment because in a typical situation ESD results from a charged human body or furniture that usually approaches the receptor equipment gradually, discharging static electricity during the process of approach. Atmospheric conditions such as humidity and air pressure may influence test results. In addition, the speed of approach can also alter test results substantially.

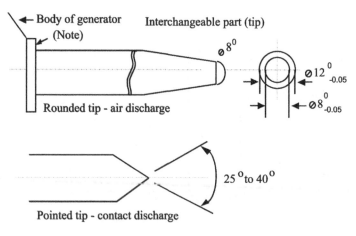

Note - The discharge switch (e.g., vacuum relay) shall be mounted as close as possible to the tip of the discharge electrode.

Figure 7.11   Discharge electrodes of ESD generator.

Thus, air discharge events are difficult to reproduce. To ensure accuracy and achieve statistically correct results, it may be necessary to conduct air discharge tests several hundred times.

The contact discharge test produces reproducible waveforms and test results; however, the waveform and amplitude of the discharge energy do not totally replicate real-life ESD as observed in nature. For purposes of obtaining a good statistical average, multiple discharges must be performed in as many locations on the EUT as possible.

Discharges to objects installed near the EUT are simulated by making contact discharges to the coupling planes. Typically, 50 discharges are made to the HCP at points 0.1 m from each side of the EUT in the most sensitive polarity. Fifty discharges may also be made to the center of only one vertical edge of the VCP, with the VCP in the positions defined in the test plan. The VCP will reradiate the magnetic field to the EUT.

7.2.2.5 *EUT Performance Criteria.*

It is important that external monitoring equipment not be used to evaluate fault conditions. This is due to the fact that additional probes, cables, and monitoring devices can cause the EUT to fail at a lower test level, providing incorrect test results. In addition, remote monitoring equipment can be susceptible to an ESD event making it difficult to ascertain if the failure was due to the EUT or monitoring instrumentation. For this reason, it is recommended to limit the number of auxiliary equipment to a minimum and to use passive loads at the end of cable assemblies (resistive loads). Only cables associated directly with the EUT need to be provided, except for fiber optic. If a failure is observed during the test and a fix is implemented, chances are that by reciprocity unintentional radiated emissions will be minimized. Appendix B provides guidance on performance criteria.

7.2.2.6 *Diagnosis and Fixes.*

When a failure occurs, the following is a recommended guideline on how to analyze a problem area. To begin with, test results must be recorded and analyzed carefully. Items of concern are operational and failure levels at all discharge locations and the type of error or malfunction associated with the failure.

Fixes include the following. Details on each suggestion are provided elsewhere [12, 13, 14]. The magnitude of the fix may be extensive:

1. Improve shielding of external cable assemblies [single to double shield, greater shield coverage/density, improved bonding of the shield to the (metallic) chassis, bonding the cable shield at both ends, replacement with coax, not bonding the shield (drain wire) to chassis ground, etc.].

2. Replace I/O connectors with enhanced metallic versions. The connector faceplate must make a low-impedance bond with the metal chassis, closing off apertures and slots that surround the connector. Bypass isolated connectors using a capacitor to chassis ground. Ensure that the angle bracket on the connector faceplate to the PCB is not connected to logic return.

3. Improve bonding of door panels, housing covers, doors, and other mechanical assemblies with conductive gaskets.

4. Provide filtering for all I/O signal lines. Filtering can include ferrite beads, T and π filters, bypass capacitors, noise suppression connector assemblies, isolation transformers, etc.

5. Improve grounding of user interface points. These points include knobs, switches, display panels, keyboard, pointing devices, and the like. Do not rely on a small drain wire for any level of performance related to draining ESD energy from a direct hit into chassis ground. Use a flat braid with a width-to-thickness ratio of 5 : 1.

6. Reduce the physical size of slots, apertures, and openings within chassis enclosures. Electrostatic discharge is typically a 1-ns event, which translates to 320 MHz.

Try to visualize how the ESD energy propagates throughout the system and not rely upon trial and error. If the path of energy travel can be determined, it becomes a simple task to incorporate cost-effective fixes.

7.2.2.7 *Concerns Related to Analyzing ESD Events.*   Humidity is only one of many parameters involved with ESD. Simulating a dry environment in a chamber that is completely lined with metal surfaces may not be sufficient for emulating a customer's environment, especially one that may be very low in humidity.

Electrostatic discharge failures at differing levels of voltage become an interesting problem due to different scenarios. The value of x is product dependent

1. An unacceptable hard ESD failure at a level of x KV or any higher level is generally related to a grounding problem. To prevent failure, improve the grounding methodology as a potential solution.

2. An unacceptable hard ESD failure at a level of x KV, without failing above this level is probably a decoupling issue in combination with a grounding problem. Careful analysis of PCB design and layout may be required.

For example, a grounding problem may include a ground wire that is too long or improperly secured or a wire diameter that is too small (recommendation is to use a braid strap or strip of copper tape). The width of the braid or tape must be very wide. A decoupling problem may involve incorrectly sized decoupling capacitors, improperly placed components, leads of mounting traces that are too long, wrong sized ferrites beads or cores, etc.

In dry conditions, not only is the charging processes enhanced but also the severity level is increased (faster rise time transitions of digital components and higher peak inrush surge current). For both reasons, it is common to see ESD problems differently around the world based on environmental conditions.

Try to debug the problem using the contact mode of testing. The effect of humidity on the electromagnetic properties of a system is generally very small. The purpose of this test is to identify if traces are susceptible to disruption or components and other circuits sensitive to ESD energy. Locally injecting pulses via direct, capacitive, inductive, or differential means has proved to be an effective way of achieving this diagnostic technique.

7.2.2.8 Alternative ESD Test Simulator.

For in situ testing related to an ESD event and when one does not have access to an ESD gun, the item used in Figure 7.12 has been found extremely useful [15]. This is a piezoelectric gas lighter modified to be an ESD gun. These inexpensive lighters require only two enhancements. One is to secure a wire to the barrel for grounding purposes: the other is to insulate the center conductor to prevent the generation of a spark, which is not good for job security reasons. The output waveform is unknown but could probably be calibrated using the test fixture detailed in IEC/EN 61000-4-2 [10]. What is achieved when using this simulator is a high-energy transient event that simulates ESD when a real simulator is unavailable.

The advantages of using a piezoelectric gas lighter are that it is low cost for a quick test and has the ability to find a weak spot in a system in minimal time away from the EMC test facility, related to an ESD event. A piezoelectric lighter is better than not having any test equipment.

Another ESD simulator is one that simulates different forms of ESD in the environment that are not covered by current standards. The electromagnetic fields from these ESD events have been shown to cause problems in electronic equipment. Characteristics of these *unusual forms of ESD* include fast rise times and multiple events over a few seconds. Examples include internal chair discharges and "jingling change in one's pocket." One can induce voltage in nearby circuits with rise times much faster than present ESD standards, which gives a bandwidth rating of up to 1 GHz. Simulation of this nature emulates real-world events.

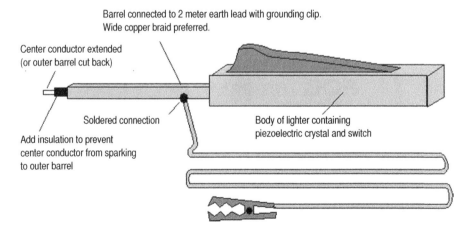

Figure 7.12 Alternative ESD test simulator (piezoelectric gas lighter) [15].

To test a product for ESD susceptibility, take the simulator and send ESD currents over the entire surface of the EUT. If the system does not react to the electromagnetic fields from the simulator, it is likely that the equipment will not respond to many types of "unusual" ESD that may be found in the field.

An example of an ESD simulator is shown in Figure 7.13 with optional mounted antennas that enhance the ESD pulse transmitted to the system.

7.2.2.9 Other Uses for ESD Simulators.
The ESD simulator can obviously be used for performing ESD testing per the standard. It is also an excellent source for simulating and troubleshooting conducted and radiated immunity problems [15]. The ESD discharge energy is an excellent broadband noise generator. If the cost of purchasing EMC test equipment is a concern and for many individuals and companies, an ESD simulator is one of the few tools that provide extensive benefit for the value of money spent to procure this device. The ESD simulator can be used to perform ESD, EFT, conducted immunity, and radiated immunity preliminary testing and/or for troubleshooting known problems. For the price of one ESD tester, up to four tests can be performed.

To utilize the ESD tester for non-ESD applications, simply tape a length of wire (~ 1 m long) to any cable or wire assembly in question and discharge the ESD tester to one end of the wire that was tape securely to the transmission line. Connect the ESD simulator return wire to the far end of the test wire. This essentially creates a short circuit between one end of the wire and the opposite end to the simulator.

To improve the coupling of this transient energy charge into a wire or cable, apply aluminum foil ½ to 1 m long to the taped wire assembly by wrapping the foil around the cable. This procedure can also be done to the AC power cord. Since the

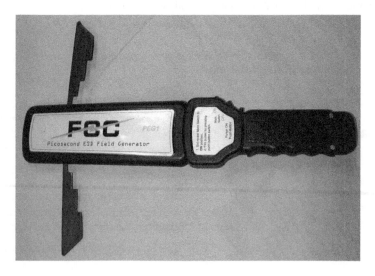

Figure 7.13 Bulk current injection probe with calibration fixture. (Photograph courtesy of Fischer Custom Communications, Inc.)

majority of radiated immunity events are cable related, using an ESD simulator can provide qualitative data for troubleshooting purposes. The significant drawback to this type of test is that one cannot simulate a radiated immunity event directly into an enclosure or chassis.

Possible problems encountered may include the following:

- Missing or incorrect grounding of connectors to chassis
- Lack of proper filtering
- Lack of AC power line filters
- Wires running too far inside the EUT prior to filtering

7.2.2.10 Sensing ESD Events within One's Environment. Within a manufacturing location or at a customer site where ESD events may be present on a frequent basis, one must be confident that the test environment does not corrupt test results. Equipment may be damaged without showing signs of problems, leading one to assume that the system is fully functional. For example, data that is supposed to be on cables are not present, since the source driver has been damaged by an ESD event. An operational system may in fact radiate significant amounts of EMI that will surely be detected by the customer when placed into service.

Simple devices are available that monitor unwanted ESD events. Even though wrist straps, antistatic chairs, conductive floors, and ionizers may be present, ESD can still radiate into a system and cause harm. A sample of one commercial device, shown in Figure 7.14, identifies relevant ESD events and their magnitude if they exist in a room.

Electrostatic discharge event monitors are placed in the areas where ESD events cannot be tolerated. Indication of an ESD event is a sure sign that ESD-preventive measures need immediate attention.

7.2.3 Power Frequency Magnetic Field Disturbance

Electronic products are often subjected to magnetic fields at AC mains frequencies, usually 50 or 60 Hz. Certain installations use 400-Hz power. Magnetic fields near

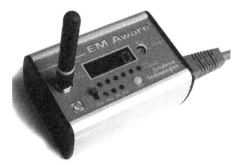

Figure 7.14   ESD environmental sensor. (Photograph courtesy of Credence Technologies.)

and within power transformers can cause significant problems with video displays, especially if they operate at the same scan frequency as the AC mains input. Hall effect sensors and other electronic products are also sensitive to magnetic fields.

Electrical products that may be sensitive to magnetic field disturbance must be tested for immunity to power frequency magnetic fields to ensure continued and reliable operation when placed in service. The European Union's EMC Directive mandates power frequency magnetic field disturbance testing for certain categories of equipment as a condition for obtaining the CE mark before shipping products to member states of the European Union. This requirement does not exist in North America or countries outside Europe.

The Basic EMC Standard for power frequency magnetic fields, EN 61000-4-8 [7], defines methods of generating consistently reproducible magnetic fields. Although higher magnetic field levels are described, compliance to the Generic Immunity Standard for residential and commercial products is 1 or 3 A/m up to 30 A/m for industrial products. While the Basic EMC Standard specifies how to perform power frequency magnetic field testing, the Generic, Product, and Product Family Standards specify test levels and performance criteria.

7.2.3.1 General Conditions. Magnetic field disturbance testing is intended to demonstrate the immunity of equipment when subjected to power frequency magnetic fields related to specific location and installation conditions and is strongly recommended for all products, even if not legally mandated. Tests must be performed on all electrical equipment that includes CRTs and Hall effect sensors. The power frequency magnetic field is generated by power frequency current in conductors or, less frequently, from other devices (e.g., leakage of fields from transformers) in the proximity of equipment [16].

As for the influence of nearby conductors, one should differentiate between (1) current under normal operating conditions, which produces a steady magnetic field with a comparatively small magnitude, and (2) current under fault conditions, which can produce comparatively high magnetic fields of short duration until the protection devices operate (a few milliseconds with fuses, a few seconds for protection relays).

The test with a steady magnetic field may apply to all types of equipment intended for public or industrial low-voltage distribution networks or for electrical plants. The test with a short-duration magnetic field related to fault conditions requires test levels that differ from those for the steady state; the highest values apply mainly to equipment to be installed in exposed places of electrical plants.

In locations such as household areas, substations, and power plants operating under normal conditions, the magnetic field produced by harmonics is negligible. In certain applications, harmonics may be present in heavy industrial areas from large power converters.

7.2.3.2 EUT Performance Criteria. For power frequency magnetic field tests, the Generic Immunity Standard requires the product continue to operate as intended during the test; however, CRT interference is allowed above 1 A/m for residential, commercial, and light industrial locations and above 3 A/m for heavy industrial environments. No degradation or loss of function is allowed below a performance level specified by the manufacturer. The performance level may be replaced by a permissible loss of performance. The product must not become unsafe under any operating condition.

7.2.3.3 *Typical Test Setup.*   The equipment to be tested can be evaluated in an open area without the need for a shielded enclosure. In addition, placement of a person next to the device under test will not affect results. The test setup comprises the following components. Precautions shall be taken to ensure that the radiated magnetic field does not interfere with instrumentation or other sensitive equipment near the test setup:

- Ground reference plane
- Equipment under test
- Induction coil
- Test generator

There are two typical setup configurations: table top (Figure 7.15) and floor standing (Figure 7.16). The test procedure is identical; only the manner in which the test is performed differs. According to IEC 61000-4-8, tests must be performed in compliance with the manufacturer's test plan, and shall specify the following:

- How the test is carried out
- Verification of the laboratory reference conditions
- Preliminary verification of the correct operation of the EUT
- An evaluation of the test results

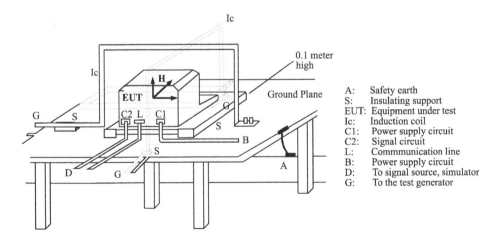

Figure 7.15   Test setup for table-top equipment.

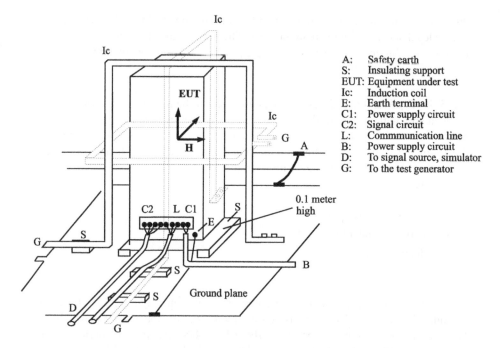

A: Safety earth
S: Insulating support
EUT: Equipment under test
Ic: Induction coil
E: Earth terminal
C1:   Power supply circuit
C2:   Signal circuit
L: Commmunication line
B: Power supply circuit
D: To signal source, simulator
G: To the test generator

Figure 7.16   Test setup for floor-standing equipment.

A ground reference plane shall be placed under the EUT with auxiliary test equipment located on top and connected with secure bonding. The ground plane shall be a nonmagnetic metal sheet (copper or aluminum) 0.25 mm thick. Other metals may be used with a minimal thickness of 0.65 mm. Overall size of the ground plane shall be 1 m × 1 m for table-top equipment. The final size will depend on the dimension of the EUT. In addition, the plane shall be connected to the safety earth system of the laboratory.

The equipment is to be configured and connected into its functional operating condition and located on the ground reference plane with a 0.1-m-thick insulating support (to ensure that the metal plane does not touch the EUT). Equipment cabinets shall be connected to safety earth ground directly to the ground reference plane through the earth terminal of the EUT.

The power supply and input and output circuits shall be connected to appropriate interfaces. Cables supplied or recommended by the manufacturer shall be provided during the test. In the absence of a recommended cable, an unshielded cable appropriate for the signals involved shall be used. All cables must be exposed to the magnetic field for at least 1 m of their length.

Back filters, if any are required, shall be inserted into the circuits that require this protection at a distance of 1 m from the EUT and connected to the ground reference plane.

An induction coil (*Ic*) having a standard dimension of 1-m per side for a rectangular coil or 1-m diameter for a circular coil is to be used for testing small equipment. For large equipment, larger dimensions may be used; however, the coil should be able to immerse the EUT in a magnetic field. The coil dimensions must give a minimum distance of coil conductors to EUT equal to one-third of the largest dimension of the EUT being tested.

7.2.3.4 *Waveform Verification.*

IEC/EN 61000-4-8 requires that the simulator output and magnetic field be periodically verified. Any oscilloscope is capable of verifying the power frequency alternating current to the coil. Monitoring the actual field requires an AC field probe and appropriate monitor. Characteristics to be verified are the output current value and total distortion factor. Both items must have an accuracy of ±2% (Figure 7.17).

7.2.3.5 *Performing the Test.*

The test shall be performed per a test plan, including verification of performance criteria detailed in the product specification sheet. The power supply, signal, and other functional electrical portions of the system shall be operated at rated values. If actual operating signals are not available, they may be simulated.

Preliminary verification of EUT performance shall be carried out prior to applying the test voltage. The test magnetic field is that applied to the EUT, immersing the unit in a magnetic field operating at the AC mains frequency.

For table-top equipment, the entire EUT is immersed by the induction coil. For the floor-standing test, the induction coil must be moved around the unit 360° in order to test the entire volume of the EUT for each orthogonal direction. This is a time-consuming test. Moving the induction coil in steps corresponding to 50% of the shortest side of the coil gives overlapping test fields. The induction coil shall

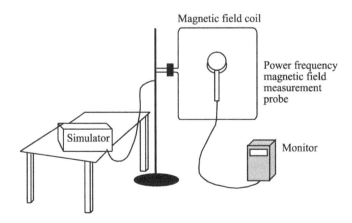

Figure 7.17 Waveform verification test setup.

then be rotated 90° in order to expose the EUT to the test field with different orientations using the same procedure as above.

REFERENCES

1. CISPR-11. *Industrial, Scientific and Medical (ISM) Radio Frequency Equipment—Radio Disturbance Characteristics—Limits and Methods of Measurement.*

2. CISPR-22. *Limits and Methods of Measurements of Radio Interference Characteristics of Information Technology.*

3. CISPR-16-1. *Specification for Radio Disturbance and Immunity Measuring Apparatus and Methods—Part 1: Radio Disturbance and Immunity Measuring Apparatus.*

4. CISPR-16-2. *Specification for Radio Disturbance and Immunity Measuring Apparatus and Methods—Part 2: Methods of Measurement of Disturbances and Immunity.*

5. Williams, T., and K. Armstrong. 2001. EMC Testing Part 1—Radiated Emissions, *EMC Compliance Journal,* February. Available: www.compliance-club.com.

6. IEC/EN 61000-4-3. 1997. *Electromagnetic Compatibility. Part 4. Testing and Measurement Techniques. Section 3. Radiated Radio Frequency Electromagnetic Field Immunity Test.*

7. IEC/EN 61000-4-8. 1994. *Electromagnetic Compatibility. Part 4. Testing and Measurement Techniques. Section 8. Power Frequency Magnetic Field Immunity Test.*

8. Williams, T. 1996. *EMC for Product Designers,* 2nd ed. Oxford: Newnes.

9. Kimmel, W., and D. Gerke. 1994. The Designer's Guide to Electromagnetic Compatibility. *EDN,* January 20.

10. IEC/EN 61000-4-2. 1999. *Electromagnetic Compatibility. Part 4. Testing and Measurement Techniques. Section 2. Electrostatic Discharge Immunity Test.* Basic EMC publication.

11. Kodali, P. 2001. *Engineering Electromagnetic Compatibility,* 2nd ed. New York: IEEE.

12. Hartal, O. 1994. *Electromagnetic Compatibility by Design.* West Conshohocken, PA: R&B Enterprises.

13. Mardiguian, M. 1992. *Electrostatic Discharge, Understand, Simulate and Fix ESD Problems.* Gainesville, VA: Interference Control Technologies.

14. Montrose, M. I. 2000. *Printed Circuit Board Design Techniques for EMC Compliance—A Handbook for Designers.* Piscataway, NJ: IEEE.

15. Williams, T., and K. Armstrong. 2001. EMC Testing Part 3—Fast Transient Burst, Surge, Electrostatic Discharge. *EMC Compliance Journal,* June. Available: www.compliance-club.com.

16. Williams, T., and K. Armstrong. 2000. *EMC for Systems and Installations.* Oxford: Newnes.

17. Williams, T., and K. Armstrong. 2002. EMC Testing Part 6—Low-Frequency Magnetic Fields (Emissions and Immunity); Mains Dips, Dropouts, Interruption, Sags, Brownouts and Swells. *EMC Compliance Journal,* February. Available: www.compliance-club.com)

18. Armstrong, K. 2002. Adding Up Emissions. Conformity Magazine, Vol. 7, No. 6 (July).

CHAPTER 8

GENERAL APPROACHES TO TROUBLESHOOTING

This chapter deals with how one should approach the process of troubleshooting for EMC, both radiated fields and conducted currents. The only way to know if a system has excessive emissions or is susceptible to an external RF event is by testing. There are numerous tests possible. Only an overview of popular types of tests is provided in this chapter. At the time of writing, there are more than 30 different immunity tests with standards written, in addition to numerous configurations and specification limits for radiated and conducted emissions. Determining the appropriate test standard required for the product and how to perform the test in a manner that is repeatable can become a formidable task.

Electromagnetic compatibility measurements are not exempt from problems during testing. Regardless of how testing is performed—developmental, formal, field, or in situ—it typically takes a few hours to days to perform a series of EMC tests. If a problem is observed, finding and fixing the problem can take weeks. What is frustrating to engineers is eventually learning that the problem observed is due to test setup, environment, or operator error and that the EUT is actually compliant.

A key aspect of regulatory compliance lies in repeatability of measured data and continued conformity of a product throughout its life cycle. In Europe, should a problem develop after certification and the product has been placed into service, the end user or one's competitor has the legal right to have any system or operational unit retested to ensure conformity exist.

A methodical approach needs to be taken so as to not overlook anything. Asking questions can provide valuable information. Some questions to ask are as follows:

Testing for EMC Compliance. By Mark I. Montrose and Edward M. Nakauchi
ISBN 0-471-43308-X © 2004 Institute of Electrical and Electronics Engineers

- What was done to change the operating characteristics of the system?
- Has any maintenance been performed?
- Who has had access to the unit?
- When was the last time maintenance was performed?
- Were any components or cables changed?
- What equipment has been installed in the vicinity of the system?

From such questions valuable information may be learned and applied to the troubleshooting process.

8.1   GENERAL SYSTEM TESTING AND TROUBLESHOOTING

Those involved in the field of EMC should be comfortable in understanding various approaches regarding testing and certification following procedures detailed in specification documents, be it domestic, international, military, or customer specific. Sending a product to a competent test facility generally means that testing will be performed professionally and with accuracy. A primary concern for the customer (manufacturer) should be the expertise level of the test engineers(s) employed and their understanding of the EMC environment, especially if a problem develops and the manufacturer is not present to resolve this issue.

Many test laboratories perform the function of testing "only" a product. If a failure is observed, such as emissions over the specification limit or a system shutdown due to an immunity event, the product is generally returned to the customer without further investigation. Many test laboratories are prohibited from performing troubleshooting or diagnostic analysis based on their accreditation status (i.e., conflict of interest or other legal reasons). When this happens, the in-house person responsible for EMC must take corrective action without further guidance or information on where the problem might be or how to incorporate a fix. Under this condition, the cost of compliance can become significant along with delaying product shipment.

Once EMC testing has been completed related to emissions and immunity and a failure is discovered, the next step is to determine what to do next and where the problem area is. Troubleshooting EMI can be frustrating, especially if one is simply trying one idea after another without really understanding the basic cause of the problem. A radiated emission is due to current flow. Therefore, one solution to a radiated emission problem is to reduce undesired RF currents by using a conducted solution. Like a doctor, one needs to understand the problem or symptom before treating the illness and prescribing proper medication. Sometimes luck enters into the process when this type of "shot-gun" approach is used. In the long term, one may miss something that will make a difference in solving the next EMI problem more efficiently. Understanding what the source is and why or how the noise couples and applying it to the next problem will make for a more efficient learning process.

8.1.1 Emission Testing

Every EMI event has three parts: a source of the energy, a coupling path, and a victim or receptor of the noise. Troubleshooting should be directed toward attempting to discover where this energy is from and removing the element that is the easiest to achieve. Questions to consider include the following:

1. What is the source or frequency of the noise?
2. Is it continuous or intermittent?
3. Does the event happen in relation to another event, such as only when a printer is printing or data transfer is in process between two systems?
4. How is the noise getting to the victim or receptor?
5. What is the error that is observed?
6. Is disruption temporary or did permanent damage occur?

Some of this information will come out during the process of identifying the problem. Identification is the first step in the troubleshooting process. What is the failure mode and what do the symptoms indicate? What piece of equipment is being affected, and if it is an immunity issue, at what frequency does the product fail? The same is true for emissions that exceed the specification limit. Knowing answers to these questions can help narrow down the search to a certain section of the system, even to a particular component on a PCB. The next step is to locate the actual noise source. This can be accomplished with an antenna or various held-hand probes of some configuration and a spectrum analyzer or receiver. In other chapters of this book there are descriptions of probes and techniques that may prove useful in locating and identifying problem areas (Chapters 4 and 9).

Another tip to keep in mind is to keep all EMI fixes in place until the problem is solved. Many times a radiated emission concern may be due to several leak points, much like a bucket with holes holding water. Plugging one hole will not keep the bucket from leaking water; it will continue to leak until all the holes are plugged. Electromagnetic interference works the same way. It will continue to leak until all major holes are plugged. Therefore, when plugging one hole and then removing it in an attempt to plug another hole, the bucket will never stop leaking water. Only after the water has been stopped is it okay to determine which modification is necessary. Remove one fix at a time and retest. If the noise level is still acceptable, then the fix applied is not stopping a major leak. If the noise level goes up significantly after removing the fix, then that fix should remain. Repeat this process to determine how many fixes need to be implemented.

Considerations Related to Radiated Emission Testing. When performing radiated emission tests, the following provides guidance on what must be considered, as many decisions may have to be made by the test engineer on the spot without the ability to consult another engineer or the test standard. Consideration must also take into account the local environment and specific aspects of the EUT, which will be different for each product tested:

1. Can an ambient scan be taken at the chosen test point with the EUT turned off or in standby mode? If it is not possible to eliminate the EUT from the ambient, assuming the EUT cannot be powered down under "any" condition, measured data will generally be inaccurate. A notation must be made in the test report to this effect. Measurement uncertainty plays a significant role for this situation.

2. A peak detector sweep is preferred with a reasonably fast scan speed, taking into account the EUT cycle time. This identifies parts of the spectrum in which the EUT emissions are more significant. Make a list of these frequencies for later investigation with greater emphasis on measurement accuracy.

3. When measuring alleged frequencies assumed to come from the EUT, consider taking quasi-peak or average measurements and compare them to the specification limit.

4. The above steps must be taken at all locations of antenna placement. It is recommended that the antenna be physically moved to all four sides of the EUT. In some situations, tests may only be performed at one or two locations due to obstruction from other systems or installation limitations. In addition, measurement at 1 m distance generally does not provide an accurate reading of the true signature profile. Antennas are designed to measure far-field emissions, and 1 m distance is still in the near field for very large equipment, especially if operating frequencies are below 100 MHz.

5. Consider the use of a correlation analyzer for "canceling" ambient signals.

Considerations Related to Conducted Emission Testing. When performing conducted emission tests, the areas of greatest concern lie in coupling of the interference to the ground reference, stray capacitance between the EUT and earth reference, and layout of cable assemblies, especially AC mains. These aspects must be documented in the test report to ensure that testing was performed correctly. It is also important that the AMN/LISN has a quality, low-impedance connection to earth reference.

When performing EMI diagnostic testing and debugging, use of a ground reference plane is highly desired. However, in situ testing sometimes does not allow a ground reference plane to be present. One must use what is available. Regarding ground reference planes, CISPR 16-2 [1] recommends the following:

1. The existing ground system at the place of installation should be used as the referenced ground. This reference system should consider the possibility of high-frequency (RF) noise. Generally, this is accomplished by connecting the EUT with wide, flat straps with a length-to-width ratio not exceeding 3 to structural conductive parts of buildings that are connected to earth ground. These include metallic water pipes, central heating pipes, lightning rods to earth ground, and concrete-reinforced steel and steel beams.

2. In general, the safety and neutral conductors of a facility may not be suitable as a reference ground, as these conductors may carry extraneous disturbance voltages that can have undefined RF impedances.

3. If no suitable reference ground is available in the surrounding area of the test object or at the place of measurement, sufficiently large conductive structures such as metal foils, metal sheets, or wire meshes set up in the proximity can be used as reference ground for measurement purposes.

8.1.2 Immunity Testing

For immunity testing, one needs to understand what the system does and how it does it. If the system resets during an immunity test, then one may need to examine the RESET line on a PCB. Many times, in attempting to resolve an immunity problem, one must force an error condition to occur. The purpose is to find the main noise entry point. This can be accomplished by disconnecting cables or loads and by shorting inputs. Using localized noise sources with hand-held probes is useful to discover entry points. For both immunity and emissions tests, isolating circuit sections or devices by breaking or shorting out signal paths can be one approach to use.

Remember to keep in mind the basic facts regarding testing and troubleshooting. Magnetic fields are caused by low impedance and changing current circuits while an electric field is caused by high impedance and changing voltage circuits. The human body is a dielectric and conductive to electric fields; it reflects and absorbs electromagnetic radiated fields. Sources are characterized by their amplitude and spectrum content (waveform) while victims or receptors are characterized by their sensitivity level and bandwidth. The terms *source* and *receptor* are relative terms. A device or circuit creating an EMI problem can easily become the receptor in another application. It would not be at all strange for one device or circuit to be both source and receptor in the same assembly.

Considerations Related to Immunity Testing. When performing immunity tests, the following are concerns if a product is to be tested in an uncontrolled environment:

1. What is the local environment in regard to transient or radiated noise based upon intended application of use?

2. What specific areas of the EUT need attention? Each host system may be different in configuration, and susceptibility levels may be totally passive.

3. Is the equipment chosen to do the test the right one? For example, which type of injection probe is to be used for conducted immunity testing?

4. Has proper calibration been performed on the injection probes and other test equipment, and what was the last date of calibration?

5. If not possible to perform an immunity test due to physical limitations or installation, is there another test option that can be used to evaluate total immunity levels?

6. Is one testing exactly to the standard or is testing being taken to a higher level to determine the exact point of system failure? For many customers, especially those in Europe, compliance with immunity requirements is more of a quality assurance program to guarantee that their purchased device will remain operational at all times. How much margin is designed into the product prior to testing?

7. Do all tests have to be performed, or only select ones based on functional design, application, and use?

8.1.3 In Situ Testing

If a product or system fails any EMC test at a qualified test facility, an investigation must be performed, usually in-house in the engineering laboratory or manufacturing facility. A controlled environment may not be available. Many companies do not own test chambers, sophisticated instrumentation, or an OATS. Also, certain systems cannot be sent to a test range due to size, weight, or need for support equipment (i.e., gas panels, water for thermal cooling, vacuum pumps, AC mains input greater than 100 A per phase, etc.). Under these conditions, it becomes difficult to perform comparison testing to ascertain conformity should changes be implemented in an effort to resolve the EMC concern. Rented test equipment is sometimes used to debug a system in-house under very uncontrolled conditions.

No test standard provides adequate guidance for in situ testing. *In situ* refers to testing a product at either the manufacturer's or user's premises or any other location that is not a controlled environment. There are concerns by many that testing in situ does not represent the compliance status of a system and that this type of testing does not meet the intent of EMC standards. CISPR 11 [2] Section 1.2 provides the following statement that reflects this concern:

> Measurement results obtained for an equipment measured in its place of use and not on a test site shall relate to that installation only, and shall not be considered representative of any other installation and so shall not be used for the purpose of statistical assessment.

For those products that require conformity testing performed in situ, use of a Technical Construction File for the EMC Directive of Europe (approved by a European Competent Body) or extensive documentation for other countries is required. CISPR 16-2 [1] states, "when a given system has been tested at three or more representative location, the results may be considered representative of all sites with similar systems for the purposes of determining compliance."

Site-specific test plans have to be developed. Decisions must be made by the EMC engineer or technician based on experience of how one is expected to perform testing and analysis under harsh condition. If in situ testing does take place, it makes more sense to carry out tests in the actual configuration and not in any artificial configuration. The real configuration may be the cause of interference when the "artificial" configuration may not. The recommendations provided herein are applicable only for actual configurations.

The definition of "sufficiently large" is vague. The area representing a ground plane should extend beyond the EUT for at least half the height distance of the system to maximize stray capacitance. It depends on what is being achieved and whether the need to minimize stray capacitance is critical to the type of test being performed. Within certain environmental locations, stray capacitance is not a concern. Determining optimal size of the ground plane may be difficult. In certain test standards, use of a ground plane is prohibited, as this ground plane is not part of the system when installed in its operational environment or place of use.

When measuring conducted emissions from products that are permanently installed, it is reasonable to use the existing bonded metal structure of the building and not create a customized ground plane reference. Both a LISN and voltage probe require a reference point. This reference is typically the boundary of the system where the power supply is connected, which is often the terminal of a power outlet or a supply transformer dedicated to the system. Transducer reference connections must be bonded using a short, wide strap to the chosen reference point. A typical green/yellow ground wire is not an adequate source of reference bonding.

The layout of the AC mains cable should be as close as possible to the configuration required during normal operation during the test. This routing can be defined as in a fixed conduit, in which case additional consideration is not necessary. When this cable routing requirement is not possible or is defined by the manufacturer as a requirement for the end user, any excess cable or coil of cable should be avoided; it should be routed as if used in a typical installation. Coiled cable provides stray capacitance between the coils and ground plane, affecting measured results. Regardless of how the AC mains cable is provided, this cable should be fixed for the duration of all tests and documented in the test report along with all other support cables and interconnects.

Considerations Related to In Situ Testing. Most products can be tested in accordance to requirements outlined in the test standard being referenced. One must keep in mind the following when performing radiated testing in situ:

1. Site attenuation between the EUT and antenna is generally unknown. CISPR 16-1 [3] details requirements for NSA to be within a defined range of ±4 dB across the usable frequency range of concern. When performing in situ tests, site attenuation can vary significantly above or below this value, thus skewing the results of the measured data with regard to a specification limit. Some signals may be well above the limit when measured but would be compliant at an OATS and vice versa. Imperfections in the ground plane, if one is provided, can be a contributing factor to measurement uncertainty.

2. The measurement distance is specified by many standards at 3 or 10 m. For many in situ applications, the antenna distance is usually at a maximum distance that may be considerably less than 3 m. Under this condition, the specification limit

needs to be corrected, assuming a linear $1/r$ relationship for the radiated signals. A problem with extrapolation is that the real signal that radiates from the EUT may not be measured accurately due to near-field placement of the antenna for a far-field event. In addition, an antenna that is physically close to a large-size unit may only see a portion of the EUT and may not reflect the actual emission profile of the system.

3. The antenna may not be capable of being manipulated to provide accurate measurement data. In situ testing generally has the antenna at one fixed height. Automatic means of changing the antenna from 1 to 3 m is not easily achieved. Changing the height of the antenna at all test locations extends the test time considerably. In addition, height limitations may not permit this change.

When using any antenna in situ, accuracy of measurement is based on the need for a ground plane between the EUT and the antenna. The purpose of the ground plane is to stabilize the reflected wave from the EUT and to minimize variations in reflectivity between tests. A ground plane ensures that the amplitude of the reflected wave is known, assuming there are no losses in the plane. If other systems are located in close proximity between the EUT and the antenna, reflections may skew measured results.

There is a significant difference for the need of a ground plane between conducted versus radiated emissions. For conducted emissions, the ground plane must be properly bonded to the facility's earth ground with a low-impedance connection to provide for a stable earth reference. When used for radiated emissions, the plane acts only as a reflector for incident waves. Its electrical connection to earth ground is irrelevant, except where safety considerations must be accounted for.

4. The antenna may not be capable of being located at multiple points around the EUT. Unlike testing a small product at an OATS on a turntable where the radiated signals can be easily maximized, the antenna must be physically moved to as many locations as possible. Under this situation, it becomes nearly impossible to state with any degree of confidence that the maximum amplitude of radiated energy has been located. Tests are generally done at 0°, 90°, 180°, and 270° in relation to the EUT. If the radiated signal is at any azimuth that differs from these four values, can one state definitely that the system is EMI compliant?

5. The EUT configuration may play a significant role during in situ testing. The system must be configured in a representative build state and operated in a typical mode of operation. This build state may not represent the final installation or end use. It is recommended that the EUT be manipulated into various configurations, if possible. Measurement requirements for in situ testing are relaxed under these conditions. Any variation of configuration depends on how flexible the equipment is and is best determined by an experienced test engineer. Generally, it is not necessary to explore the effects of each mode of operation.

6. Are all immunity tests required for the product depending on installed application and environment? For example, if a semiconductor manufacturing tool is installed in a controlled environment with filter AC mains that are surge protected and monitored at all times through uninterruptible power sources, are surge testing and voltage variation/sag/brownout tests required?

8.2 POTENTIAL PROBLEMS DURING TESTING AND TROUBLESHOOTING

When performing testing or troubleshooting, emissions or immunity problems may arise. If one is unaware of the following concerns, the time it takes to complete testing can be extended significantly or measured results are skewed and inaccurate [4]:

1. *Setup and Auxiliary Equipment.* When performing any test, how does one know if the data taken are really from the EUT, the auxiliary equipment, or other sources? In order to validate conformity with various regulations, reproducibility is a primary concern should the product be audited by a third party. Auxiliary systems may be different between test units or at an installation site. The EUT may be compliant with one vendor's product but not with another's, although the auxiliary device is compliant with sufficient margins when tested under normal modes of operation. It is important to recognize that not all configurations are identical and separate units already EMC compliant may end up noncompliant.

To ensure that it is possible to reproduce measured data and to demonstrate due diligence when performing tests, detailed documentation must be precise. Digital cameras are making the process of photographing test setups quick and inexpensive. The placement of all cables and distance spacing between units can be illustrated. For example, was instrumentation installed and used in the manner described by both the test specification and the manufacturer?

2. *Validation of Measurement Data.* When taking data, how accurate is the measured results? Did background noise, or ambients, cause false readings to be recorded? Did an intermittent signal arise when a wideband spectrum of signals being measured happened without one's knowledge, thus missing this signal of interest? Are correction factors accurate and have they been applied correctly? Questions we must ask include:

- Was the signal measured an ambient or not, and is the ambient level of the test environment at least 6 dB or more below the signature profile of the EUT or the specification limit being referenced?
- Are all transducers (antennas and probes) working within their range of calibration? Damage to an antenna can significantly change the calibration factor, causing an incorrect reading to be recorded. A faulty or weak coaxial connector can skew measured results significantly.
- When recording a signal that appears to be over the specification limit by a significant amount, was the EUT turned off to verify if the signal was real or not? If the signal was real, where did the signal come from, the EUT or auxiliary support? Was power cycling performed on all units?
- Does measurement instrumentation have sufficient dynamic range for intended use? Is an amplifier or preselector required for the spectrum analyzer? Conversely, will the signal overdriving the input to the analyzer create gain

compression or intermodulation products, which in turn will cause a false reading?

• Will background noise or insufficient sensitivity of the receiver/analyzer cause a measured EMI level to barely emerge over a broadband cluster of signals? Is this background or broadband noise greater in amplitude than the existing narrowband emission that we wish to measure?

3. *Instrumentation-Related Problems.* Are all antennas and probes in working order and are they calibrated? Is the spectrum analyzer and related support equipment capable of taking accurate data? Coaxial cables have a tendency to deteriorate over time if exposed to environmental elements such as rain, snow, and sunlight (ultraviolet). Deterioration may include cracking along with a higher insertion loss and higher values of VSWR. Areas of concern include the following:

• Is the spectrum analyzer, receiver, or any other immunity test equipment calibrated and fully functional? A system may have a calibration sticker that indicates the unit is still current within a typical 1-year time frame of a standard calibration process, but something could have caused the unit to become defective and the calibration sticker provides a false sense of security. Is vertical linearity out of specification? If so, verify with external extenuator attenuators and ensure that the signal being measured does not overload the front end of the analyzer or receiver.

• Was the proper value of resolution bandwidth and detector selected? If an incorrect resolution bandwidth is used, both signal and ambient can add together and provide a false reading.

• Are out-of-band signals causing a false reading to be measured? Signals that are harmonics of the AC mains (for conducted test, 50/60/400 Hz) and local broadcast stations (radiated tests) can affect a measured signal without being displayed on the screen of the analyzer or receiver. Broadband signals (generally not a concern) can affect narrowband signals (the signal of interest) by combining power densities, causing a larger amplitude of the signal to be observed.

• Is the scan rate too fast? If so, then narrowband signals may be missed. This is especially true for conducted emissions if the resolution bandwidth is not properly selected.

• When cable lengths are excessively long, generally greater than a quarter wavelength, significant errors can develop, especially when performing radiated or conducted immunity tests. There can be an impedance mismatch between the RF generator and injection device (antenna or probe). This mismatch is identified as VSWR. The VSWR is the ratio of reflected power to forward power. The higher the mismatch, the greater the power loss from the amplifier to the transducer. Reflected noise can cause serious damage to the amplifier in addition to having the need to provide more RF power to achieve the desired field strength at a particular frequency. The VSWR is a greater concern at the lower frequency range being tested. Never use a T junction in the coax during radiated and conducted immunity tests as these junctions result in significant VSWR from impedance mismatches provided by the use of an additional connector.

4. *Environmental-Related Concerns.* External noise or ambients can cause significant problems or false readings, especially for emissions when tests are performed at a test site that is not ideal. When testing is conducted in an anechoic chamber, this concern is minimized.

Immunity tests are generally not affected by the environment under normal operating conditions and installation. If impossible to conduct an emission test in a chamber, either radiated testing using antennas or conducted testing using LISNs, alternate means of measurement need to be performed. One possible means of doing an alternate measurement is to use a current probe on interconnect cables for radiated emissions or a voltage probe for conducted emissions. The difficult part in isolating environmental concerns is understanding the test environment and the criterion for making accurate measurements. To help determine environmental concerns related to radiated emissions, perform the following (Figure 8.1) [4]:

- Turn off the EUT and all auxiliary and support equipment.
- Remove the coax from the transducer (antenna or probe) and provide a 50-Ω terminator to the coax. The input to the receiver is now through a 50-Ω "alleged" perfect transmission line. This will allow one to ascertain if the signal measured is from the environment, especially for radiated noise. If a signal is observed on the analyzer/receiver, the coax is defective and environmental noise will corrupt the signal of interest being measured.

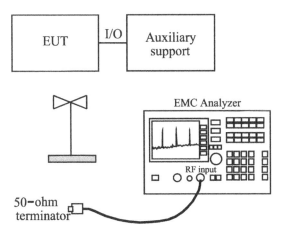

Figure 8.1 Testing for environmental noise—radiated emissions.

- If noise is still observed on the coax and nothing can be done to change the test environment or equipment setup, numerous ferrite clamps placed on the coax throughout the length of the cable may be needed.

To test for environmental noise related to AC mains, the following is recommended (Figure 8.2) [4]:

- Turn off the EUT and all auxiliary and support equipment.
- Remove the coax from the LISN and provide a 50-Ω terminator to the coax. The input to the receiver is now through a 50-Ω alleged perfect transmission line.
- Terminate the outer barrel of the coax connector to a ground plane that is also bonded to all measurement instrumentation. If the ground part of the terminator is left floating, inaccurate data may be observed. If bonded to a different ground plane, the potential difference between the two planes can be sufficient to cause measurement error. Bonding everything together will allow one to ascertain if the signal measured is from the AC mains provided by the facility.
- If signals are observed on the analyzer/receiver, the coax is defective and environmental noise will corrupt the signal of interest. If nothing can be done to change the test environment or equipment setup, multiple ferrite clamps placed on the coax throughout the length of the cable may be needed. Add these ferrite clamps also to all AC mains cable of support instrumentation, but "not" the EUT side of the LISN.

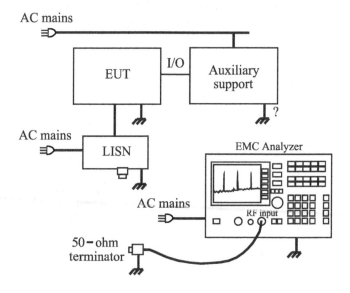

Figure 8.2 Testing for environmental noise—conducted emissions.

8.3 TESTING AND TROUBLESHOOTING CONCERNS

The most difficult part of being an EMC engineer is having a product once compliant become noncompliant in the field or in production or, worse, cause a problem with other electrical equipment—either emissions or immunity. This problem may not have been discovered during original testing, or the user made a change that he or she believed was minor. An example is adding a new adapter board into a personal computer after removing adapter plates and not replacing them, or not tightening the top cover back onto the chassis to ensure compression of gaskets, if provided.

To illustrate the complexity of a coupling concern, Figure 8.3 is provided:

- *Antenna coupling* refers to the effects that any antenna has based on distance spacing between the system and antenna. Included in the coupling equations are reflections from a reference plane and other environmental factors.
- *Power line coupling* is when the AC and/or DC portion of the power supply is disturbed by an external source of RF energy, such as switching noise from digital components traveling back into the power supply or radiating into the switching transformers. This coupling is then observed elsewhere in the system or out of the power cord.
- *Common ground impedance coupling* refers to a circuit that has two or more operational components sharing a common return line. An example is a PCB containing both analog and digital circuits with isolated analog power sharing a common DC return plane or circuits at different reference potentials tied together, such as interconnect cables.
- *Field-to-air coupling* implies that a radiating source will induce a common-mode voltage into a loop circuit formed by components communicating with each other using a common reference plane. Once this common-mode noise

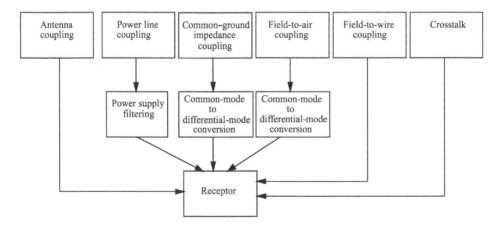

Figure 8.3 Complexity of EMC coupling.

has been absorbed, differential-mode currents will be developed. Differential-mode currents will propagate throughout the assembly, as they should, until they develop common-mode currents at I/O interconnects or other parts of the PCB.

- *Field-to-wire coupling* is present when two different parallel circuits provide mutual capacitance and mutual inductance to a propagating field. The circuit that picks up this RF voltage and current will be observed by the component most susceptible to disruption. Parallel circuits refer to differential-mode signal transfer where one signal propagates down a transmission line to a load and a return path exists to support differential-mode data transfer.

- *Crosstalk* refers to energy transfer between transmission lines. The higher the frequency of operation, along with faster edge rate transitions, the greater the concern that crosstalk will be present.

8.3.1 Systematic Approach for Emission Testing and Troubleshooting

When performing emission testing, the process is straightforward using the procedure illustrated in Figure 8.4 [4]. There are three basic levels of testing: engineering evaluation, prequalification, and formal certification. Each has unique process steps. The easiest one to perform is engineering evaluation. Here one can spend considerable time in learning about potential problems and incorporate solutions prior to production. Once production has started, time to market and product cost will increase.

Emission testing does not require one to monitor the EUT for possible malfunction, as is the situation for immunity testing. The equipment just needs to be operational with all subsystems working. Accuracy of performance is not a concern. Of all emission tests, conducted emissions testing is the easiest to perform along with taking far less time than radiated emissions testing, and is generally a computerized process. This is due to a simple test setup and less items to investigate should a problem arise. When performing emission tests, simple diagnostic probes and sensors can be used early in the design process.

If a problem develops during emission testing, the following provides guidance on how to approach the troubleshooting aspect:

(a) Determine if troubleshooting will be performed at the test facility, at the prequalification location, or in-house on the bench.
(b) Determine what kind of instrumentation, probes, and support equipment will be required.
(c) Determine if the system to be investigated will represent a production unit or if the unit to be retested is a prototype.

Engineering prototypes are best evaluated in the laboratory using simplified instrumentation and equipment. Prequalification equipment is best used at a test facility that will be either the formal qualification site or a location that has been charac-

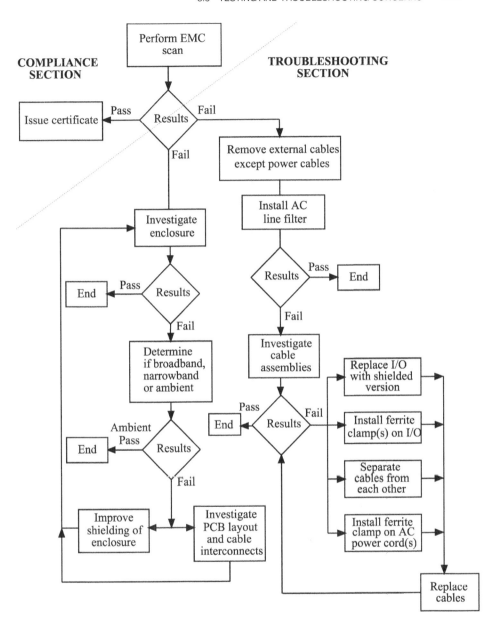

Figure 8.4 Flow chart for emissions testing and troubleshooting.

terized as being accurate for pretesting purposes. Formal tests must be done at a site that is registered or assessed as a competent test facility.

Emission testing can be performed quickly. It is recommended that conducted emissions occur before radiated emissions. It is easy to locate a problem area as well as detect field coupling to power cables that may have an effect on radiated emissions and/or immunity. Since conducted emission tests are carried out on AC mains, according to the standard (at least most standards), and in a much smaller bandwidth than the radiated emission tests, it makes sense to carry out radiated emission testing before conducted emission testing, even during the precompliance stage.

After conducted emission testing, using the concepts from Figure 8.4, radiated emissions testing is performed. This is more time consuming as the test engineer must manually search the entire frequency spectrum for any radiated emissions above or near a specification limit using personal judgment on whether the signal is from the EUT or ambient. It is impossible to use automation for radiated qualification testing unless the OATS is free of ambient noise. When tests are performed in an anechoic chamber, the frequencies observed are generally those that will probably have the greatest area of concern when repeating the test at an OATS.

Instrumentation commonly used for emissions diagnostic includes clamp-on probes, near-field and closed-field probes, and other simplified diagnostic techniques and tools detailed later in this chapter. Commercial and home-made products can be used to help isolate and debug a noncompliant system.

8.3.2   Systematic Approach for Immunity Testing and Troubleshooting

After emissions testing is completed, immunity comes next with the probability of system-wide damage due to the severity of certain tests that have to be performed. Conducted immunity is easier to do than radiated immunity since a shielded enclosure or anechoic chamber is not required. Testing on the bench allows one to isolate a victim circuit or components, validate filtering or shielding that has been implemented, and debug software interrupt routines. When performing immunity tests, the following is a suggested order of performance:

1. Voltage variation, sag, and dropout
2. Magnetic power line frequency
3. Electrical fast transient/burst
4. Conducted immunity
5. Radiated immunity
6. Electrostatic discharge
7. Surge

When attempting to isolate an immunity event, conducted immunity tests will suffice for radiated means in the frequency range of 10 kHz–230 MHz. This frequency range is only an approximation and is based on standard commercial instrumentation. Current clamps and injection devices that are placed over cable interconnects can cause a failure to occur with logic circuits. If circuits can be disrupted by an immunity current, the same cable can radiate undesired EMI.

When using bulk injection, a CW or pulse stimulus is desired, as these signals are the only ones that are feasible for troubleshooting and diagnostic work outside an anechoic chamber. We are able to track, isolate, and harden victim components and circuits or verify if filters, shielding, and other EMI mitigation components are working as desired. Verification of performance includes software with error detection and recovery algorithms.

For immunity tests, it is mandatory (or desired) for the EUT to be fully functional and to be set in an automatic mode of operation with software running an infinite loop program. This allows the test engineer to determine if a failure or glitch happened during the test. Monitoring the EUT outside an anechoic chamber is easy. When located in a test environment where one does not have visual access to the EUT or cannot be in physical proximity due to potential biological harm, use of remote video monitoring equipment with zoom features becomes mandatory. When using remote monitoring equipment, the interconnect cable should be fiber optic as regular coax or instrumentation cables may be susceptible to radiated fields and themselves be disrupted.

When monitoring a system for failures, the following may be observed:

1. Error message indicating a system failure or incorrect processing of data on a display monitor
2. A summary report printed out after the test noting abnormal modes of operation
3. Reset or system failure (shutdown)
4. Smoke emanating from the power supply or other circuits or the system bursting into flames (generally from surge testing)
5. Alarms activated

During testing, it is recommended that minimal use of auxiliary support equipment be provided. The purpose of this analysis is to evaluate the EUT and not third-party assemblies. Providing passive loads for all interconnect cables running in a loop-back mode is acceptable, if not preferred, for most immunity tests.

8.3.3 Systematic Approach to Detecting and Locating Problems

One of the more important elements to consider when doing forensic analysis on a compliant product becoming noncompliant is keeping one's composure. By remaining calm, one can think in a logical manner and go directly to the emission source instead of immediately trying to apply a fix using copper tape or ferrite clamps. Temporary fixes are just that, temporary. The problem signal will usually reemerge in the future from another leak point. A second visit to a customer site does not do well for public relations or future business opportunities.

Understanding EMC theory, even in basic form, provides insight into how RF energy is developed and propagated (Chapter 2). Look at all elements that can cause problems—high-speed digital switching components, cable interconnects, and mechanical assemblies. Investigate each area one at a time. In most cases a problem frequency can be solved with one component on the offending signal, yet many patches are implemented by inexperienced engineers who experiment with everything they can think of. This approach provides no benefit while raising the cost of the rework.

The following must be considered when doing formal analysis [4]:

Before the Investigation

1. Is the problem continuous or intermittent? If intermittent, can the event be recreated? Continuous emissions are generally easier to solve for obvious reasons. Intermittent problems can take considerably longer to detect and repair.
2. Does the problem exist at only a specific time of day and, if so, where are nearby transmitters located that can be the source of the problem?
3. If transmitters are nearby, are they commercial broadcast or private networks? What is the transmitter power and antenna gain along with frequency of operation? How far away are these transmitters from systems being disrupted? If the equipment being evaluated is causing harmful radiated emissions, an inverse investigation must occur using these guidelines.
4. Where could the possible coupling paths be located?
5. Is the problem radiated or conducted. If conducted, how many interconnect cables are provided and are they shielded?
6. Anticipate the problem and bring appropriate electrical and mechanical parts. This includes instrumentation and rework supplies.

During the Investigation Cycle (Figure 8.5)

1. Perform a visual inspection. Look for the following:

   - Proper placement and installation of power line filters along with the quality of bonding to the enclosure, if required
   - Use of data line filters (both common mode and differential mode), including ferrite clamps and their implementation
   - Bonds between metal covers that are clean of debris, paint, and other insulating coating and securely tightened as required
   - Grounding methodology (i.e., bonding through poorly designed screw mating system and using round bus wire instead of flat braid for bonding purposes)
   - Use of plastic versus metal connector housings
   - Shielded versus unshielded cable assemblies
   - Top covers not screwed securely to the main enclosure

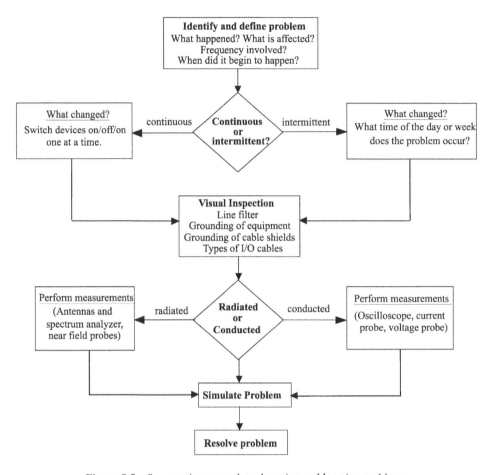

Figure 8.5 Systematic approach to detecting and locating problems.

- Missing or damaged gaskets
- Worn out or damaged cable interconnects and housings
- Rust on metal enclosure seams and other components

2. Investigate cable routing and component placement. Are cables routed in tight bundles with slow-speed control lines mixed in with high-speed data signals and clocks?

3. Are all cable assemblies routed in metal conduit properly grounded at both ends? Crosstalk may be a cause of system failure.

4. Is the device located in an area with other systems, including legacy products, or is the unit isolated from external RF noise? How many systems share the same AC mains branch circuit?

5. Is the system located in an area where personal communication devices are used, such as cell phone, pagers, hand-held computers, and wireless transmitters?

6. Is the system located within a reasonable distance from broadcast stations, transmitting towers (AM/FM/television/satellite), and radar installations (e.g., an airport or military facility)? Neon signs and industrial plants with welding equipment also pose potential RF threats.

Once the investigation process has been completed, the next step is to repair or replace the defective subsystem whether this is a PCB, cable assembly, or installation of new gaskets. This assumes that the source of the problem has been identified. If elimination of the energy source cannot happen for various reasons, the next area of approach is to remove the coupling path. Removing the coupling path may involve changing interface cables, enhancing the enclosure assembly to lower the transfer impedance between covers, replacing plastic housings with metal ones, changing the grounding method of the unit, or applying an uninterruptible power source (UPS) for AC mains problems, among other possible fixes.

Anticipate what the problem may be between testing starts and bring proper equipment to perform the job professionally. Proper equipment means instrumentation, probes, replacement (shielded) cables, ferrite devices, filter capacitors, gaskets, conductive tape, and so on. It does not look professional for a debug engineer to say, "I forgot my sniffer probes. I will be back tomorrow." Make a list of every item that may be required for troubleshooting and keep this list available when the need arises to create a portable debug kit to take on the road or to a test site. The same is true for in-house diagnostic work. Have all the equipment and supplies readily available.

The following (partial) recommended list is for "travel" or "debug" purposes [5]. The items that are assumed available include measurement instrumentation such as spectrum analyzers or receivers and laptop computers with appropriate software:

- Equipment to record data: digital camera, plotters, and printers
- Measuring tape to determine distance between antenna and EUT in addition to length of cables and enclosure seam slots
- Small hand tools: various types of screwdrivers, wire strippers/cutters, pliers, insulating tape
- Rolls of flexible conductive fabric, copper tape, and aluminum foil
- Various collapsible antennas appropriate for intended use along with a light-weight tripod
- Assortment of coaxial cables, connectors, attenuators, and terminators
- Assortment of field probes
- Coaxial single-pole, double-throw (SPDT) switch for easy comparison between antennas and probes
- RF preamplifier and optional comb generator (to ascertain ambient levels)

- Voltage probe with alligator clip
- Current probes with different frequency ranges of operation and inside core diameters
- Clamp-on ferrite cores in various sizes, shapes, and impedances along with various capacitors in all sizes, shapes, and values
- Selection of line filters (signal and power), shielding material (gaskets, elastomers, etc.), and connector backshells

8.3.4 Minimum Requirements for Performing EMC Tests

In order to perform testing, and troubleshooting if necessary, the following is recommended for emissions:

1. *Test Facility or Location with Minimal Ambient Noise.* If background noise is excessive, attempting to characterize a signature profile or isolate problem frequencies may be difficult, if not impossible, to achieve. Sources of ambient noise include fluorescent lights, switching networks, broadcast station antennas, and machine tools.

2. *Proper Instrumentation Calibrated and in Good Working Order.* The equipment includes the following:

    a. *Test Receiver or Spectrum Analyzer.* The test instrument must be capable of operating over the frequency range of concern with sufficient sensitivity. Sensitivity means that the tool must be able to measure low-amplitude emissions with accuracy and distinguish ambient noise from an offending signal. In addition, proper detectors and bandwidth settings are required.

    b. *Low-Noise Wideband Preamplifier.* In certain environments it may be difficult to isolate an offending signal from the ambient or determine the peak amplitude of the signal. An amplifier (typically +26 dB in gain) makes for ease in locating and measuring signals. If an amplifier is used, the gain value must be deducted from the measured amplitude.

    c. *LISNs or CDNs.* These provide for interfacing test instruments to cables and wire assemblies directly.

    d. *Antennas (Biconical, Log Periodic, Horn, etc.).* As with test instruments, antennas will operate only within a certain range of frequencies. For each frequency, antenna factors will affect measured results. The antenna factor must be taken into consideration when calculating actual measured field strengths.

    e. *Current and Voltage Probes (Receive and Bulk Injection).* These are excellent measurement tools, but like antennas they have a limited frequency range of operation and current-carrying capability. Calibration factors play a significant role in determining accuracy of the measured signal.

   f. *High-Quality Coaxial and Interconnect Cables.* A significant amount of time and effort can be wasted if poor cables exist. Worn-out or weather-beaten cables can cause nonreproducible data and other glitches. The quality of the braid, along with coverage percentage, is critical for accurate test results.

3. *Transducers (Antennas, Probes, Clamps, etc.).* The right transducer is required to measure electric, magnetic, and electromagnetic fields. Will the measurement be conducted or radiated?

4. *Noise-Free Power Mains.* If unwanted RF energy enters a product through the power mains cable, additional problems may develop which are difficult to determine. Isolation of the EUT from the environment provides for optimal measurement capabilities.

5. Ground Plane. Most EMC tests require use of a ground plane. Depending on size, the ground plane can be located on a table top or the floor, extending beyond the physical dimensions of the EUT. This ground plane provides for a stable well-defined environment for the EUT relative to routing of cables and peripherals and helps ensure repeatability of measurement between test locations.

For immunity tests the following are minimal requirements:

1. RF Ambient Environment. The ambient where testing is to be performed is generally not a concern. The injected RF energy into a EUT will generally be magnitudes greater in amplitude. Most systems operate well in a standard environment except under unusual conditions. Where the real problem lies is in interference to other systems located in close proximity to the test laboratory. Radiated fields may not cause harm to the EUT but may cause significant damage to other equipment.

2. Availability of a Ground Plane. A ground plane is required for reproducibility of injected noise. Almost all immunity tests require a ground plane that is securely bonded to both test instrumentation and earth ground. This is especially important for ESD and surge testing.

3. Filtered or Isolated Mains Power. As with emissions, filtered mains power is required. Noise injected into AC mains or other cables can be transferred into the facilities' mains, causing disruption to other electrical equipment on the same branch circuit. Use of a LISN/CDN will minimize this potential damage.

8.4   REPEATABILITY OF SYSTEM TESTING

A key aspect of regulatory compliance lies in repeatability of measured data and continued conformity of a product throughout its life cycle. In Europe, should a problem develop after certification and the product has been placed in service, the end user or one's competitor has the legal right to have any system or operational unit retested to ensure conformity exist.

When regulatory agencies retest a product to ensure conformity, a certain amount of leeway is sometimes provided. A system can be several decibels over the specification limit and still be considered compliant. Action taken by the authority is only for those products that are significantly out of specification or clearly not designed to meet the EMC Directive, that is, a false Declaration of Conformity. When this situation occurs, drastic measures can be taken by the authorities.

Why does audit testing allow radiated emissions to be over the specification limit, especially for frequencies above 1 MHz? This is due to measurement uncertainty between test laboratories and lack of repeatability. To ensure conformity, use of a "golden product" reduces measurement error and improves repeatability for *all* EMC tests; emissions and immunity.

Repeatability of emission and immunity tests can sometimes be confounding when conforming to full-compliance test methods at the same test facility using the same test engineer every time. When an outdoor test location (OATS) is certified during site attenuation measurements, the accuracy between laboratories can vary by as much as ±6 dB. It is for this reason that a product is permitted to be over the specification limit by a few decibels when being audited. If the difference in measured data exceeds 6 dB, then several possibilities may explain this discrepancy:

1. Instrumentation may be out of calibration or defective without the operator knowing it.
2. The site attenuation at the OATS is either very high or low at a particular frequency.
3. The site attenuation at the OATS never complied with legal limits in the first place.
4. The test engineer (or technician) may have made an error during testing (not maximizing cable placement, rotating the table a full 360°, changing the height of the antenna the entire distance, etc.).
5. The test engineer at the laboratory may not have been having a good day (i.e., lack of sleep) and missed important emissions by accident or forgot to quasi-peak a signal.
6. The test setup and layout of the EUT, especially cable routings, are different between tests.
7. Incorrect data may have been entered into the data sheet by accident.
8. Environmental conditions on the day of the test may skew measured results (lightning strikes, power line noise spikes, cars driving by with transmitters on, etc.).
9. The manufacturer may have incorporated special modifications in the system to pass the test. These special fixes were "not" incorporated into production.
10. Junction temperatures in circuit devices can change spectral performance.

Accredited test laboratories perform site attenuation measurements using reference noise generators and calibrated antennas. If used properly, site attenuation data should be accurate. One must compare the results of a particular site's attenuation data with those of other test sites to ascertain if a difference exists at only one frequency or across the entire spectrum. A system may be compliant at one laboratory but noncompliant at another. One disadvantage of relying on site attenuation lies in the types of fields used to calibrate the range. One does not use calibrated antennas for both electric and magnetic fields at all frequencies. The EUT will emit both fields. For this reason, a detailed setup and layout of the EUT, its software version, associated cables, and auxiliary equipment must be documented and provided in the test report. Moving a cable by even a small amount can cause significant differences in a radiated emissions or immunity measurement. Different ways of bundling long cables can affect emissions by more than 20 dB.

Using a known good product ("golden system") is one way to minimize test site errors and ensure continued conformity. This unit is then sealed so there is no possibility of being modified—hardware, software, mechanical, or electrical. Extensive documentation (photographs or line art drawing) showing cable placement must be archived with the system. Some companies put a "Warranty Void if Open" sticker on all covers of the unit to ensure that no changes can be made. Careful handling of the unit must be performed every time the system is used to validate a test laboratory's site attenuation or to ensure accuracy of measured data for repeatability purposes. This golden system must remain archived for the life of the product, be it 10 months or 10 years. Some companies paint test cables a unique color to identify the need for archival purposes.

Where the test method used is different from a standard test procedure, for example, measuring radiated emissions using closed-field probes or a current clamp instead of a spectrum analyzer, the golden system allows this alternative method to be improved until it gives a reasonable *correlation* with the accredited test results. With closed-field or current probes, there should be close correlation between the golden system and the device under test. The signature profile of the golden system becomes a master reference document for a particular type of measurement.

When testing is performed on an infrequent basis, verification of the site should be performed before conformity testing. If the results correlate, the test facility is acceptable and certification of a new product can occur. If testing is performed on a daily basis, the golden system should be brought out at regular intervals to verify that site attenuation has not changed or that instrumentation has not been damaged, such as an antenna getting bent or a crack in the braid of the coaxial cable due to environmental issues, misuse, and handling.

Use of a golden system will diminish with time, as newer technologies will have significant effects on the signature profile of a EUT, including a higher frequency range of emissions as well as changes in the shielding effectiveness of the unit's enclosure.

8.5   UNEXPECTED PROBLEMS AFTER PRODUCTION HAS BEGUN

A concern for any manufacturer is what to do when a problem develops in the field after a product has been sold or permanently installed. This can turn into a serious political, social, and/or professional nightmare; the goal is to keep customers happy while not being subjected to potential liability lawsuits. The equipment must be repaired in the field or upgraded with new features that include EMC mitigation components. If it is impossible to modify the product internally, external enhancements must be incorporated or the physical environment modified, which is more difficult to achieve.

Problems may develop after the fact for a variety of reasons:

1. Installations may contain both new and legacy products. Legacy products are those manufactured and placed in service long before the EMC Directive or legal requirement became mandated. These older products may be significant sources of EMI in addition to being susceptible to the ambient environment.

   Many newer products are using RF generators for the treatment of material (i.e., silicon wafer processing in semiconductor manufacturing equipment, medical diagnostic equipment, etc.) or for communication purposes (cellular phones, pagers, wireless networks, global positioning, etc). These newer products with intentionally generated fields can be harmful to legacy equipment located nearby.

2. The EMC environment may have changed. With the proliferation of wireless communications and broadcast stations, equipment that was once compliant is now susceptible to disruption.

3. Certain products cannot be upgraded or modified due to sensitive components or features. This includes instrumentation and sensors.

4. Inadequate cable installations by contractors who may have substituted shielded cable with unshielded cable by accident.

When approaching a problem in the field, the engineer must not take anything for granted. Careful analysis of the environment and playing the role of forensic detective become an efficient mode of operation. Areas to consider when evaluating problems lie not in immediately fixing the EUT but in determining what caused the system to have a problem worthy of a service call. Items to consider or questions to ask are as follows:

1. Where could the harmful signal be coming from and what equipment is capable of generating a propagating field that is of sufficient amplitude to cause harm?

2. Where in the frequency spectrum is the problem observed? Are there local broadcast stations (radio or television), cellular telephone towers, and new construction facilities containing high-technology equipment?

3. What are the possible coupling paths—radiated or conducted? Are communication cables routed above ground on tall poles or underground? Are wire trays within the building containing numerous bundled cables of Ethernet or other telecommunication systems exposed to the environment? How many systems are connected to the same AC mains distribution branch?

4. Where is the exact location of the system(s) being affected and its relationship to other electrical equipment? Is the building located in an environment with minimal ambients? Is the system located near high-energy generating devices such as elevators and wiring closets?

8.6 CREATIVE APPROACHES TO TROUBLESHOOTING (CASE STUDIES)

In trying to solve an EMC event, one must be creative. Sometimes the standard method of analysis proves to be incorrect due to parasitics and other design features that are not readily noticeable. Simulation may not identify problem areas—source of energy development or transmission media/path. Examples on how one can approach a problem are provided below based on real-life products.

1. *Conducted Emission Situation.* An avionic motor controller box failed to meet a required MIL-STD-461 conducted emission level. There was very little space available for a line filter. This element was the last item to be incorporated during the design process. A paper design demonstrated that a conventional π- or T-filter would require too large (physically) an inductor. The problem frequency range was from 20 kHz to about 800 kHz as well as a high current requirement (40 A average with 100 A peak). The effectiveness of a filter is not only determined by individual component values but is also dependent upon the ratio of the component values to each other as determined from filter design equations. Since it was not possible to have a large inductor, a lot of capacitance was substituted for inductance. A modified value of inductance was chosen (25 μH) with total capacitance of 3400 μF. The filter design worked and the unit successfully met the MIL-STD-461 conducted emission test. This is a good example of thinking "outside the box" and going back to the fundamental design equations to determine or understand what effects various parameters have on a filter design. As long as there is good engineering judgment behind a decision, one should not be swayed by unconventional results or design practice. In this case, test engineers commented on how they had never seen a filter design with that much capacitance, but because of packaging constraints, this unique design apparently was the right one to solve the problem.

2. *Another Conducted Emission Situation.* In attempting to get a switching power supply to meet FCC Class B levels, no matter what size common mode inductor was used, the power supply could not meet the limits. Using a larger common-mode inductor had the opposite effect of generating more noise on the power lines. Analysis indicated that excessive interwinding capacitance between the coils of the inductor was the possible cause of failure. As the number of turns on the core in-

creased, both inductance *and* interwinding capacitance increased, allowing more RF energy to bypass the inductor. A larger core was a solution, but as usual, there was insufficient space for a larger device. However, there was room for a second, smaller size inductor. According to basic circuit theory, capacitors in series behave as resistors in parallel. A smaller value capacitor in series with a larger capacitor results in an equivalent circuit with total capacitance that is smaller in value of either capacitor individually. By placing a smaller size inductor (5 μH) with a small interwinding capacitance in series with the larger common-mode inductor (in the millihenry range along with its large interwinding capacitance) resulted in a reduced emission level, thus meeting the Class B requirement.

3. *Radiated Emission Situation.* A commercial bread-slicing machine was being EMI tested and failed radiated emissions from 30 MHz to about 200 MHz. The cause of the emission was traced to the controller's PCB by use of a near-field "sniffer" probe. The board was a common two-sided assembly with traces on both top and bottom. The company did not want to incur the expense of going to a multilayer board; it had already ordered several thousands of these boards that were beginning to be assembled. The problem was due to bad layout and trace routing. No time was available for a second revision of the PCB. Fortunately, the failure was by only about 3–5 dB. A solution was to make a discrete ground plane from copper laminate and attach it to the bottom of the board with some edge clips. This reduced the noise by 6–8 dB, yielding a "passing" criterion of 2–3 dB.

4. *Another Radiated Emission Situation.* A system was being tested for radiated emissions with a failure at 2.4 GHz, the data rate frequency for fiber-optic transceivers. A problem frequency can be translated into a physical dimension by the equation $\lambda = C/f$ as starting point for analysis. Any aperture or wire antenna/transmission line that approximates this physical dimension would be the first area of concern related to locating the radiating source, as this dimension equates to being an efficient radiator for a particular frequency. The other dimension for antenna efficiency would be $\frac{1}{2}, \frac{1}{4}, \frac{1}{8}, \frac{1}{16}, \ldots$ of a wavelength. This was verified by tracing the signal with a near-field loop probe to the fiber-optic connector front-panel mounting area. The wavelength for this frequency is about 5 in., and therefore, the half-length is 2.5 in. The leakage was located to be around the end of a mounting plate that holds and mounts the fiber-optic connector to the front panel. There is a rectangular slot in the front panel 2 in. in length and ½ in. wide, which are divisible into the number 5. The mounting bracket was made from metal and had a conductive coating as did the front panel. However, at these higher frequencies, the quality of the grounding or bonding becomes critical. A gasket made from fabric over foam was cut out to fit into this area to improve grounding and thus reduce the resistivity of the connection. The signal at 2.4 GHz dropped by 8 dB.

REFERENCES

1. CISPR 16-2. 1996. *Specification for Radio Disturbance and Immunity Measuring Apparatus and Methods: Part 2: Methods of Measurement of Disturbances and Immunity.*

2. CISPR 11. 1997+A1:1999. *Industrial, Scientific and Medical (ISM) Radio Frequency Equipment—Radio Disturbance Characteristics—Limits and Methods of Measurement.*
3. CISPR 16-1. 1993. *Specification for Radio Disturbance and Immunity Measuring Apparatus and Methods: Part 1: Radio Disturbance and Immunity Measuring Apparatus.*
4. Mardiguian, M. 2000. *EMI Troubleshooting Techniques,* New York: McGraw-Hill.
5. Hoffman, H. R. 1995. Experiences in Making On-Site EMC Measurements. Paper presented at the Eleventh International Symposium on EMC, March 7–9, Zurich, pp. 35–40.

CHAPTER 9

ON-SITE TROUBLESHOOTING TECHNIQUES

At some point during product development, every system must undergo a variety of tests. This book focuses on EMC, although electrical product safety, environmental, and functional testing must also be performed. Earlier chapters in this book provided details on the need to test a system prior to production. These chapters discussed basic EMC theory and application, the use of probes, commonly used instrumentation, and test facilities. Conducted and radiated field testing, both emission and immunity, was also presented. In addition, a systematic approach toward achieving EMC was detailed in the previous chapter.

When an EMC test is performed, there is usually some kind of failure on the first pass. Preliminary testing allows one to understand how the system operates under various environmental and installation conditions along with the magnitude of the failure. From here, finding the problem and fixing it becomes the next task. This chapter provides insight into locating areas of concern using a variety of methods. Every engineer has a favorite troubleshooting technique. In this chapter, popular techniques are provided to stimulate thinking on behalf of the engineer related to the simplicity of troubleshooting products. More techniques are available than those provided herein.

When identifying radiated and/or conducted electromagnetic fields, the right tool is required for measurement purposes, especially if a problem develops, be it emissions or immunity. Commercial instrumentation, near-field probes, and diagnostic equipment may be too expensive for some engineers and their company. For troubleshooting purposes, in lieu of commercial products, home-made probes and alternate test equipment provide significant benefit. Appendix A illustrates how to

Testing for EMC Compliance. By Mark I. Montrose and Edward M. Nakauchi
ISBN 0-471-43308-X © 2004 Institute of Electrical and Electronics Engineers

build a variety of probes for little or no cost used for identifying, debugging, or troubleshooting EMC events.

For diagnostic or troubleshooting home-made probes work as well if not better than commercial products at a significantly lower cost. The reason is that commercial probes are designed to have certain operational parameters and response characteristics. In addition, a calibration factor chart of the probe's response is sometimes included in the purchase price. Calibration charts are valuable only when trying to determine the exact magnitude of a signal source. When debugging a product, the primary items of concern are (1) where the energy is being developed and (2) how I can minimize or eliminate the problem. Formal qualification testing requires a certified or assessed test site with the ability to perform measurement uncertainty calculations. If a product fails EMI testing, someone has to quickly identify the problem, take appropriate action, and incorporate a fix.

This chapter illustrates how to identify various RF fields. Once identified, an examination on various debug and troubleshooting techniques is presented. The majority of troubleshooting techniques are conventional. Some may be unorthodox; however, all techniques presented have a record of debugging EMI problems quickly with minimal cost. The key element to remember is that once a problem is isolated a solution must be incorporated. Most solutions are only quick fixes and are not designed for large-scale production. Sometimes, a simple fix on a test unit can be made permanent at minimal cost to the finished product.

It must be pointed out that by no means are the techniques mentioned in this chapter the only ones available. It is hoped by the authors that the techniques and suggestions presented not only act as guidelines for the reader but also are an inspiration for developing the reader's own techniques.

9.1   QUICK FIXES AND SOLUTIONS

This section is intended to give a brief overview of typical components and devices that are currently available to reduce or resolve an EMC event, whether at a range during qualification tests or at a customer's site after installation to resolve a non-compliant situation.

Many engineers and companies maintain a stockpile of EMI suppression components. Vendors are more than happy to contribute to your inventory of sample parts. A variety of, for example, ferrite components (clamps, beads, torroidal, etc.), conductive tape (metal or fabric), aluminum foil, various types of gaskets, power line filters, signal line filters, and copper braid are some elements to include in a laboratory EMI kit. It should be mentioned that sometimes there is nothing one can do except redesign or re-layout a PCB. This is a last-resort effort at significant cost, even if the re-layout includes moving only a few parts around, splitting planes, or adding a discrete component (ferrite beads, series-terminating resistor, or extra decoupling capacitors).

What happens when it is not possible to return the unit for repair, upgrade, or rework? The repair technician is asked to solve a miracle yesterday without making changes or causing downtime. The problem must be identified immediately and fixes incorporated that permit the unit to work within its intended environment. For an experienced engineer, this is "another-day-at-the-office" situation—isolating the problem area and incorporating a fix using only the tools in his or her possession, a few EMI rework components, and common sense.

9.1.1   Conducted Solutions

For conducted problems , there must be a conductive path involved where current is flowing. From a solution's standpoint, one needs to place a high-impedance component in series with the noise path, reduce current flow, or add a low-impedance filter in parallel to divert the current to 0 V (ground). These types of devices fall under the generic name of "filter." Examples are AC power lines, ferrite beads, ferrite clamp-on capacitors, opto-isolators, transformers, and transient protectors.

When specific components are used in series with the noise source, they block the undesired energy. Inductors will be low-pass filters, whereas capacitors are high-pass filters. When filters are used in parallel across a noise source, they divert the energy. Other parameters must be considered when choosing a suppression component that will fix the problem. Frequency, circuit impedances, component resonance effects, and physical location are important considerations. The location is also important because it will affect the quality of grounding, especially when used in a shunt or diverting mode. The concept is that the lower impedance path of a shunt filter will divert the current away from the receptor or victim circuit. If the impedance path of the shunt filter is too high, then very little of the current will be diverted. The use of bonding straps is critically dependent upon impedance as well. The opposite is true for series impedance components. High impedance is desired, but not so high that normal circuit operation is affected. Ferrite devices are therefore not very useful for low-frequency applications. Inductors are excellent at lower frequencies, such as those below 1 MHz. When dealing with higher frequencies, remember that small inductance in the circuit or a transmission line can become a high impedance that may affect operation, and that large capacitance can become a low impedance at a specific resonant frequency.

Another important consideration in choosing appropriate suppression component(s) is to determine whether the noise current is common mode or differential mode. Filter techniques, configuration, placement, and effectiveness are dependent upon proper implementation. For a differential-mode current flow, a capacitor must be placed across or between the two lines carrying the current. For a common-mode current flow, capacitors must be used on every line connected *only* to chassis ground. The same is true for using ferrites. For a differential-mode current flow, a single ferrite device can be placed in series in the transmission line. For common-mode current flow, both wires must be "filtered." Nowadays, filtered connectors are available in all types of configurations such that these components can be added as a retrofit solution.

When a conducted problem is observed, the first area to investigate is each cable assembly. A field patch is only temporary but allows one to isolate where the problem is and permits one to investigate further. Field patches include the following devices [1]:

1. *Ferrite Material.* Ferrites play a significant role in the removal of undesired EMI. Ferrites are produced in every shape, configuration, constructional material, and means of application conceivable. Ferrites are produced from various mixtures composed of iron, nickel, manganese, zinc, or magnesium oxides in a binder. By varying the composition of these different ferrite materials, the losses or attenuation of the ferrite can be controlled over selected frequency ranges or bands. The resultant material is a hard and brittle substance.

The major considerations as to which material is required for a particular application depends upon frequency of concern, bandwidth of performance, and impedance required. Ferrites feature the magnetic property known as permeability, which is a frequency-dependent parameter. The amount of attenuation or suppression is also dependent upon the permeability of the material, the size of the ferrite, and the frequency of the energy to be suppressed. The impedance of the ferrite is composed of an inductive part and a resistive part, as shown in the simplified illustration of Figure 9.1. The inductive reactance $L(f)$ of the circuit contributes to the insertion loss at lower frequencies while the lossy resistive $R(f)$ part attenuates higher frequency signals. The inductive portion of the material can be raised by increasing the number of turns of wire through the ferrite core base. Normally, it is not recommended to exceed more than three or four turns due to increasing the interwinding capacitance $C(f)$, thus lowering the resonant frequency of operation. The impedance of a ferrite is described by the equation

$$Z = R(f) + j2\pi f L(f) \tag{9.1}$$

Since the equivalent circuit is a parallel RLC circuit, the impedance is more like a $Z = R(f) \parallel j2\pi f L(f)$ equation.

The insertion loss of a ferrite device, in decibels, is obtained by the equation

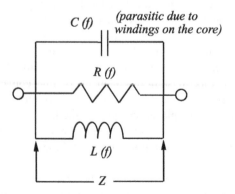

Figure 9.1   Simple circuit model for a ferrite.

$$\text{Attenuation} = 20 \log\left(\frac{E_L}{E_{LB}}\right) \tag{9.2}$$

where E_L is the voltage across the load without the ferrite in the signal path and E_{LB} is the voltage across the load after the ferrite has been inserted in the circuit. Therefore, attenuation is related to the ferrite's impedance as well as the source and load impedance of the circuit (Figure 9.2).

When using ferrites, care must be given to the alternating or direct current through the device. As with any magnetic material, performance will be degraded if too much current exist. Too much current, or magnetizing force, can exceed the *saturation* point of the substrate. This is the point where the magnetic flux will stop increasing its impedance value. Once the saturation point has been reached, the ferrite core will offer minimal attenuation to the undesired signal being filtered. For further insights on uses and applications of ferrites, refer to the many excellent catalogs and application notes that manufacturers of ferrite provide.

Applications of ferrites include noise suppression on PCBs to reduce coupling of undesired RF energy (noise) to other circuits and/or to confine desired energy required for operation to a defined region on the board. Ferrites can also be used to reduce or dampen self-oscillation in circuits caused by fast transient switching events

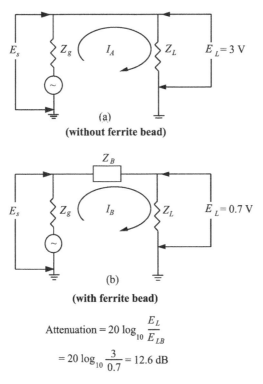

Figure 9.2 Attenuation calculation of a ferrite device.

and distributed parasitics (inductive and capacitive) found PCB layouts and chassis enclosures.

An example of self-oscillation can be found in power supply designs. Ferrites can be used as snubbers to reduce high levels of RF noise generated by switching diodes. Noise transients generated by switching diodes are a major source of not only radiated noise but also conducted energy. For this application, ferrites are a possible solution for improving EFT immunity problems. Ferrites can be used by themselves or in conjunction with capacitors to filter power bus noise generated by high-speed logic devices. Ferrites can also be used to filter high-frequency noise from I/O cables such as an output driver to a keyboard. The interconnect cable can act as a radiator of RF energy; therefore, high-frequency RF noise should not be allowed to flow in the I/O cable. A ferrite bead placed between the I/O driver and the external cable will attenuate high-frequency noise while not disturbing operational characteristics of the desired signal. This same concept can be applied to any external I/O cable assembly that carries or has high-frequency energy coupled onto it from digital components or external immunity fields. Video cables are another source of EMI radiation that is caused by RF spectral currents on the internal conductors. Again, placing ferrite components on these conductors will attenuate the high-frequency noise and thus lower radiation.

Most EMC laboratories contain a wide selection of ferrite beads and clamps for retrofit purposes. A small sample of ferrite products is shown in Figure 9.3. Every EMC engineer and technician should have sample kits from different manufacturers in their possession at all time. For the majority of troubleshooting problems, clamping a ferrite bead onto a cable may quickly isolate the problem area and may be the "only" remedy to solve a product before shipment of the product.

2. *Signal and Power Line Filters.* Signal and power line filters are used to minimize or prevent propagation of harmful RF energy either entering or leaving a system, generally in the conductive mode: common mode or differential mode. Filters consist of four basic modes of operation (Figure 9.4) and come in various configurations:

- Low pass—Rejects undesired RF energy above a desired set point, passing frequencies below this point with little or no attenuation. The AC line filters

Figure 9.3 Various configurations and types of ferrite material. (Photograph courtesy of Steward.)

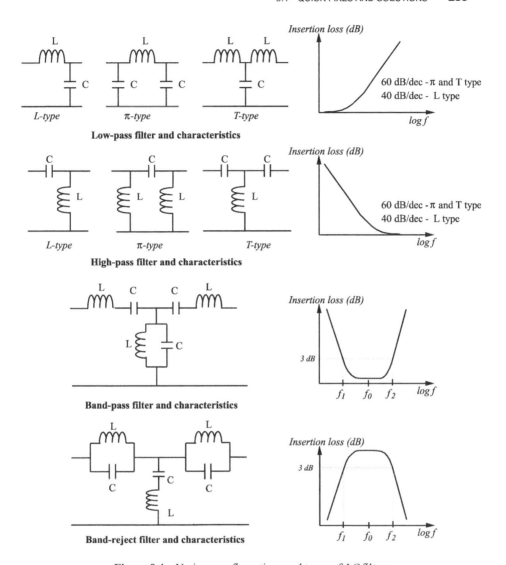

Figure 9.4 Various configurations and types of *LC* filters.

are typically of the low-pass variety. This filter is the primary type of device used in circuits and assemblies.

- High pass—Rejects undesired RF energy below a desired set point, passing frequencies above this point with little or no attenuation.

- Bandpass—Passes a range of desired frequencies with little or no attenuation, rejecting frequencies outside this specific range.

- Band reject—Rejects a range of frequencies within a particular frequency band of operation while passing all other frequencies outside this band.

Essentially, a filter is made up of two basic types of components—capacitors and inductors. The simplest type is called a first-order filter consisting of a single reactive component. Capacitive reactance X_C will decrease with an increase in frequency. Inductive reactance X_{LI} will increase with an increase in frequency [See Eq. (9.3)]. Capacitors shunt noise current away from a load while inductors block or reduce the noise in the transmission line. Generally, these single-component filters are not very useful as their attenuation only increases at a rate of 6 or 20 dB/decade.

$$X_C = \frac{1}{2\pi f C} \qquad X_L = 2\pi f L \qquad (9.3)$$

To achieve greater attenuation, a second or higher order filter consisting of two reactive components or more is required. This filter provides 12 dB/octave or 40 dB/decade attenuation. The value of the inductive or capacitive components is determined by the impedance of the source and load and the highest frequency to be passed (i.e., cutoff frequency). This two-element filter is sometimes referred to as an L filter. The design of a second or higher order filter requires careful analysis. Filter resonances and ringing must be considered and involve a design characteristic called damping factor, sometimes symbolized as the Greek letter zeta (ζ), which describes gain and the time response of the filter. A third-order filter, of course, consists of three or more reactive elements yielding an ideal attenuation of 18 dB/octave or 60 dB/decade. These types of filters are sometimes referred to as pi (π) or T filters. For each additional component, the rate of attenuation is increased by 6 dB/octave or 20 dB/decade. The disadvantage of a larger filter is that cost also increases as well as physical size. The third-order filter is among the most popular topologies of filters used and should not be overlooked when attempting to solve an EMC problem.

One aspect of filter design is impedance mismatch. The first filter element nearest the source or load end should be selected to provide the highest mismatch at EMI frequencies possible. Typically, this means that if the source or load impedance is low ($<100\ \Omega$), then the first filter element should be an inductive component. Conversely, if the source or load impedance is high ($>100\ \Omega$), the first filter element should be capacitive. This provides the designer an extremely efficient design with the least number of stages or components.

Differential-mode filtering involves placing capacitors between signal lines and/or an inductor in series with either the high or low side of the line. Common-mode filtering involves capacitors to ground and/or a common-mode inductor in series with both sides of the line or lines. A common-mode inductor does not affect differential-mode currents except for whatever imperfect coupling exists (i.e., leakage inductance). It is best to split the inductor evenly on both sides of the transmission line to maintain balance in the circuit. This is important for common-mode rejection within the circuit.

Mutual inductance will maximize the impedance to common-mode noise. Because of different modes of propagation, it is important to determine which type of noise current exists so that proper filtering can be implemented for maximum efficiency and cost.

It can appear deceptively easy to design filters. Unfortunately, many other factors must be considered, such as peak load current, saturation of inductors, elevated temperatures, and output impedance. The output impedance of the filter can be a critical item when incorporated within switching power supplies. The negative input impedance of a switching power supply can oscillate in conjunction with the output impedance of the line filter. An important design criterion known as Middlebrook must be considered. The Middlebrook criterion states that the output impedance of the filter must be less than the reflected load impedance of the switching power supply.

Various types of commercial filter configurations are presented in Figure 9.5.

9.1.2 Radiated Solutions

When a radiated problem is observed, first cables and interconnects should be checked, then the enclosure and top covers. A field patch is only temporary but allows one to isolate where the problem is and permits one to investigate further. Field patches include the following [1]:

Signal line filters   Power entry modules

Power line filters   Typical inside construction of a line filter

Figure 9.5 Various configurations and types of signal and power line filters. (Photographs courtesy of Corcom.)

Shielding Materials

1. *Aluminum Foil.* Aluminum foil is used to prevent RF-propagating fields from traveling into free space or can be used as a conductor similar to that of wire. This is the best friend of many EMC engineers and technicians and is usually purchased in bulk quantities. It is recommended that 2-mil-thick foil be used (heavy duty), as it does not tear as easy as the regular 1-mil-thick variety. Aluminum foil is easy to apply and can be removed quickly. One can manipulate the foil into any configuration imaginable. Aluminum foil is excellent only for electric fields but provides no benefit for magnetic fields as the permeability value of the material is too low and the thickness too small. Regardless of which side is used, shinny or dull, do not be concerned. The shiny side of the foil only works when used for cooking or thermal heat transfer. Aluminum foil is metal. Either side affects electromagnetic fields equally.

2. *Copper Foil or Tape.* Copper foil or tape is the next best friend to an EMC engineer, just after aluminum foil, for shielding purposes. The availability of this material may be difficult to acquire when an emergency develops, as only certain manufacturers and distributors carry multiple configurations of this material in stock. A significant difference between copper tape and aluminum foil lies in conductivity. Copper is a more efficient conductor, which is why it is the preferred choice of material for conductors. The pressure-sensitive adhesive material on the back of the tape or foil will be either conductive or nonconductive. The majority of tape purchased comes with nonconductive adhesive. This can lead to bad assumptions regarding low-impedance bonding. When one has located a weak spot in a cable shield or enclosure and no change in radiated emissions is observed after application of tape, an assumption can be made that the tape provided no benefit when, in reality, the nonconductive adhesive prevented the tape from making a nice, tight bond connection to ground.

Two types of copper-shielding material are available, laminar sheet and embossed. When using embossed copper tape, scraping the tape with the back of a fingernail, pencil eraser, or similar item will cause the copper boss to break through the nonconductive adhesive for improved performance. Copper tape is sometimes incorporated as a permanent part of an assembly, such as the inside of a connector housing prior to molding, helping ensure a low-impedance connection to chassis ground and to shield the wires from radiating to the environment. Another application for tape is to seal seams and joints of sheet metal assemblies, although gaskets are the preferred material. Foil is great for debugging purposes but not for production use except for certain applications. Copper tape is always placed directly on the leak point, unlike aluminum foil, which is easier to wrap around an entire assembly without identifying the exact leak point.

Caution: The edges of copper tape and foil are razor sharp! If a finger is used to apply the tape to any structure and if the finger rubs against the edge of the tape, the razor sharp edge will slice the finger wide open. The need for a first-aid kit in the laboratory is mandatory whenever copper tape is used.

3. *Zippertubing.** This product, after being applied to a radiating cable, provides a shielded assembly that can be zippered shut without requiring any physical re-

*Zippertubing is a tradename of Zippertubing Corporation.

work to the cable. This is easy to implement if one is attempting to reduce a magnetic field problem. Aluminum or copper tape/foils will have no effect on low-frequency magnetic fields, generally less than 10–20 kHz. Materials with a high permeability value must be used. These shielded cable jackets are available with a steel mesh inside an insulating jacket along with a zipper assembly.

4. *Ground Straps or Heavy-Gauge Wire.* Many products are enclosed in sheet metal enclosures; however, use of plastic housings is commonly found in disposable consumer products. A plastic enclosure reduces weight and can be manufactured in any shape and configuration conceivable. The outer enclosure may be considered a fire enclosure (product safety concerns). For cosmetic and functional purposes, conductive paint or plating is sometimes applied to the plastic. For metal enclosures, plating (alodine, nickel chromate, zinc phosphate, anodizing, etc.) is usually applied. When different pieces of metal are mated to each other, poor bonding may develop, resulting in increased radiated emissions or lowering the immunity level of the system. To ensure low-impedance bonding between assemblies, ground straps or heavy-gauge wire is used. This bonding method is excellent for low-frequency applications but may not work well above 1 MHz.

A benefit of using ground straps or heavy-gauge wire to secure panels and doors to the primary AC mains chassis connection deals not with EMC but with product safety. A metal chassis can become electrically live, especially with voltage above ± 42.4 V AC or 60 V DC, resulting in potential electrical shock. This shock can result in serious injury or death. Bonding all portions of metal enclosures with a low-impedance connection, using straps with a width-to-length ratio of 1 : 5 will minimize electrical shock hazard. These straps are only required if the panels do not make hard electrical contact with other panels. The 1 : 5 ratio is important at high frequencies but has little effect at power line frequencies (i.e., 50/60/400 Hz) where safety grounding is a concern. A ratio of 1 : 3 is preferred, with 1 : 5 being the *absolute minimum.*

If one tries to use straps for EMI bonding purposes, it must be noted that the strap may be highly inductive, causing a voltage potential difference between panels. Any common-mode eddy currents flowing in the chassis may have the opportunity to become RF live and thus radiate undesired EM fields.

5. *Gaskets.* The use of gaskets plays a significant role in preventing radiated fields from leaving an enclosure. Many books describe various types of gaskets and how to properly select them: advantages and disadvantage, comparison of application between gasket types, and the like. The References at the end of this chapter provide a partial list of books that discuss this topic. A brief summary follows:

(a) *Common types of gasket:* beryllium copper, spiral wrap, tin-plated copper, wire oriented in elastomer, monel mesh, extruded elastomers, cloth over foam, cloth tape, spring rubber, thermoset plastics, quick-dry liquids, etc.

(b) *Common means of application:* pressure-sensitive adhesive—either conductive or nonconductive—U-shaped knife edges, mounting studs, conductive glue, spring clips, etc.

(c) *Common location of use:* between panels of enclosures, PCBs and metal enclosures, grounds of PCBs to a chassis frame, shield partition between functional assemblies, securing assemblies in productions, and common debugging applications.

A solution to radiated problems is found by making all joints or seams of adjoining metal pieces continuous from a continuity standpoint. If one cannot measure continuity between metal pieces, then this will become a radiating aperture for RF currents. One can use aluminum foil or conductive tape to cover the seam for engineering purposes. This is where a gasket can be used. However, it must be kept in mind that all conductive materials attempt to provide a low-impedance joint and that low-resistance contact must be available on the metal seams. In other words, decorative paint or anodized finish on the surfaces to be joined must be cleaned of any insulating finish before application of the gasket. A protective conductive finish such as tin or some type of chromate must be used instead.

Figure 9.6 illustrates samples of various types of gaskets available. It is up to the design engineer to select the material appropriate for intended application. Gaskets are used to maintain shielding effectiveness by proper seam treatment. It is the effect of seam discontinuities, in general, which accounts for most of the leakages in

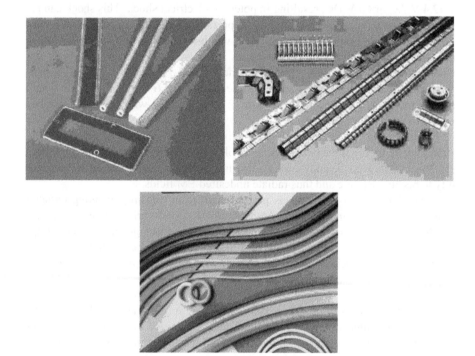

Figure 9.6 Various configurations and types of EMI gaskets. (Photographs courtesy of Laird Technologies.)

an enclosure design. Gaskets maintain conductive contact across mating surfaces. With a constant conductive contact along long seams of metal, it may become possible to permit a larger distance spacing between fasteners. Screw spacing is determined by the stiffness of the enclosure material and compression force required for the selected gasket. If the gasket has a large variation in conductivity, such as elastomers, screws must be tightened (torque) to a specific force by a torque wrench or the mating surfaces must have a positive mechanical stop to provide uniform compression and hence consistent shielding effectiveness.

Regardless of gasket type, important factors that must be considered during the selection process are RF impedance, shielding effectiveness, material compatibility, corrosion control, compression force, compressibility, compression range, compression set, and the environment. For RF impedance control, high conductivity and low inductance are desired. Is should not be a surprise that BeCu (beryllium–copper) has the highest conductivity. For low-frequency (<10–100 kHz) magnetic field shielding, the permeability must be much greater than 1 for optimal performance. This is why most gaskets, regardless of their composition or type, normally do not provide effective low-frequency magnetic field shielding. High-permeability materials must be used, which are very expensive.

Corrosion is also a concern because it leads to reduced shielding effectiveness as it causes the gasket material to become nonconductive. The major contributors to this problem are surface contact area, material dissimilarity, and the presence of an electrolyte. Galvanic compatible materials are those that are within 0.25 V of each other. For commercial applications where the environment is controlled, the range can be increased up to 0.5–0.6 V. If large contact voltages are present, the more anodic material will eventually be destroyed. To prevent this problem, either the gasket material or mating surface or both will need to be plated with a material or finish that is compatible with the base material.

Compression is the force required to achieve maximum shielding effectiveness. The combination of compression and gasket material characteristics determines compression set. Compression set requires one to apply a higher compression force after each opening in order to maintain a constant shielding effectiveness value. After many openings and closings, the effects of compression set starts to be noticed, resulting in permanent deformation of the gasket, which continues until gasket resiliency is lost, resulting in the need for replacing the material.

Finally, selection and location of the gasket material chosen for a given application must take into consideration damage to the constructional contents of the gasket. If a gasket is damaged, one may end up with tiny broken wire strands or conductive flakes and particles that can be distributed throughout the system through natural and/or forced air cooling.

9.1.3 Crosstalk Solutions

Crosstalk can be difficult to locate and potentially very expensive to fix. References 2 and 3 provide a detailed list of solutions in addition to a technical discussion. Generally, crosstalk is found within PCBs and cable assemblies. One way to resolve problems is with a new layout of the PCB and possible relocation of components and cable assemblies. It takes a considerable amount of time to correct a PCB design flaw in addition to the expense of purchasing new assemblies and reperforming functional testing.

Crosstalk is observed in cable assemblies when many bundles or wires are located in close proximity. The easiest solution is to provide a shield over the susceptible cable or separate the wires. Routing cables physically adjacent to metal enclosures also provide benefit of minimizing field coupling.

To prevent crosstalk within a PCB, suggested design and layout techniques are listed below. Crosstalk will sometimes increase with a wider trace width. This is not true if the separation distance is held constant because of the ratio of self and mutual capacitance being held at a fixed ratio. If the ratio is not fixed, mutual capacitance C_m will increase. With parallel traces, the longer the trace route, the greater the mutual inductance L_m. An increase in impedance, along with mutual capacitance, will increase with faster rise times of a signal transition, thus exacerbating crosstalk.

Design and layout techniques for PCBs include the following [3]:

1. Group logic families according to functionality. Keep the bus structure tightly controlled.
2. Minimize physical distance between components.
3. Minimize parallel routed trace lengths.
4. Locate components away from I/O interconnects and other areas susceptible to data corruption and coupling.
5. Provide proper terminations on impedance-controlled traces or traces rich in harmonic energy.
6. Avoid routing of traces parallel to each other. Provide sufficient separation between traces to minimize inductive coupling.
7. Route adjacent layers (microstrip or stripline) orthogonal. This prevents capacitive coupling between the planes.
8. Reduce signal-to-ground reference distance separation.
9. Reduce trace impedance and signal drive level.
10. Isolate routing layers that must be routed in the same axis by a solid planar structure (typical of backplane stackup assignments).
11. Partition or isolate high-noise emitters (clocks, I/O, high-speed interconnects, etc.) onto different layers within the PCB stackup assignment.

9.2   SIMPLIFIED TROUBLESHOOTING TECHNIQUES

This section will present simplified troubleshooting techniques when trying to locate unwanted radiated emissions. For certain applications, radiated immunity can be investigated using these tools but rarely do these diagnostic techniques work for immunity events. The item to keep in mind is that these techniques work best with a spectrum analyzer and can be used not only in the engineering laboratory but also at a test site. What we are interested in is whether radiated energy from the EUT is affected. The element of interest is not the *magnitude* of the emission but the fact that a disruption of a propagating field occurs. When field disruption happens, one is able to locate the leak point to a certain degree. After one performs this high-level troubleshooting, probes and diagnostic tools become useful.

9.2.1 The "Plain Wave and Standing Wave" Technique

Of all diagnostic techniques, using the bare hand is the least expensive but highly efficient means of locating where a leak point of radiated EMI from enclosures exists. This is an uncalibrated technique. The human body not only has a specific self-resonant frequency and impedance but also contains parasitic capacitance. This parasitic capacitance is the result of metallic structures both on and in the body (jewelry and metal plates/screws from surgery).

Radiated fields are affected by parasitic capacitance to nearby metallic structures. Parasitic capacitance permits the transference of radiated fields from a driven source to 0 V potential—the same configuration as a dipole antenna with a voltage-driven element and a return. The body is at 0 V potential while the EUT is at voltage potential; thus an efficient transmitter for RF fields is available between the system and human body (or domestic animals, which are an appropriate but poor substitute).

By placing the body between the antenna and EUT during EMI testing or by waving the hand between assemblies and components, significant disruption of radiated fields may occur. Placing a hand over or on an enclosure seam or on top of a PCB may cause a disruption in field propagation. We are interested in the effects that a hand has on the system, not the actual value of attenuation provided. It is for this reason one is not permitted to stand on the turntable next to the EUT during a qualification test. The body affects field propagation, especially when one stands between the system and the antenna.

9.2.2 The "Disabling-the-System" Technique

When troubleshooting EMC problems, one of the first goals is to determine the source of the noise before locating the coupling mechanism. One approach is to disconnect periphery devices or accessories, various subassemblies, circuit boards, and finally individual functions on a PCB. At the system level, disconnecting interconnect cables one at a time will aid in defining which cable is the primary radiator. If all cables are disconnected and radiation still exists, the source is likely not a cable but from the enclosure.

Powering down or disabling each element of a system one at a time will aid in determining from where the source of the energy is coming. Sometimes in performing these steps the system will not be operational, which is acceptable at this stage of analysis. Troubleshooting does not mean the system has to be 100% functional. Disabling parts of a system is a valuable tool to quickly isolate a particular signal, especially if more than one part of the PCB has the same clock frequency and harmonics. If there are ten 33-MHz clock traces and disabling only one trace removes the undesired energy, then the other nine traces are fine and only one trace needs attention.

If radiating energy disappears when removing a particular cable from a PCB, an assumption can be made that the radiating mechanism is associated with two potential areas (source and/or load) including a potentially bad cable shield. From these actions, it is possible to determine which device, subassembly, or component is the offending source and the coupling or radiating mechanism.

These same basic steps can be applied when attempting noise isolation at the component level. Begin by disconnecting internal cables to isolate which subassembly or which PCB is the potential source of the problem. Integrated circuit components can be isolated by unplugging, unsoldering, or lifting particular pins, disabling through software control, cutting traces between devices or circuit functions, and the like. Suspected noise paths can also be isolated by removing series components such as resistors. These steps can be performed in conjunction with near-field probes and a spectrum analyzer to see what happens, similar to when a cable, trace, or device is disconnected.

9.2.3   The "Cable Disconnection" Technique

If one has a radiated emission problem, turning off power to individual devices or circuit sections will help isolate the source. Disconnecting cables may be necessary. These cables could also be internal if the unit is a self-contained system. Keep disconnecting cables one at a time until a drop in radiated emissions is noticed. The last disconnected cable, when a drop in emission is observed, is probably the propagating path. Apply filtering or shielding accordingly. Reconnect previously removed cables one at time and repeat the process until all the cables are reattached. If the noise level decreases by only a few decibels at a time (i.e., 3–6 dB) during the disconnection process, this usually implies that all the cables are probably contributing about equally and a solution must be applied to all or that one must probe the system in greater depth to locate the actual internal source. Hand-held probes and various other techniques discussed in this chapter can then be utilized to perform this task. Even after locating the source internally, some amount of filtering or shielding may still be necessary. If the problem persists after disconnecting all cables, then the source may be the AC power cord itself. If this is the case or if it is known that a conducted emission problem exists, determine if the noise is common mode or a differential mode and apply an appropriate filter solution. If filtering or shielding the power cord does not reduce radiated emission, then the emission is probably emanating directly from the case or enclosure. From a general standpoint, low-frequency (i.e., <200-MHz) radiated emissions or immunity problems are typically cable related due to the cable's physical length; the lower the frequency, the longer the wavelength (an inverse relationship).

Wavelength is a physical length representation of a frequency. Wires or structures become an efficient radiating source when their physical dimensions approach a quarter or half wavelength. The longest physical structures in a device or system are usually cables, which means they radiate at the lower frequencies. Higher frequency problems (i.e., >200 MHz) are typically enclosure related because the physical dimensions of slots are approaching a quarter or half wavelength for these higher frequencies (shorter wavelengths). Again using hand-held probes and a spectrum analyzer, one can locate, for example, leakage areas such as a display window or ventilation slots or a leaky joint between enclosure sections.

Apply fixes to the enclosure to reduce RF leakage or modify internal circuit boards to reduce the noise source. The noise source could be a clock oscillator circuit. After reducing the RF noise from the enclosure or PCB, remember to replace all the disconnected cables to verify that the cables are now not a secondary source of radiation.

In performing near-field probing, keep in mind that the levels one observes on the spectrum analyzer cannot be correlated to readings obtained at an OATS. The near-field probe readings are just a relative indicator. It should be used to prioritize where or what is the largest source of noise or leakage. Disconnecting or unplugging devices and/or cables either internally or externally can lead to systems not operating quite the same or even not operating at all. If the problem appears to be due to a clock and its harmonics, then functional modes may not be an issue. The only requirement is to have power turned on and in standby mode. For other types of situations, one may require the services of a programmer to make software modifications or to slightly modify a PCB. The radiated emission profile may change depending upon the functional mode of the system. The concept here is to isolate one device or one section at a time to determine the noise source.

What is important is not only the location of the source but also whether the method of coupling is a true radiation problem, common impedance coupling, or a near-field electric or magnetic field coupling problem. Determining the coupling mechanism is a great time-saver in that an appropriate type of solution can be implemented with little waste of time. A simple example of this technique would be coupling of noise from a low-frequency, high-current conductor. High currents would indicate a larger magnetic field than an electric field; therefore, to prevent coupling the shield must consist of a magnetic-based material. A common-impedance coupling of RF transfer will not be solved by any amount of shielding. Rerouting or further analyzing current paths is indicated for this type of problem.

9.2.4   The "Sticky Finger" Debugging Tool

Here is a technique that must be used cautiously. It involves changing the RF characteristics of the noise source by the "loading effect" of one's finger. As described in the previous section, isolation can be performed by removing components, lifting pins of integrated circuits (ICs), or cutting traces. When one performs this task, it will not take long to discover how time consuming the disabling-the-system technique can be. Utilizing one's finger can eliminate some of the pain and efforts of the prior technique. Time can be spent fixing, not isolating, the problem area. By using any finger, touching various pins on a suspected IC component or circuit on a PCB can cause a change in radiated emissions. A wet finger has better dielectric properties and can result in greater radiated effects when a spectrum analyzer is used to observe the radiating waveform using probes or an antenna (Figure 9.7).

If one touches a pin on a component, a significant change in amplitude and/or a possibly frequency shift in the offending noise may be observed on the spectrum analyzer.

> *Caution 1.* Before sticking your finger on any component or pins, make sure that it safe to do so (the circuit must be low voltage DC, known as SELV—Safety Extra Low Voltage). An SELV circuit implies that the voltage under normal or a single fault condition will never exceed 42.4 V AC (60 V DC) and is considered safe to touch with no harm. Once a particularly sensitive component or circuit board area is found, one can use the technique described next to better isolate the noise source down to a particular component pin level.
>
> *Caution 2.* The human body sometimes contains a high electric charge that may cause an ESD event to occur. Care must be taken to discharge oneself before making physical contact with any electrical circuit. The body can contain charges into the upper amperage range and thousands of volts. For most silicon-based components, direct injection of 500 V is sufficient to cause permanent damage.

9.2.5   The "Sharpen-Your-Pencil" Tool

For those squeamish about sticking fingers onto a PCB, especially when not sure of voltage levels present, this technique might be more appropriate. It is also used to further define which particular pin on a component might be the offending source after using the previous troubleshooting techniques to narrow the search down to a particular component.

Figure 9.7   Sticky finger debugging tool.

As in the case of the sticky finger technique, the sharpen-your-pencil technique produces a loading effect on a component pin that will change radiated emission characteristics (Figure 9.8). Pencils contain lead, which injects capacitive loading on a signal trace. Adding capacitance to a transmission line will affect the time-domain signature characteristic, which in turn affects EMI when observed using sniffer probes or antennas and a spectrum analyzer. Like the previous technique, one can quickly locate a potential noise source without sophistical instrumentation.

9.2.6 Coolant Spray Tool

Many systems use more than one clock source that may operate at the same frequency. For this situation, it can be extremely difficult to determine which oscillator is the worst offender. If one has access to a correlation analyzer, the problem may be easy to solve. However, in lieu of not having such an analyzer, there are several simple methods to remove or eliminate multiple clock sources operating at the same time on a PCB. One method is to simply remove or unplug cards or oscillators one at a time. The problem with this technique is that normal circuit functioning will usually cease to operate. When it is not feasible to unplug or disconnect oscillators, another approach is to shift the clock frequency and monitor this shift with a spectrum analyzer. The span and bandwidth are set to minimum values in order to observe this frequency shift. It is acceptable to have an out-of-calibration mode on the analyzer since one is concerned with only relative amplitude levels. Even the best thermally compensated oscillator will shift in frequency when hit by a sudden temperature change. Obtain a can of cold spray or coolant. Aim the application tube to one oscillator at a time. Observe the frequency in question and record a reference level. After spraying, watch the display. A shifted frequency should appear. The frequency may last for a few hundred cycles. If the shifted frequency is lower in amplitude from the stable frequency, then the oscillator just sprayed is probably not

Figure 9.8 Pencil debugging tool.

the major source. If the shifted frequency is higher in amplitude than the reference frequency, then this oscillator is likely the offender.

9.2.7 The "Piece-of-Wire" Approach

Another economical troubleshooting tool is an ordinary piece of wire. Touch one end of wire 2–3 ft (60–90 cm) long directly to a suspected piece of metal or near a suspected gap, seam, or aperture in an enclosure. Gauge or diameter is not important. Stretch the wire out. Carefully observe the spectrum analyzer for any increase in radiated emissions. The wire will work as a radiating antenna to propagate RF currents present on the enclosure. If there is an increase in emissions, the area on the chassis touched by the wire has an RF potential present. Adding a piece of wire presents high-RF impedance across the joint, creating localized RF field coupling to the wire.

A piece of wire can sometimes be used to short out an aperture, joint, or seam leak point. A short circuit can produce a reduction in radiated emission due to a low-impedance connection that the wire has created across the aperture being investigated. Depending upon the frequency of the noise source, a screwdriver can also be used in lieu of a wire.

Touch the blade of the screwdriver to the suspected metal piece or across a seam, gap, or joint. To enhance this troubleshooting technique for greater accuracy, place the screwdriver "and" wire through the center of a current probe. Monitor the current probe output with the analyzer. Many times, it becomes difficult to observe radiated emission changes with an antenna or probe since, by their physical presence, our bodies can affect radiated fields. If there is an RF potential or field coupling to the wire, there will also be a current flow in the wire or screwdriver blade at offending frequencies. By using the screwdriver to short out the aperture seam, the field potential on the wire is minimized. A reduction in RF current is sometimes easier to observe on a spectrum analyzer than monitoring radiated emissions level.

9.2.8 Radio Control Race Car Diagnostic Sensor*

This simple diagnostic sensor provides significant value when debugging problems with test facilities and installations used for EMC analysis. Amplitude accuracy is generally not required when performing debugging and diagnostic work.

When debugging or locating the presence of radiated emissions, the primary area of concern lies in finding the exact point from where the undesired source of energy is coming. For large installations, it may be difficult, if not impossible, to determine which subassembly module is causing the energy to exist. This concept is valid from the smallest of products, such as a calculator, to industrial-level systems, such as naval aircraft carriers or space shuttles. If a propagating field has the ability to penetrate through an opening or aperture in a metal chassis or enclosure, we have both a susceptibility and emissions problem. In other words, this sensor evaluates shielding effectiveness.

*Section 9.2.8. courtesy Amy Pinchuk, InField Scientific Inc., Montreal, Quebec, Canada

For those who are attempting to determine if an installation has leakage problem, the use of a child's toy as a diagnostic sensor can be used. This sensor is a radio-controlled (RC) car available worldwide. This inexpensive toy comes in all sizes from very small to monster truck (child to adult size—who says engineering has to be all work and no play). This tool is shown during actual use in Figure 9.9.

To use this diagnostic sensor, for example, locate the car inside an anechoic or shield room enclosure and activate the radio control. It is recommended that a second person be available, as one needs to operate the control and the other to watch the car in soundproof facilities. If the car starts to move or makes noise, then the enclosure has a shielding integrity problem. Sometimes, the location of the leak point can be discovered with minimal effort. The same technique is true in reverse—with the car outside of the chamber and the operator inside.

For smaller systems, a very small car can be placed inside a chassis enclosure that is assumed EMI tight. Remove PCBs or other assemblies that may interfere with the physical size of the car being placed inside the box. This is where a smaller size car is useful. The object is to not have the car physically move but to activate the motor (engine). If one hears a "vroom . . . vroom," it can be assumed that the design of the enclosure is inadequate for its intended application.

One advantage of using this sensor is not to perform official EMC analysis but to demonstrate to management that a product design is inadequate. Have the program manager or vice president operate the controls of the car. If the car moves at 30 MHz, how will EMI performance be at 300 MHz (10th harmonic)? By using this toy to demonstrate shielding effectiveness (negligible cost compared to commercial EMC diagnostic tools), one presents unarguable conclusions. A red Formula-One

Figure 9.9 Radio-controlled race car diagnostic sensor. (Photographs courtesy of Amy Pinchuk, InField Scientific Inc.)

race car (27 MHz) works best because the color not only impresses management but also captures people's attention. Everyone enjoys operating the car, which further illustrates the need to improve shielding effectiveness.

It is amazing how a small-scale RF signal penetrates a supposedly secure shielded enclosure. From experience, this has proven to be far more convincing to management than shielding effectiveness data acquired at an EMI laboratory. It is particularly useful for convincing others of the need to implement maintenance procedures such as replacing broken finger stock and missing gaskets.

Applications of this diagnostic sensor has been to demonstrate vulnerability of a Tempest shielded chamber and the need to convince the captain of a navy vessel that the proposed location of a new 1-kW high-frequency antenna would cause interference onto the bridge (contrary to the captain's belief that the bridge was well shielded based on weeks of mathematical analysis).

If anything interferes with the seal of a gasket, such as nonconductive paint, oil from fingers, an aperture opening, a cable entry, a thin fabric netting, or a nonconductive (non-EMC) gasket, shielding integrity is lost and the car can be heard racing around inside a metal enclosure. This diagnostic tool is extremely effective for convincing others about the need for EMC control devices such as metal EMI connectors and gaskets. For many system designs, the best shields are defeated by paint and open doors. A car bashing around in an enclosure is a diagnostic tool that should not be discounted.

9.2.9 The "Tin Can Wireless Antenna" for Signals above 1 GHz

This simple antenna design allows for measurement of RF signals above 1 GHz or where an expensive horn antenna is required. The wireless network and public communication systems that operate at 1.25- and 2.4-GHz applications (i.e., Bluetooth, IEEE 802.11, etc.) need an antenna that has high gain and a narrow beamwidth. The tin-can antenna has been used extensively in the wireless community.*

This "very" inexpensive antenna is made from a 40-oz can of beef stew. A tin-can design is a waveguide antenna, or Cantenna, and may be the best diagnostic tool to not only measure IEEE 802.11 signals but also test for immunity concerns in the frequency range of interest. The design reuses a food can made out of steel or tin, not aluminum. Details on construction of this sophisticated antenna are found in Appendix A (Figure 9.10).

Once the antenna has been constructed, connect the can to the wireless control card or access point on the system. This connection requires a special cable commonly called a "pig tail." One end of the cable will have an *N-series* male connector while the other end will have a connector appropriate to the control card or access point. In other words, this is a coax cable with two different types of connectors on both ends.

This antenna has linear polarization. That means that rotating the tin-can antenna on the tripod will affect the signal strength. First, place the cable connection straight

*The antenna is in the public domain. Material is available on the Internet on numerous websites.

Figure 9.10 Tin-can wireless antenna.

down. To improve signal strength for both transmit and receive, experiment with rotating the can while watching the signal strength meter on the host system to achieve optimal performance. If for communication purposes, after making a wireless connection between two systems, determine the maximum distance of operation, which can be considerable. For measurement of RF radiation from a device, locate the antenna at a distance that permits measurement of the signal. Calibration is not required as the item of interest is the magnitude of change within signal strength when a change is made in the configuration and design of the product

9.3 TESTING AND TROUBLESHOOTING USING PROBES

The material in the simplified troubleshooting techniques section does not work all the time for various reasons. Results are very preliminary and are used only to ascertain if a problem exists and potential areas of noise generation. For more accurate testing and troubleshooting, the techniques described in this section are recommended.

9.3.1 Using Probes for Immunity Testing and Troubleshooting

Instrumentation. Typical laboratory benchtop equipment such as digital multimeters, oscilloscopes, and logic analyzers provide minimal benefit when attempting to isolate immunity events or disruptions and are difficult to use for immunity testing. These instruments are measurement tools designed for use in the time domain, whereas immunity problems are generally a frequency-domain event. However, transient immunity is better evaluated as a time-domain event.

When trying to analyze immunity concerns using time-domain probes or analyzers, it can be difficult to distinguish what is an immunity disruption, ambient event, or EMI. Time-domain probes have different impedances than propagating electromagnetic fields (377 Ω in free space) or a transmission line (PCB trace), which can vary from 40 to 150 Ω, not including the impedance of both the source driver and the receiver. In addition, time-domain probes can act as injection points into high-impedance circuits, causing a disturbance and disrupting circuit operation, thus making immunity events harder to isolate.

Should these types of instruments be used, special probes are available that will minimize loading effects in transmission lines. Types of measurements possible are voltage, current, and local field strengths. These probes connect to instruments using a fiber-optic interconnect and are useful for bandwidths exceeding 1 GHz.

Localized Emission Testing—Use of Near-Field Probes. Every product is unique in its design. A solution for one product may not be acceptable to another even if it has the same problem. Electromagnetic compatibility can be generalized into simple concepts. Typical areas to consider include looking inside an enclosure to find long transmission lines acting as antennas, such as wires and cable assemblies or traces on PCBs, especially if they are carrying high-frequency signals. Internal components on a PCB that cause EMI includes microprocessors, memory arrays, DMA and I/O buffers, internal cables or harnesses interconnects, video display processors/screens, and power supplies. Enclosures generally contain seams, joints, apertures (holes such as ventilation slots), gaps, penetrations of cables, connector shells, and surface or contact resistivity of gasket material, especially for higher frequencies (i.e., >1 GHz).

In general, radiated emission frequencies (e.g., <100–200 MHz) are usually due to cables or wires while higher frequencies (>200 MHz) are typically related to leakage from the enclosure. Remember that wires or cable do not necessarily have to generate or contain RF energy. Any antenna can radiate RF that has been coupled into/onto that assembly.

When using near-field probes, slowly move the probe physically close to the surface of the enclosure. Continue scanning over the enclosure surface concentrating near suspected seams, joints, openings, or I/O interconnects and their attached cables. Calculate wavelength from frequency ($\lambda = C/f$) and physically measure all seams and potential antenna structures that are roughly either a half-wavelength or quarter-wavelength dimension based on this equation. Write down the relative levels from each "source" using a constant measurement distance from each radiating source. The highest level provides a starting point as to the potential location of the enclosure leak point. If the highest reading is from a wire or cable, use a current clamp to measure the noise current on it. The radiated field is directly proportional to this current level. If one needs a 6-dB reduction, current in the cable must be reduced by 6 dB. Because the source of the RF noise is usually generated internal to the enclosure, further near-field probing may be performed on inside wires, traces, or components on various PCBs. If excessive radiation is measured directly on the PCB, this may indicate poor layout, trace routing, or decoupling.

With near-field probing, one will generally observe RF noise only when physically close to the source. It must be kept in mind that near-field readings are dependent upon the nature of the source, its geometry, and its properties. There will be no correlation between near-field and far-field measurement. Near-field measurements are a means of identifying the source of the noise. Small movements of a near-field probe will usually result in relatively large changes in the amplitude. The usefulness of the near-field probe is to show where the emission is generated.

Localized Immunity Testing. When performing troubleshooting for an immunity event, one must use tools available, either in a test laboratory or at a customer site. Sometimes basic instrumentation is available. Chapter 5 details various probes that are effective in locating problem areas, mainly close-field magnetic and electric field probes. When commercially available probes are not available, home-made probes should be sufficient. Appendix A provides details on building probes that work for almost all applications: emissions and immunity.

Several areas of concern must be noted when using probes, either purchased or home made. The primary concern lies with the ability of the probe to be robust enough to survive potential abuse by the user. If the probe survives operational use, other concerns for localized immunity testing include the following:

1. What kinds of circuit or devices are to be analyzed?
2. Is the unit to be tested a prototype or a production unit?
3. If there are failures during formal testing, will probes help identify the problem area quickly?
4. When ensuring conformity of a product during production, will probes provide sufficient information to determine if a possible problem will develop in the field?

Creating Transient Events. Most closed-field and current probes detailed in Chapter 5 can be connected directly to the output of a fast transient/burst generator, creating a localized magnetic or electric field. Depending on the type of probe and signal provided, magnetic and/or electric fields can be generated simultaneously that correspond to the injected waveform. A simple home mode probe that can be used to inject transient energy into a cable is shown in Figure 9.11 [4].

The procedure for using these simple probes is similar to that used for detecting localized sources of emissions. The probe is carefully moved over the portion of the system suspected of being the problem area. Usually, the probe has to be right next to the device or touching it. This procedure is used until sensitive areas are located. Home-made and commercial probes work well in providing a radiated energy source to a circuit. A disadvantage of doing direct injection is that components could possibly be damaged from high voltage levels. With magnetic or electric field probes, the radiated field will only cause operational concerns. If the energy source is transmitted through cables, the inherent capacitance and inductance of the wire may attenuate the signal to the point where it becomes impossible to locate the problem area.

In addition, a small loop probe can be used as a noise source for locating immunity problems on, for example, cables, PCBs, and enclosures. This technique uses the probe as a noise source along with a quality signal generator. The output of the signal generator connects to the probe. The loop probe then injects a mag-

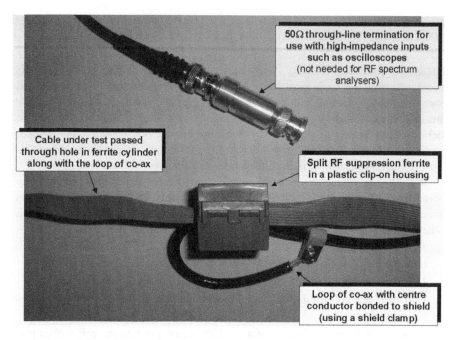

Figure 9.11 Simple probe circuit for injection of a transient pulse into a cable. (Photograph courtesy of *EMC Compliance Journal* [4].)

netic field by inductive coupling to a wire or device by simply moving the probe over the suspected circuit. Caution must be used to not supply too much power to the loop probe. Either place a 50-Ω terminating resistor in series with the probe, in which case the generated field will be lower, or just physically touch the probe to verify that it is not becoming too hot from power dissipation. One can make loop probes in a variety of sizes to accommodate cables and seams (1–2 in. in diameter) or smaller ones (<1 in. in diameter) to use on traces or circuit components.

In most cases, the output of a signal generator is adequate, especially for troubleshooting sensitive analog-type circuits. Occasionally a low-output-power amplifier (1–5 W) may be necessary when using this technique for digital components. If one is careful enough, direct injection from the signal generator itself through a capacitor onto the suspected trace or device can also be performed. The capacitor is necessary to block any DC voltage from causing component damage.

For troubleshooting problems related to power line surges, do not attempt to perform this type of testing in the field. Use only an approved test facility with safety features provided. The easiest way to determine if a failure exists from a surge event is to not resurge a good unit hoping to find the problem area but to open up the power supply assembly (or equivalent circuit) and look for physical damage. A surge pulse usually causes a component to self-destruct or become scorched by high

Figure 9.12 Measurement of cable currents due to EMI coupling; modem phone line and serial cable. (Photograph courtesy of Doug Smith.)

temperatures. Sometimes, one can visually determine the problem area. At this stage, refer to schematics and actual PCB layout artwork to determine if proper creepage and clearance distances were implemented and the component selected was rated for a high transient event and used correctly for its intended application.

9.3.2 Differential Measurement of RF Currents on Cables and Interconnects

Immunity problems in electronic equipment caused by conducted EMI noise on cables can be difficult to diagnose, especially if operational problems appear to be intermittent in nature. Sources of noise include EFTs, ESD, and radio transmitters. A simple procedure using a matched set of current probes can be used to troubleshoot immunity problems in equipment.* By determining the path of the resulting noise currents in equipment, one has made progress toward finding the source of the problem.

The procedure uses a pair of matched current probes to find which equipment cables are carrying RF current and the direction of the current flow. In Figure 9.12, high-frequency energy from a radiated ESD event is coupled onto the telephone cable (upper right cable). The resulting RF current is monitored by a current probe. The question is, which cable is carrying undesired RF current and in which direction? Current arriving on one cable must exit on the other cable or radiate from the equipment itself.

A second current probe is placed on the cables one at a time to determine the role of each cable in creating a path for the ESD pulse. Figure 9.13 shows the result of placing a second current probe on the serial cable (the left cable in Figure 9.12).

*Details of this troubleshooting technique are derived from a *Technical Tidbit* article (August 2001) provided at www.emcesd.com courtesy of Doug Smith.

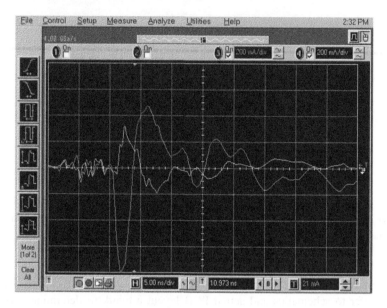

Figure 9.13 ESD currents on external cables; phone line and serial port. (Photograph courtesy of Doug Smith.)

The larger sized cable is the phone cable and the smaller cable is the serial cable. The current probes are placed on the two cables in the same orientation with respect to the equipment. The expected current entering the phone line should be of opposite polarity as the current on the serial line. The first peak of the current on the phone cable is –800 mA. The serial cable carries a little less than half of that current, about +300 mA, in a direction away from the equipment. There is a slight time delay of about 1 ns between the two currents, representing the time it takes for the current to enter the equipment and then exit on the other cable. The small amount of noise to the left of the current peaks is due to radiation directly through the air from the ESD source.

Figure 9.14 shows the second current probe moved from the serial cable to the power cable. The result is shown in Figure 9.15.

Figure 9.15 illustrates that the power cable conducts a little over +150 mA away from the equipment. Twice as much current is being conducted on the serial cable as on the power cable. This information allows a designer to pay attention to the circuitry around the phone and serial cables. Hopefully, there is no critical circuitry between the connectors associated with these cables and the PCB that can be disrupted by undesired ESD currents.

Often, the cable or cables carrying noise into the equipment are not known. For this situation, a pair of current probes can be used to find the relative direction and magnitude of cables carrying currents into or out of the system. Usually, the offending current is primarily coming in on one cable and exiting on one or more of the other cables.

Figure 9.14 Measurement of cable currents due to an ESD event; modem phone line and power cable. (Photograph courtesy of Doug Smith.)

This example also points out a useful design rule. To minimize effects of external conducted noise on equipment, keep connectors as physically close to each other as the system design permits. Doing so minimizes chances that RF currents entering one cable and exiting on another will not pass though critical circuitry located in the area of the interconnect.

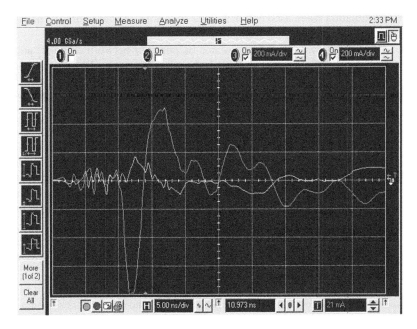

Figure 9.15 ESD currents on modem cables; phone line and power cable. (Photograph courtesy of Doug Smith.)

9.3.3 Switching Power Supply Effects on Common-Mode Conducted Noise

Solving intermittent problems that occur in equipment can use significant engineering resources. An important source of intermittent system problems is common-mode noise current that is rarely covered by power supply specifications or EMC tests yet represents a major source of system problems. It can take quite a while for a series of 50-kHz spikes to line up with a vulnerable state in a system making for an apparent intermittent problem. The cause of a problem due to switching power supply noise is often difficult to find because of its hidden, unlikely nature.[4]

Switching power supplies are capable of generating various types of noise: radiated and/or conducted. Any EMI mode can disrupt system operation (signal corruption), even if operational circuits are not powered from the supply in question. One area of concern is from common-mode currents flowing between DC isolated outputs or between an isolated output and the power supply's input ground connection. Common-mode currents can reach substantial amplitudes, causing an $L(dI/dt)$ voltage drop along a conductor connecting the input and output returns. This amplitude level may exceed 1 V/cm!

Often, power supplies are tested fully loaded. However, input AC mains return (frame ground) and the output 0-V reference (return) are not connected together during testing. In most electronic systems, both input and output returns (AC mains and 0-V reference) *are* connected, sometimes at a single point and sometimes at multiple points. To properly test and evaluate a power supply on the bench and for accurate simulation when used within a metal enclosure, the configuration of Figure 9.16 is discussed.

Noise currents flowing between isolated output(s) or between output(s) and input ground are due to parasitic coupling in the power supply and can vary substantially between units. Even a small, seemingly innocent design change can cause a 10-fold increase in noise currents. Figure 9.17 shows one way to measure the $L(di/dt)$ voltage drop due to noise currents.

The input and output returns are connected with a plain piece of wire. The $L(di/dt)$ voltage drop in the interconnect wire can be estimated by mutual induction from a square magnetic loop probe made from a piece of stiff wire (see Appendix A for details on this probe). The output of the loop will be $M(di/dt)$. This magnetic induction represents a lower bound limit for the voltage drop caused by the long interconnect wire. Ensure the oscilloscope end of the coaxial cable from the loop probe is terminated in the cable's characteristic impedance, usually 50 Ω. The load connected to the power supply should be varied throughout all output current ranges. Some power supplies may be noisy with low load currents while others have considerably more noise at higher load levels.

Another method to measure the voltage drop within a wire is detailed in Figure 9.18 using a balanced coaxial probe. Again, vary the load current on each output of

---

[4]Details of this troubleshooting technique are derived from a *Technical Tidbit* article (February 2001) provided at www.emcesd.com, courtesy of Doug Smith.

Figure 9.16   Switching power supply with input/output grounds connected together. (Photograph courtesy of Doug Smith.)

the power supply. There are numerous diagnostic techniques available to measure common-mode currents; however, those shown here are easy and sufficient in most cases.

When using a loop probe, approximately how much RF energy is considered significant? One volt induced across 1 cm of wire is too high. Eventually a problem due to the noise current will surface. Often, this occurs when an otherwise innocuous change is made in a nearby circuit and then the system no longer works. A lim-

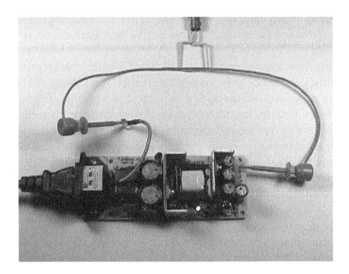

Figure 9.17   Measuring common noise ground currents produced by a switching supply (magnetic loop probe).

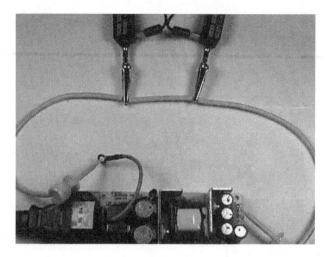

Figure 9.18 Measuring common noise ground currents from a switching supply (balanced coaxial probe). (Photograph courtesy of Doug Smith.)

it of 50 mV/cm will suffice for many circuits. Systems with optical transceivers and low-level analog circuits may require the amplitude level of the magnetic field to be much lower.

9.3.4 Discrete Component Diagnostic Tool

A combination of components such as a ferrite bead in series with a resistor or an inductor in series with a capacitor works extremely well in helping reduce RF energy within a transmission line or PCB trace. The best filter circuit is one that is placed in series.

Another excellent filter element is a ferrite bead. Ferrites exhibit primarily resistive characteristics to signal lines. At low frequencies, inductive performance dominates. Ferrite material exhibits very low impedance at a specific frequency of operation. At higher frequencies, ferrite beads exhibit resistive characteristics, and hence they have high (resistive) impedance. The vector angle formed between the resistive and reactive portion of the ferrite is identified as a "tangent delta" and is important only for phase-sensitive circuits. Due to the RL characteristics of the ferrite (which perform as a parallel $R//L$ circuit), the component is a low-Q device and will NOT resonate, unlike normal inductors that contain parasitic capacitance, forming an $L\|C$ circuit. An LC circuit will exhibit a reactive resonance, which is exactly what we do not want.

How does one place discrete components in series with a trace or pin on a PCB? Cut open the suspected trace with a sharp knife and scrape away soldermask. Place one end of the filter assembly on one side of the cut. Place the other end of the filter combination on the other end of the cut. These series elements will bridge across the gap that was made when cutting the trace. Using this technique and a combination of ferrite bead and a resistor and/or capacitor can significantly reduce the RF energy content within a transmission line, providing for quick identification of a suspected problem area. For differential-mode noise, use of a shunt element such as a capacitor may be required. For this situation, scrape away the solder mask from the transmission line and nearest ground reference and connect the capacitor between these two areas.

For bypassing or shunting effects, place an oscilloscope probe tip on the suspected pin or trace and the other end of the probe on the nearest signal ground connection. This type of probe configuration consists of a series inductor/capacitor circuit as mentioned earlier or a single capacitor. The probe can quickly be used to optimize the location for placement of a bypass component. One can also touch a capacitive probe tip along every pin or pin-to-trace to see what effects occur. Using this technique may make the circuit nonfunctional, which is acceptable, since what is being achieved is location of a fault area containing unwanted RF energy, not system operation.

9.3.5 Tweezers Probe

A unique and highly efficient simple diagnostic probe can be created with an item usually found in the home or laboratory [5]. Creation of other unique probes and applications are unlimited and is left for one's imagination. The idea presented hopefully provides a catalyst for creativity on one's behalf.

The tweezers probe consists of the following:

1. Metal tweezers
2. Suppression components (e.g., capacitors, ferrite beads, resistors) that can be used to touch a suspected pin or trace connected to the end of the tweezers
3. Optional switch

For keeping leads as short as possible and for ease of manipulation, the modified tweezer probe is a highly efficient diagnostic probe. It is easy to locate a problem area using only one hand. Tweezers are inexpensive and readily available. The tweezers must be split apart at the joining end. For inexpensive (poor-quality) tweezers, this should be a matter of simply pulling them apart with minimal effort. For higher quality tweezers, one may have to drill the spot weld at the joined end in order to pull the two sides apart. After breaking the tweezers in two, place an insulator between the split halves. This insulator can be a piece of plastic or wood (Figures 9.19 and 20). The two ends are now reconnected using epoxy glue to the insulator. Be careful to leave exposed the end tip of each half (epoxy-free area) for soldering discrete components, if desired. As shown in the figure, a switch can also be added to allow band switching among several frequency ranges. One can connect a shunt capacitor between the bottom half of the tweezers or a series resistor, depending on application of use.

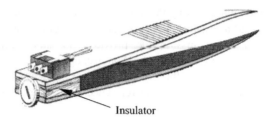

Figure 9.19   Tweezer sniffer probe assembly.

Figure 9.20   Fully manufactured tweezer probe assembly. (Switch is provided for switching in discrete components.)

9.3.6   Miniature High-Discrimination Probe

The miniature high-discrimination probe is an *H*-field probe used for picking up radiated emissions. It is extremely small and offers shielding for enhanced electric field discrimination. The small size offers a spatial resolution of about 1 mm [6].

The probe incorporates a 50-Ω-termination resistor to minimize reflections and contains approximately a 10-turn coil located in a small shielded tube (Figure 9.21). The probe's output is proportional to the rate of change of the magnetic field being measured, indicating nearby magnetic field flux. The 50-Ω terminating resistor is not critical but will aid in reducing distortion to the waveform. If desired, an external 50-Ω-termination resistor can be used.

The probe is sensitive to magnetic fields located along the probe's axis. For a single conductor or trace, the probe's response will be greatest on either side of the

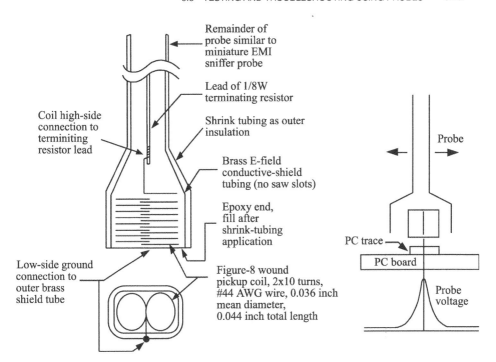

Figure 9.21 Miniature high-discrimination probe construction details.

transmission line. If the RF current return is located in an adjacent trace or conductor, a sharp peak will be found when the probe is placed between the two lines. The response will be a null when over either conductor. The response to a trace over a ground plane or with the return trace directly underneath the top trace will be the same as for the single trace.

When using this probe, a two-channel oscilloscope is recommended. One channel is used to look at the noise that is to be located and the second channel to view the probe's output voltage. A trigger signal can stabilize the waveforms. Move the probe around the board until a probe output waveform is found that is synchronized with the noise transient. This provides correlation between the transient present and the potential noise source. The shape of the waveforms may not be identical but should have a strong resemblance, especially at the physical edges of the trace. The probe can be used to locate sources of di/dt related transients such as diode recovery problems in power supplies, leakage fields from transformers and inductors, and poorly bypassed digital devices (Figure 9.22).

9.3.7 Using Current Probe as Substitute for Radiated Emission Testing

Many times, it is more convenient to perform troubleshooting tasks or investigations in the engineering laboratory where easy access to support equipment is avail-

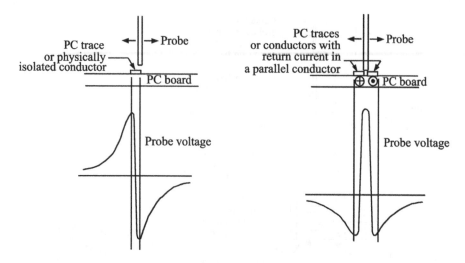

Figure 9.22   Typical high-discrimination probe output waveforms.

able (people and hardware). When a formal test laboratory is not available, a simple bench technique for identifying and reducing noise from equipment under evaluation is highly desired. This diagnostic technique is based on the premise that most radiated problems are due to common-mode currents flowing within cables, especially in the frequency range of 100–300 MHz.

Radiated emissions in the higher portion of the frequency spectrum are usually a result of enclosure leakages. For this condition and using near-field probes described in earlier sections, one can determine the actual value of the radiated emission. For referencing, always use diagnostic probes at the same distance before and after modifications to determine relative improvement levels.

Using Eq. (9.4) [12], we can calculate expected radiated field levels by measuring the actual amount of common-mode current within the transmission line:

$$E = \frac{(12.6 \times 10^{-7})fIL}{r} \quad \text{(V/m)} \tag{9.4}$$

where f = frequency of operation (Hz)
    I = measured current in transmission line (A)
    L = length of cable (m)
    r = distance between source and antenna (m)

EXAMPLE

For a 1-m cable at 50 MHz, the measured common-mode voltage expected to be observed at a test site with 5 μA current in the cable is calculated to be

$$E = \frac{(12.6 \times 10^{-7})(50 \times 10^{6})(5 \times 10^{-6})(1)}{3}$$

$$= 105 \ \mu\text{V/m}$$

The specification limit for FCC Class A at 3 m is 90 μV. For FCC Class B, the limit is 100 μV. For this example, the unit fails the FCC Class B limits.

Note: This answer is approximate. Between the physical cable and the probe is air, or a dielectric. There will be a small dielectric loss in the measured signal. This loss is generally small enough to not be a concern.

For each frequency of interest, we can calculate radiated field strength levels. Instead of performing radiated emission tests at an OATS, troubleshooting can be performed on a laboratory bench with just a current probe measuring the noise current on a cable. As usual, it is wise to perform a null test by measuring the ambient with the current probe by itself without being energized by cable currents. Depending upon the current value measured, a low-noise preamplifier may or may not be necessary.

9.3.8 Enclosure Resonances and Shielding Effectiveness

All metal structures, when used as a chassis enclosure, have a self-resonant frequency. When internal RF sources such as oscillators and their harmonics coincide with structural resonances, the chassis can become excited by this energy and may "amplify" the noise. Chassis resonances can be predetermined during the prototyping stage. A spectrum analyzer (preferably with an internal tracking generator), signal generator, and one loop or pin probe are required for resonance testing. To test the shielding effectiveness (characteristics) of the enclosure, two loop probes are required (Figure 9.23). A directional coupler is an additional piece of equipment re-

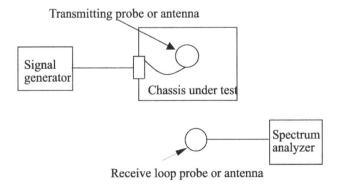

Figure 9.23   Measuring shielding effectiveness of a chassis.

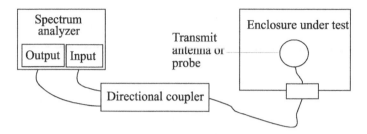

Figure 9.24 Measuring resonances of an enclosure.

quired when attempting to measure resonant points. The second pickup loop probe used during shielding effectiveness testing can be substituted by the normal OATS antenna if located at an outdoor test site.

Use a bulkhead feed-through connector such as BNC or TNC to bring the signal to one of the loop probes placed inside the sealed enclosure. Placement should be considered carefully. The location where high-speed electronic circuits are to be placed should be given highest priority. The amount of current drive to the probe can be determined by monitoring the external portion of the enclosure with a second probe (make sure the chassis is open for this verification test prior to sealing up the box). Chassis leaks can be found and corrected as necessary using the tracking or signal generator driving the probe internal to the unit. For higher frequencies in the spectrum, an electric field antenna can be substituted for the loop probe (antenna). This can be as simple as the exposed center conductor of a coax cable.

A null test should be performed to verify that only radiated fields from the transmit antenna are being received. Replace the transmitting loop with 50 Ω termination and repeat the test with the same output signal level from the signal generator. Nothing should be received by the receive antenna or probe.

The setup is almost the same when testing for chassis resonances. The loop probe reflects the energy from the signal generator. However, as the signal approaches a resonant frequency, energy should be absorbed. This will cause a dip in the analyzer's trace or display (Figure 9.24).

9.4 ALTERNATE TROUBLESHOOTING TECHNIQUES

9.4.1 Using Oscilloscope to Debug Signal Integrity Waveforms and Radiated Emissions

Designers with access to an oscilloscope and appropriate probes can utilize an oscilloscope to investigate systems and PCBs for radiated emissions. For systems that fail testing, they can be used to make qualitative checks that have some relevance to radiated emission performance.

An oscilloscope must have a bandwidth appropriate for the range of frequencies to be investigated. A 60-MHz oscilloscope will not show detailed information on frequencies above 60 MHz. In addition, a 10X probe will not work above 500 MHz if the probe configuration is incorrect, especially if a 100-mm-long ground lead is clipped to a reference point [7]. Appendix A illustrates how to make high-frequency measurement probes.

Before using an oscilloscope to examine radiated emissions, one must know how to translate time-domain signals to the frequency domain. Scopes with FFT analysis functions are useful, especially if one is familiar with knowing that a higher rate of changes in voltage or current (dV/dt and dI/dt) means greater threats of radiated emissions. Spurious ringing on a waveform identifies a great deal about the resonant frequency of a circuit and which frequencies are likely to cause problems.

A useful diagnostic technique to investigate signal integrity on a PCB is to follow a digital signal (i.e., clock trace) from its source to the load and observe waveform degeneration. A properly designed PCB with low emissions will always maintain optimal signal integrity along the length of its trace route. When a waveform degrades significantly, ringing on the trace generally indicates likely emissions are present. One problem when using an oscilloscope probe is the impedance of the probe tip. If the impedance of the probe is approximately the same as the circuit, serious signal degradation can occur. This degradation will cause the waveform of the signal to alter, thus affecting signal integrity within the transmission line in addition to masking the true EMI characteristics present. A high-impedance probe minimizes effects of circuit loading.

One of the few benefits of oscilloscopes over spectrum analyzers is that scopes can be triggered from a clock or other waveform using a standard voltage probe and a second channel. If the measured signal is not sharp and clean, one needs only to change the triggering point on the oscilloscope. Select the trigger point to be from a different clock or signal source. This technique allows one to ascertain which trace is contributing the most to the propagation of unwanted noise.

Voltage probing suffers from one significant problem: The loop formed by the probe's signal and ground wire acts as an unshielded loop antenna picking up radiated noise from its local environment. Most oscilloscopes have very poor common-mode rejection. It can sometimes be difficult to determine what one is measuring— a DC voltage signal or AC radiated emissions due to induced coupling into the scope's ground clip.

Transducers such as closed-field probes, current probes, and antennas have no metallic connection to the circuit being evaluated, usually a PCB trace. These transducers do not suffer from problems occurring from a long ground clip wire commonly found on oscilloscope probes. Closed-field current probes do not measure signal waveforms as efficiently as a voltage probe.

Common-mode voltage can be detected using an electric field or pin probe connected to the chassis or 0-V reference. Connecting a 10X probe to chassis or 0-V reference without connecting its ground lead can also detect common-mode voltage. The problem of AC mains frequency "hum" can be solved with a high-pass filter (such as a low-value series capacitor provided at the tip of the probe).

Using existing oscilloscope probes is clearly the lowest cost method for an engineer during development of a product or when performing emission analysis. However, a lot of skill in reading waveforms to achieve any quantitative correlation with standard EMC tests is required.

When using an oscilloscope to measure waveforms, one must recognize that measurement error may occur (the small difference between two measured values). An easy test is available to determine measurement error. This test is to short the oscilloscope probe ground to the oscilloscope probe tip. Now move the probe near the area being measured. The measurement environment, as observed by the loop of the ground wire to the center pin (acting as a single-turn loop antenna), may indicate radiated noise generated in the loop antenna. If the radiated noise is nearly the same level as direct connection to the circuit, the loop area of the ground lead may be the source of error. The test environment is made up of the probe body (in some expensive probes), the probe interconnect cable, and the oscilloscope environment.

Items to remember when using an oscilloscope probe to measure waveforms or EMI include the following:

1. Keep the probe ground wire as short as possible as this minimizes formation of a loop antenna from picking up unwanted noise.
2. Use an integrated damping resistor internal to the probe as this prevents damage to the oscilloscope as well as minimizes ringing that may be present.
3. Always consider the impact of the probe's impedance on the circuit performance as this prevents waveform degradation and ensures optimal signal integrity of the transmission line.

Any EMI sniffer or E- or H-field probe can be connected to an oscilloscope if 50-Ω termination is applied at the scope's input by either selecting the 50-Ω-termination option or providing an external 50-Ω through terminator to the input connector. Standard high-impedance voltage probes can also be connected to signal and data conductors, *not* AC mains (see warning below). Voltage probes can measure both common- and differential-mode noise within circuits. Depending on the common- and differential-mode impedance of the test environment, making comparisons to actual test levels is difficult unless a CDN (coupling/decoupling network) is used to provide a proper fixed impedance value.

The following diagnostic test requires the use of two probes and a two-channel oscilloscope. For common-mode measurements all conductors in a signal or data cable, including signal returns, should be probed and their channels summed in phase $(A + B)$. Differential-mode measurements on conductors should be probed and their channels subtracted $(A - B)$. The ground leads for the probes should not be connected to anything (left floating). Common-mode measurements tend to suffer from high levels of AC mains frequency. This can be reduced by using high-pass filters with good rejection at the AC mains frequency. Attenuation for higher frequencies to be measured is minimized. Sometimes, ferrite chokes on the probe or mains lead to the oscilloscope can help reduce high-frequency common-mode interference.

When connecting a scope probe to a LISN or CDN output connector or to an AC mains voltage probe, high levels of AC mains frequency can once again be a problem. The high-pass filter mentioned above can be very useful to minimize this low-frequency problem.

Safety warning when using oscilloscopes: The equipment safety earth connection must *NEVER* be removed. Many engineers will isolate the ground plug of the AC mains cord to prevent a ground loop from occurring between the EUT and scope. If a ground loop problem exists, do not disconnect the ground wire; use a ferrite choke or add series impedance into the ground circuit. In addition, isolating transformers, differential amplifiers, and similar devices works well in isolating ground loops. Normal scope probes are not rated for AC mains use. If a scope probe is connected to AC mains, lethal shock hazard can exist in addition to damage of the instrument.

In addition, scope probe ground leads must never be connected to AC mains neutral. Although the neutral is connected to earth at some point in the distribution network, any voltage potential drop across the impedance of the ground wire due to heavy currents flowing along the length of the neutral plus three-phase imbalances will allow the neutral line to develop a voltage potential. This voltage drop across 1 Ω impedance, typical of most power cables, is capable of driving current through the probe ground wire and into the oscilloscope, causing damage. Many installations have voltage on the neutral lines of up to 60 V, which will make the ground wire on the probe burn up in seconds, causing a possible fire hazard.

9.4.2 Using Inexpensive Receivers for Emissions Testing

For low-cost development, diagnosis, and quality assurance, the cheapest and easiest test instrument to use is the standard AM/FM radio. The cheaper the unit, the better the performance. Portable domestic radio and TV receivers can be used to get some idea of how badly a product is doing but only in the broadcast bands in which they are designed to operate.

A disadvantage of using a standard AM/FM/cassette/CD radio is that the output of the radio only provides an indication of the *modulation* present through the speaker. For EMC analysis, we are interested in both the frequency and magnitude of the signal. When we tune in to the harmonic of a digital clock, detection is possible by the way it *squelches* the background noise, or reduces the amplitude of legitimate radio transmissions. If any other signal is heard from clock harmonics, it should be a faint buzz at the mains frequency due to ripple on the clock's DC supply.

Although it is sometimes possible to perform relative comparisons with a basic radio, a signal strength indicator provides significant value. The output of this signal strength indicator indicates the magnitude of the emissions. It is possible to add a field strength meter to a standard radio, if technical competence exists.

Commercial AM/FM radio receivers are available with multiple frequency ranges or a broad spectrum of reception capability. This includes amateur radio, hand-held receivers with continuous coverage from 100 kHz to 2 GHz, signal strength meters, automatic scanners for signals throughout the spectrum, walkie-talkies, and the like, all at reasonable prices. Military-grade radios are also available for minimal cost. A commercial receiver is shown in Figure 9.25 and operates on two AA batteries and covers (with some gaps) the frequency range 100–1.3 GHz. It is shown with both a semirigid coax cable loop and ferrite current probe with RG-141 coax, discussed below.

The FCC restrictions on consumer equipment mandate gaps in spectrum coverage. The commercial radio shown in Figure 9.25, for example, cannot tune (among other frequencies) to the 622.8-MHz oscillator frequency often used in telecommunication equipment. "Unblocked" receivers are available upon proof of use by government and telecommunication authorities. In addition to having almost a semi-spectrum-analyzer display, the single-sideband mode of these radios allows the user to distinguish from among several closely spaced sources by their sound and to eliminate ambient broadcast signals as problems when probing.

Home-made antennas for commercial radios are easy to build. Two types are shown in Figure 9.26. These are small loop antennas. Small loop antennas need not be efficient, and in many cases, the smaller the size, the better. A small shielded loop antenna made from subminiature coax can be used to follow signals on a single trace, even when it lies under other traces. Small loop antennas may be assembled quickly from coax cable. Figure 9.26 includes loop antennas made from semirigid coax and RG-141.

Combining a small shielded loop with a snap-on ferrite bead results in a current probe. This is useful for tracing emissions to a particular wire, cable, or conductor

Figure 9.25 Using a radio receiver for EMI troubleshooting. (Photograph courtesy of Cortland Richmond.)

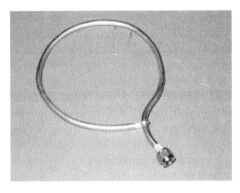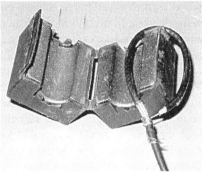

Figure 9.26 Troubleshooting antenna for a radio receiver. (Photographs courtesy of Cortland Richmond.)

as it excludes fields from conductors not passing through the ferrite. Smaller ferrites allow probing of more closely spaced conductors. Larger ferrites can accept more than one conductor, which allows distinguishing common-mode from differential-mode currents. Differential-mode currents cancel if both source and return are passed through the ferrite in the same direction. When source and return cables are passed through the ferrite in opposite directions, common-mode currents will cancel out.

Receivers not specifically designed for EMC analysis will not have quasi-peak or average detectors and their meters will not be linear. Signal strength meters that respond to the peak of the RF signal and have an attack time of under 0.1 ms are extremely useful. To calibrate a signal strength meter, a known transmitter's output power or a previously compliant system can be used to roughly calibrate the meter. What is desired is a comparison between a known good product and a product under investigation. If the reading of the signal strength meter decreases or the volume becomes lower, indications are that a reduction of EMI energy has occurred. This knowledge is valuable during the precompliance stage of development and analysis.

9.4.3 Using Amateur Radio Transmitter for Immunity Testing

There are two types of transmitters available for troubleshooting radiated immunity problems. The first is the hand-held radio transmitter. The field strength of this transmitter is approximated by Eq. (9.5). Note that this generalized equation only provides value for estimating the field strength of a transmitted signal. Many other field parameters are required to accurately solve this equation. Details on this complex equation are beyond the scope of this book:

$$E = \frac{\sqrt{30 \times \text{ERP}}}{D} \qquad \text{(V/m)} \qquad (9.5)$$

where EPR is the effective radiated output power of the transmitter in watts and D is the distance from the transmitter to the antenna in meters. This equation is an *approximation and not accurate* at very close range (i.e., one-sixth of a wavelength at a particular frequency of interest). The most common radiated immunity limit for most commercial or light industrial equipment is between 3 and 10 V/m. With a 1-W power transmitter the field level is calculated to be about 5 V/m at 1 m distance. Remember, the field level will increase greatly when the antenna is closer than one-sixth of a wavelength of the transmitting frequency. Maximum coupling occurs with the radio antenna held parallel to the cable, wire, or trace. For seams or apertures in an enclosure, holding the antenna perpendicular to the seam in the longest dimension will maximize coupling.

Establish a reference level prior to any modification by measuring the distance from the transmitter to the suspected area and noting this dimension. After installing the necessary modification, repeat the test to determine if the modification is effective. Next, determine the distance spacing before a failure happens again. This distance spacing will identify the effective immunity level and other potential items susceptible to an immunity event.

Another type of transmitter that can also be used is a mobile or base station transmitter connected to a dummy load. With a dummy load and *not* an antenna, anyone can "use" this transmitter since it will not be transmitting over the airwaves. This allows the antenna cable to be longer and the cable can then be tape secured for a length of about 1 m to a suspected problem cable. This technique can be utilized also for troubleshooting radiated or conducted immunity problems.

The shortcoming with using amateur radio transmitters is the limited frequency range in which they are available. However, the frequency ranges of these radios are those most commonly used in environments where commercial equipment is installed, and which would present the greatest harm to other electrical systems located nearby.

9.4.4 Radiated Problem Masked as Conducted Emission Problem

Conducted emissions can result from radiated coupling of internal power supply/filter wires or traces from a PCB as well as the power supply/filter components themselves. The radiated noise will appear as conducted emissions. During troubleshooting, if the response to conducted emissions is not filter component changes, one should suspect that something is wrong and that the possibility of radiated coupling around the filter occurred. One may spend hours attempting to locate the presumed source (e.g., rectifying diodes) when there is none to be found internal to the suspected power supply. This same basic concept applies to filter designs.

Instead of disconnecting the entire power supply from the circuit, simply disconnect the output side of the supply and connect an external filter assembly. Any emissions shown in the conducted emission scan are entirely due to radiated coupling.

A separate external power supply can quickly determine this phenomenon, avoid confusion, and save precious time when trying to locate a problem area. The power supply and filter is left installed in the system; however, the output wires from the power supply/filter are disconnected and the AC main is removed from the input of the LISN. A second, separate power supply/filter is then connected to the intended load (Figure 9.27). Conducted emissions scans can then be performed with the disconnected output wires of the original power supply both open and shorted. The LISN will measure any noise on the input wires to the power supply, even when disconnected from AC mains or inactive.

Caution: Verify that the AC input source is disconnected from the LISN prior to the test since the output of the original power supply is to be shorted together, causing a potential shock hazard.

Any noise measured in the resulting scans indicates a radiated coupling problem. This test may also identify potential problems in the filter layout, such as improper location of capacitors, inductive components, or poorly designed internal wiring.

9.4.5 Determining Whether Conducted Emission Noise Is Differential Mode or Common Mode

In evaluating conducted emission problems, one needs to determine whether EMI is differential mode or common mode. From an earlier presentation, there are different filtering techniques depending upon the type of noise current present. Series inductors in line with the signal (high) side of a transmission line and line-to-line capacitors ("X" type) are effective for differential-mode noise currents. For common-mode energy, line-to-ground capacitors ("Y" type) and common-mode inductors are effective. Line-to-ground capacitors require special design considerations as they can cause excessive leakage currents from a fault condition within an

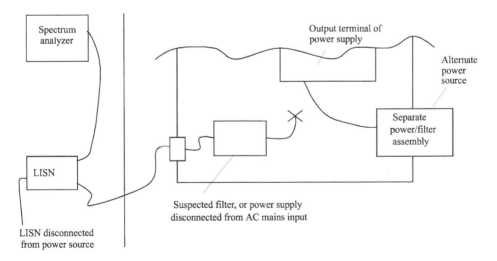

Figure 9.27 Testing a power supply for radiated emissions observed as conducted emissions.

electrical assembly (power supply or PCB) to chassis enclosure causing a shock hazard. Leakage currents are a product safety concern. Internal to common-mode filters, inductors become the main suppression component that provides benefit. A method for determining the type of noise, differential or common mode, is shown in Figure 9.28.

Place the signal or high side of the transmission line through a current clamp and obtain a reference reading. The exact value is not important. After obtaining a reference level, place the return or low-side conductor also inside the current clamp together with the high side. If the level increases by approximately 6 dB, then the noise is predominantly common mode. If the reading decreases to very low levels, the noise is differential mode. This is because common-mode noise adds between transmission lines while differential-mode noise cancels. With the determination of the type of noise present, appropriate filter circuitry can be added or values modified.

9.4.6 Another Use of EFT/B Generator

The imagination is unlimited for the engineer when it comes to finding other uses for EMI test equipment. Hot swapping is prevalent in today's mainframes and

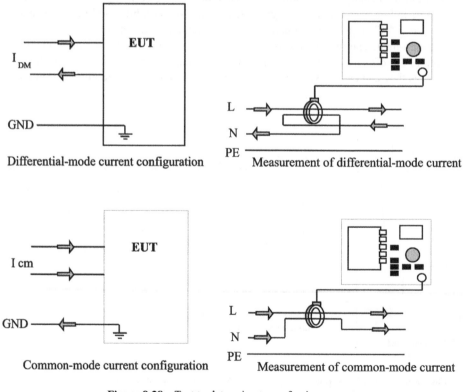

Figure 9.28 Test to determine type of noise current.

cardcage assemblies. It is very common that, when plugging in PCBs with power applied, switching transients are generated onto the power bus structure. These transients can travel along the distribution path and possibly affect other assemblies connected to the power system. The EFT generator performs an immunity-type test to simulate hot swapping. There are no regulatory compliance tests or requirements for this procedure. The reason to perform this test is to improve the quality of the product.

Simply connect the output of the EFT generator to the power bus through a high-voltage capacitor. The value is not critical. For most applications 0.01 μF is sufficient. The amplitude of the transient is left up to the person doing the test—500–1000 V are typical values. Monitor functional operation of all PCBs within the backplane assembly for potential failures due to transient noise spikes.

9.4.7 Signal Integrity Observations

When attempting to determine noise problems, do not overlook observing propagating signals in the time domain using an oscilloscope and/or a spectrum analyzer.

The coupling method and the nature of the waveform are described by the signature profile of the propagating energy within a transmission line. If the RF energy is a derivative of the voltage level present, the coupling mechanism is capacitive. If the noise is a derivative of the current present, the coupling mechanism is inductive.

If it is possible to reverse the direction of the suspected source signal through a trace and the polarity of the noise in the victim trace reverses, the coupling mechanism is inductive. No change in polarity of the noise in the victim circuit implies that the coupling mechanism is capacitive. Ringing observed on an unterminated signal line implies that there is an impedance mismatch with a low-impedance source driver. A monotonic edge implies a mismatch with a high impedance source driver.

Varying the impedance at the load in victim circuitry can also aid in determining the type of coupling mechanism. If increasing load impedance produces a voltage drop in the line, the coupling mechanism is magnetic. If decreasing the load impedance produces a decrease in voltage, coupling is capacitive. Many times, it is more convenient to disconnect the load (open circuit) at the source end. If the voltage across the victim's load decreases, then coupling is magnetic. If it increases, coupling is capacitive.

When deciding on the use of filters on signal lines, the type to select is dependent upon whether it is common mode or differential mode. Determination is based on whether coupling is magnetic (inductive) or electric (capacitive). For capacitive coupling, reducing the high-frequency content of the source spectrum is preferred by adding a bypass capacitor at the source end or at the victim's load end as well as adding a shield to either the source or victim's circuits. If coupling is magnetic, reduce the high-frequency content by increasing impedance in the source circuit (i.e., adding a ferrite bead in series).

9.5 SYSTEM-LEVEL TROUBLESHOOTING

This section provides guidance on various methods of isolating the problem area if a system fails an EMC test. It is important to distinguish between the system setup and external environment before attempting low-level diagnostic troubleshooting.

9.5.1 Switching Power Supplies—Measuring Magnetic Field Coupling

Solving intermittent problems is difficult, resulting in wasted engineering resources. One source of intermittent system-level problem is magnetic field emissions from a switching power supply.* It takes considerable time for a series of 50-kHz spikes to become phase shifted together to create a vulnerable condition causing an intermittent problem to develop during a particular machine state cycle. Often the cause is associated with switching power supply noise, which may be extremely difficult to diagnose. Magnetic storage devices such as tape and disk drives can be especially sensitive to the magnetic fields generated in a switching power supply. This is in addition to small analog signal disruption such as that in optical interconnects and phase-locked loop circuits.

Switching power supplies are capable of generating various types of noise: radiated and/or conducted. Any EMI mode can disrupt system operation (signal corruption), even if operational circuits are not powered from a particular power supply.

The strength of the magnetic fields radiated from a switching power supply can be substantial. Magnetic fields can induce voltage coupling of over 1 V/cm due to $M \, dI/dt$ induction on conductors located near the power supply, which is significant enough to corrupt signals in those conductors. Magnetic fields can emanate from magnetic devices such as switching transformers, inductors, rectifiers, diodes, or other devices that are required to switch high levels of current.

The magnetic field strength emitted from a switching power supply is not correlated with the cost or apparent quality in the design of a particular supply. The inexpensive unit in Figure 9.29 is very quiet EMI-wise, whereas other expensive high-quality supplies can be noisy related to magnetic field disturbance. Some power supply vendors provide a specification on magnetic field emissions in terms of magnetic field strength propagated from the unit. An equivalent approach to determining radiated magnetic field strength is to use a loop antenna, such as a 2-cm round or square probe (details provided in Appendix A). Ensure that the coax cable to the analyzer from the loop probe is approximately 50 Ω. A receiver with a bandwidth of several megahertz is usually adequate for this test. The loop probe is passed over the surface of the supply in addition to all interconnect cables and other circuits. The peak output voltage is recorded by appropriate instrumentation.

Figure 9.30 shows an example of this type of loop probe, used to determine magnetic field emissions from the power supply. The test should be done with a range

*Details of this troubleshooting technique are derived from a *Technical Tidbit* article (February 2001) at www.emcesd.com, courtesy of Doug Smith.

Figure 9.29 Typical small-size switching power supply. (Photograph courtesy of Doug Smith.)

of loads on the output terminals. Due to the efficiency of this particular power supply, along with control circuitry, different radiated emission levels will be noted between no load, partial load, and full load.

One advantage of using this probe and diagnostic technique, instead of relying on vendor information, is that the result of this test provides a quantified analysis on the amount of possible interference that may be picked up by nearby circuits in terms of induced voltage.

Another advantage of using a self-made loop probe is cost versus that of an expensive calibrated field probe, which is not necessary for this level of analysis. Either shielded or unshielded loops can be used. The electric field shielding of loop probes does not work well except for uniform fields, or at least electric fields that are symmetric around the centerline of the loop, a condition not likely to be met for this type of testing.

When using a loop probe, approximately how much energy measured is considered significant? For a 2-cm loop probe, signals that exceed 50 mV are generally a cause for concern. Systems with optical to electrical signal conversion and

Figure 9.30 Measuring magnetic field emissions from a switching power supply. (Photograph courtesy of Doug Smith.)

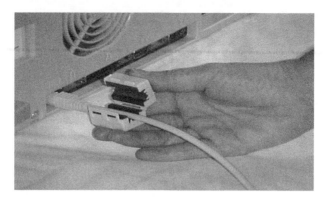

Figure 9.31 Installing a ferrite core on a cable. (Photograph courtesy of Doug Smith.)

low-level analog circuits may require the amplitude level of the magnetic field to be much lower.

9.5.2 Potential Problems When Using Ferrite Cores—Increase in Radiated Emissions

Clamp-on ferrite cores are commonly installed on cables for inserting common-mode impedance (Figure 9.31). This is done to reduce common-mode currents that cause unwanted emissions and potential failure to meet international emission regulations such as CISPR 22 and FCC Part 15.

Ferrite cores may reduce a certain portion of common-mode current on the shield of cables. Of primary interest is reducing the *magnitude* of the common-mode current to levels that are acceptable. It is impossible to remove all undesired radiated energy present. Undesired RF currents can cause unwanted emissions and potential failure to international regulatory requirements.

There are times when a ferrite core can increase radiated emission levels at a particular frequency.* Many EMC engineers have noticed that the emissions at a few frequencies go up with the installation of a ferrite core in some applications. The cores usually used for this purpose are made with a lossy material. The common-mode impedance of the core has a significant resistive component. Normally, one would not expect that a resonance is caused by the core tuning the cable due to the resistive part of its common-mode impedance.

Inserting a resistive common-mode impedance on a cable can indeed cause a tuning/resonance effect to occur along with the potential for increased emissions. One way this can happen is illustrated in Figure 9.32.

In Figure 9.32, two systems are connected by a single interconnect cable. Assume equipment enclosures are metal and sufficiently large to be part of a wave-

*Details of this troubleshooting technique are derived from a *Technical Tidbit* article (December 1999) provided at www.emcesd.com, courtesy of Doug Smith.

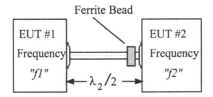

Note: The ferrite bead located on EUT #2 can cause
common-mode current at EUT #1 if "$f1=2f2$. "

Figure 9.32 Two pieces of equipment connected via an interconnect cable.

length at a particular frequency of interest. For instance, a personal computer enclosure meets this criterion at 150 MHz.

Assume both EUT 1 and 2 have internal circuitry such that both can potentially
radiate undesired RF energy. The operating frequency for EUT 1 is f_1 (in megahertz). The operating frequency at EUT 2 is $2(f_1) = f_2$ (also in megahertz). These
units transfer signals between each other. The interconnect cable is a half wavelength for f_2 and a quarter wavelength for f_1.

A half-wavelength wire will propagate a low-impedance signal from one EUT
to the other without difficulty. A quarter-wavelength transmission line will propagate a high-impedance signal from one system to a low-impedance load at the other end of the wire. A quarter-wavelength wire makes a good antenna where the
chassis of the equipment to which the wire is fastened becomes the other half of
the antenna. This is because the quarter-wavelength wire, with its end in free
space, presents a low impedance at the equipment end. This allows high-frequency currents to flow into the wire and radiate. It only takes on the order of 20 μA
to be potential radiated emissions problem. If the far end of the quarter-wavelength wire is connected to a metal chassis instead of ending in free space, the
boundary condition at the end of the wire (free space) is not met and the impedance looking into the wire from the driven end becomes higher. This results in
less common-mode current flowing and the propagating field decreases in amplitude.

The opposite situation occurs for a *half*-wavelength wire. If one end of the wire
is unterminated (open), a high impedance is present at the other end. A low impedance at one end of the cable will cause the impedance presented at the other end to
be lower.

With this discussion, how does a ferrite clamp increase radiated emissions? At
EUT 2, the impedance will be low at frequency f_2 if the enclosure of EUT 1 presents a low-impedance load (likely) at the end of the half-wavelength cable. If EUT
1 presents a low-impedance value to EUT 2 at the end of the *half*-wavelength cable,
current flow from EUT 2 occurs, resulting in radiated emissions. Because of the
low impedance presented to EUT 2, current at frequency f_2 will flow and radiated
emission results.

Looking into the cable from EUT 1, the likely low impedance presented by the enclosure of EUT 2 will be reflected back to EUT 1 as a high impedance because the cable is a quarter of a wavelength at f_1. The source of potential emission in EUT 1 cannot drive current into the high impedance at f_1 so emissions at f_1 will be low.

During testing, emissions are measured coming from EUT 2 at frequency f_2. To solve this problem, several ferrite cores are installed on the cable at EUT 2. By inserting common-mode impedance in the cable at this location, common-mode current at frequency f_2 from EUT 2 is reduced as the energy is sent to EUT 1. However, the installation of the ferrite clamp at EUT 2 raises the common-mode impedance at the EUT 2 end of the cable when observed from EUT 1. Because the cable is a quarter wavelength at f_1, the lower common-mode impedance at EUT 1 allows for greater drive current into the cable. This additional current results in increased radiated emission coming from EUT 1 at frequency f_1. Therefore, a new problem has developed that was not there before!

In this situation, putting additional common-mode resistance at the EUT 2 end of the cable caused the interconnect cable to be an efficient antenna at EUT 1 by isolating the enclosure EUT 2 from the end of the cable.

9.5.3 Measuring Shielding Effectiveness of Materials and Enclosures

When relying on shielding, one must determine optimal material to use. For enclosures, conductive paint, finger stock, wire mesh, cloth over foam, conductive elastomers, foil laminates, and copper tape are only a sample of the material available. Depending on application, a particular product is selected for a specific use. Two big questions to ask are "How much shielding do I need in decibels" and "how much attenuation occurs as a result of the shielding material?" If the material is poorly chosen or incorrectly applied, no benefit will be observed at a higher manufacturing cost. When choosing shielding material, it becomes important to validate datasheet specification from vendor values to ascertain the quality of the material chosen for the enclosure design.

A simple diagnostic test to determine shielding effectiveness of materials and enclosures is available to the engineer.* Figure 9.33 shows two shielded magnetic loops placed on opposite sides of a sample of shielding material. In this figure, the material is conductive cloth. One loop probe is driven by a signal source. The other loop is connected to a measuring instrument such as an oscilloscope or spectrum analyzer. The magnetic flux that penetrates the shielding material and passes through the second loop generates a signal in that loop. The output of the second loop gives a measure of the shielding effectiveness of the material to an incident magnetic field perpendicular to the shielding material.

When testing the shielding effectiveness of metal enclosures, Figure 9.34 illustrates how this is achieved. IEEE STD 299 specifies the "acceptable" manner of shielding effectiveness of an enclosure. Loop probes are moved around the enclo-

---

*Details of this troubleshooting technique are derived from a *Technical Tidbit* article (August 2000) provided at www.emcesd.com, courtesy of Doug Smith.

Figure 9.33 Measuring shielding effectiveness. (Photograph courtesy of Doug Smith.)

sure 360° from top to bottom and side to side. Careful examination of leakage at seams and I/O cable interconnects must be made, as this is where the majority of enclosure leaks occur, in addition to the door assembly.

The shielding effectiveness value for the magnetic *H*-field is not necessarily equal to the shielding effectiveness for electric *E*-field. Shielding effectiveness measurements must be done separately for each field, depending on frequency.

Normally the shielding effectiveness of a material or enclosure is determined by measuring plane waves in the far field (at a significant distance from the radi-

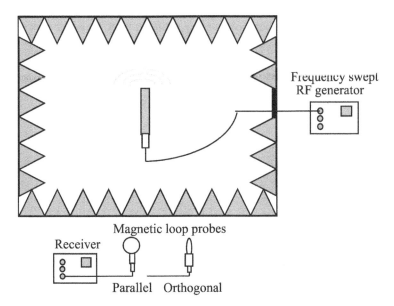

Figure 9.34 Measuring shielding effectiveness of an enclosure. (Remove any anechoic material to represent a plain metal enclosure.)

ating source). It is not easy to extrapolate near-field measured data to that of the far field, which is where most of the undesired energy exists. In this case, the magnetic field of the plane wave is parallel to the shielding material. Measurements using this technique are often expensive or difficult to perform in a development laboratory. In addition, if a product has one compartment shielded within the assembly that is separated from other compartments, such as the RF section of a cell phone, the shielding material may not be far enough away physically from the radiating source for the plane-wave assumption to be valid. For this condition, the shielding material is illuminated with a magnetic field from currents flowing in nearby circuitry. Under these conditions, the measurement pictured in Figure 9.33 is a far more accurate indicator of the real shielding effectiveness of the material.

Figure 9.35 shows two loops positioned close to each other (Figure 9.33 without the shield material shown). The loops should be of the shielded magnetic version (Appendix A), positioned just far enough apart for the proposed shielding material to be inserted. This sets up an environment for making a baseline reading of shielding effectiveness.

Loop 1 is excited from a driven source, generating a magnetic field. Two magnetic field lines are illustrated in Figure 9.35. The magnetic flux that passes through the second loop generates a voltage according to Faraday's law. Usually the frequency of the source is swept and the output plotted with respect to frequency. This is easy to do using a spectrum analyzer with a tracking generator. Loop 1 is connected to the output of the tracking generator and loop 2 is connected to the spectrum analyzer input.

Shielding material is now placed between the two loops (Figure 9.36). It is important to maintain the initial loop spacing of Figure 9.35. Magnetic flux will try to

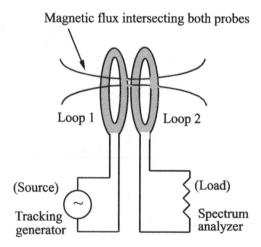

Figure 9.35 Transmission model between two magnetic loop antennas.

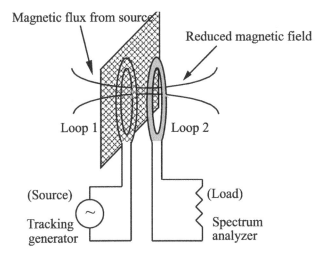

Figure 9.36 Transmission model between two magnetic loop antennas with shielding provided.

penetrate the shield, causing eddy currents to flow around the material's edges. These eddy currents cancel parts of the magnetic field. This cancellation produces reduced magnetic field intensity within the test environment. The output voltage of the second loop is plotted with respect to frequency to show how well the shielding material works over the frequency range of interest. Smaller loops work at higher frequencies and can resolve differences of shielding effectiveness over smaller distances. The results of this test are often surprising.

Shielding material rated in the 60–100-dB range for plane-wave illumination often has much lower shielding effectiveness, sometimes in the range of 10–20 dB. If the material is to be used within a small product, specifying plane-wave shielding effectiveness may not be a good choice. Measuring magnetic field shielding effectiveness will closely relate actual performance when used within a real product. For most shielding materials, knowing the application and how it relates to design specifications is important if a desired level of circuit performance is to be achieved.

9.5.4 Measuring Effects of High-Frequency Noise Currents in Equipment

A square loop probe made from stiff wire (such as a bent paper clip) can be used as a "current probe" to trace currents flowing inside equipment and the resulting voltage drops due to EMI that may be present on the outside portion of a metal enclosure, such as an ESD event or a strong radiated field from a transmitter. The key to successful use of the probe is to reduce unwanted coupling of the undesired interference into the loop probe and to prevent the presence of the probe from affecting

Figure 9.37 Loop probe used to sense current on a PCB. (Photograph courtesy of Doug Smith.)

the EUT during the measurement process. Figure 9.37 shows a typical setup using a personal computer as an example.*

A BNC feed-through barrel connector is installed onto a metal adapter bracket of a personal computer. To ensure high levels of shielding effectiveness (prevent RF leaks from occurring between the bracket and chassis), copper tape is used to seal the opening between bracket and enclosure. Skin effect will then keep currents induced on the loop cable on the outside of the enclosure.

Between the barrel connector and the loop probe, a short piece of coax is used. The probe is then moved about the box or circuit boards to locate a source of radiated emissions. A loop probe is highly directional with a sharp *null* in output voltage when RF current is perpendicular to the probe. When RF current is parallel to the probe, a *peak* in output current occurs. Depending on the probe orientation, one is able to locate problem areas quickly.

To verify accuracy of the loop output voltage, a null measurement needs to be performed [7]. Several examples of performing a null measurement are as follows:

1. Suspend the loop probe a short distance above the area being measured. Due to mutual inductance between the probe and circuit, the output voltage will drop quickly as the probe is moved away from the radiating source. This procedure allows one to determine if the fields being measured are leaking through slots and seams in the enclosure. As long as the output is much smaller than when the loop is at a distance from the suspected radiating source, one can assume the source of the radiating energy has been detected.

2. When the loop probe output is large when suspended away from a suspected circuit or area, including the enclosure, fields internal to the system may themselves be a problem. An output of 5 V under this condition means other

---

*Details of this troubleshooting technique are derived from a *Technical Tidbit* article (November 2000) provided at www.emcesd.com, courtesy of Doug Smith.

conductors inside the enclosure are picking up similar voltages and are now subject to RF corruption.

3. Replace the loop probe with a 50-Ω terminator at the end of the internal coaxial cable. Any voltage that is measured with this terminator represents measurement error of the test circuit and instrumentation.

Figure 9.38 illustrates the output voltage of the loop probe shown in Figure 9.37 when an ESD event is applied to the outside of the enclosure. The loop output represents a lower bound for the voltage drop on the ground plane between the two corners of the square probe.

The loop output voltage attained a value of almost 4 V peak. Since the loop is approximately 2.5 cm on a side, the voltage drop on the board is at least 4 V peak over a distance of 2.5 cm! An individual device would not see this much voltage per 2.5 cm of signal path because of mutual inductance between paths and the nearby ground/power plane. However, a 4-V drop along 2.5 cm of a printed wiring board trace is not to be ignored.

Figure 9.39 illustrates the loop output from the same location when an improved grounding change was implemented on the PCB. The peak voltage induced in the loop is approximately 1.6 V, a significant reduction. With improved bonding in a ground circuit, the reduction in ground bounce should not permit the ESD event to disrupt operation of the PCB.

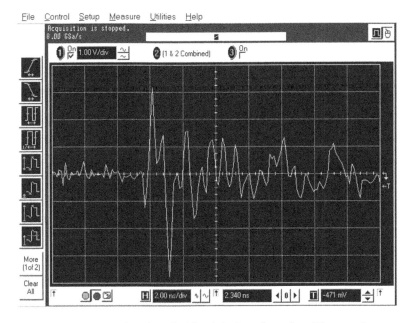

Figure 9.38 Loop output from board with original configuration. (Photograph courtesy of Doug Smith.)

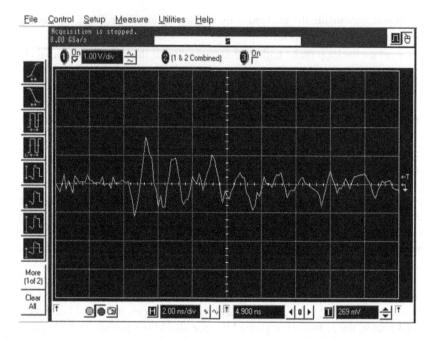

Figure 9.39   Loop output from board with modified grounding arrangement. (Photograph courtesy of Doug Smith.)

From these examples, one can conclude that not only is it possible to make measurements inside equipment subjected to severe interference, such as ESD on the outside, but also it can be done with a relatively simple measurement technique.

9.5.5   Measuring Noise Voltage across Seams in Enclosures

Seams and slots in equipment enclosures can result in emissions leakage that affects radiated emissions compliance.* Often EMC personnel will use magnetic loops to find such leakage. This technique covers an alternate way to investigate radiation from seams and slots using a voltage measurement procedure that can be roughly correlated to the emissions.

Figure 9.40 shows a commercial high-bandwidth 1.8-GHz differential probe used to measure the voltage across a seam in a chassis. This probe is detailed in Chapter 5.

With proper interpretation, this measurement can give an estimation of the emissions potential of this seam. If the dimensions of the seam and the surrounding metal are a substantial fraction of a wavelength at a frequency of interest, one could consider the metal as an antenna that will be driven by the voltage across the seam.

*Details of this troubleshooting technique are derived from a *Technical Tidbit* article (November 2002) provided at www.emcesd.com, courtesy of Doug Smith.

Figure 9.40 Voltage measurement across seam using a differential probe. (Photograph courtesy of Doug Smith.)

A dipole that is tuned to a half wavelength at a frequency of interest has a low driving point input impedance (~70 Ω). Under this condition, it takes only about 15 μA of current to cause a potential Class A (industrial) emissions compliance problem. The voltage input to the dipole necessary to drive a current of 15 μA into 70 Ω is only about 1 mV.

The metal may have a much broader spectrum over which it can radiate than a dipole. However, if we use the same numbers as for a dipole (we are only looking for an estimate), then a voltage across the seam on the order of a few millivolts (due to currents or fields inside the enclosure) might be a problem for emissions. It is this current that results in radiation in the far field where emissions are measured. The question is whether a voltage across the seam will result in a current that will, in turn, flow on the outside surface of the metal enclosure that is sufficient to result in emission problems.

There are a few points to consider:

1. Is the measured voltage due to a source internal to the equipment? Turn off the equipment to find out the source of emissions. This is identified as a null measurement.

2. Is the measured voltage due to voltage across the seam, as opposed to the common-mode response that is measured by the probe? Short the probe tips together and then to each side of the seam, one side at a time. This is another null measurement. The measured results should be much lower than the measured voltage across the slit. The result of this null experiment, shorting the probe tips, will also indicate if a magnetic field in the area is inducing a voltage into the probe tip loop in accordance with Faraday's law. In either case, if the shorted probe response is similar to the measured voltage across the slot, or seam, it may not be possible to make this voltage measurement.

3. If the measured voltage is really across the seam, the final question to answer is, Does that voltage result in current flow? Without current flow, there would not likely be significant emission. For an enclosure whose metal is a substantial fraction of a wavelength, the answer is probably yes, there will be current flow and emission. One way to check is to put a 75-Ω resistor across the seam. If the impedance across the seam is low, the voltage will not change appreciably, perhaps only a few decibels. If the impedance is high, the addition of the resistor will change the result substantially. This could happen if standing waves make the seam a high-impedance node. A low impedance across the seam can be due to the metal acting as a low-impedance antenna, thus permitting RF current to flow. This is not the only possibility. A low impedance across the seam coupled with a few millivolts of voltage across the seam probably means that there will be significant emission potential.

Caution: To get the best results from the measurement discussed herein, the differential probe should have an input impedance of at least a few hundred ohms at the frequency of interest. Many active differential probes have input impedance too low, such as 20 Ω, when used with small probe tips.

Figure 9.41 shows an oscilloscope measurement across a slot on an operating piece of equipment. Given the scope scale of 10 mV (full scale), any signal that can be displayed would represent a possible compliance problem. There is a strong frequency component at about 500 MHz having a peak value of about 4 mV. There is also a smaller component at about 4 GHz. Given the scope frequency response was on the order of 1.5 GHz, the 4-GHz component is really much larger than shown and did, in fact, cause an emissions problem. Filling the slot with an EMI gasket reduced the scope reading to "flat line," indicating that the emissions due to the slot would be significantly reduced.

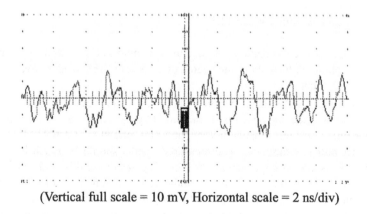

(Vertical full scale = 10 mV, Horizontal scale = 2 ns/div)

Figure 9.41   Voltage measured across a slot in an equipment enclosure. (Photograph courtesy of Doug Smith.)

Although not a substitute for an emissions test, a measurement of this nature as described may increase the confidence factor of passing an emissions test while the equipment is still in the development laboratory. Laboratory bench tests such as this one can lower development costs and speed equipment to final approval.

9.6 AMBIENT CANCELLATION OR SUPPRESSION

Ambient cancellation means the *process* of removing ambient signals by numerical methods. It is a technique for suppressing ambient noise from a desired measurement point. This is accomplished based upon an interference cancellation technique. Two channels or inputs are measured simultaneously in time and frequency. One input is the desired signal plus the interfering signal(s) or ambient(s). The second input is a *reference* antenna that measures only the interfering signal. This reference signal is then used to suppress, or "cancel," the undesired signal from the desired signal (Figure 9.42).

A primary antenna measures both ambient and EUT emissions. A reference antenna measures ambient signals and a much lower level of EUT emissions. This is due to the fact that the reference antenna is much further away from the EUT and the primary antenna. See Section 3.4 for details. The two time/frequency-synchronized signals are then processed to "subtract" out ambient signals from EUT signals. The feedback loop then compensates for (1) phase, (2) amplitude, and (3) multipath. The positioning of the two antennas is critical to the optimization and accuracy of the measurement. The reference antenna should measure –20 dB lower than any EUT signals observed on the primary antenna. This translates into the reference antenna being a distance ratio of about 10-fold further away the selected measuring distance.

If the measuring distance is 10 m from the EUT, the reference antenna must be at least 100 m away. The functional requirements for ambient cancellation are as follows.

1. Ambient reference measurements are needed.
2. The reference antenna must be 10X distance ratio to yield a maximum of 0.9 dB error in measurement.
3. Both amplitude and phase compensation are needed.

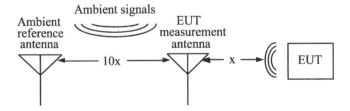

Figure 9.42 Ambient cancellation setup.

4. High level of accuracy is required even for the most modest of cancellation—amplitude accuracy of 0.1 dB and phase accuracy of 5° result in 20–25 dB cancellation.

5. Independent amplitude and phase compensation.

6. Multipath effects require additional degrees of freedom.

7. Fast response time of compensation circuitry.

Ambient suppression is different from ambient cancellation. When using a spectrum analyzer, one may test a product that is fully functional and then power down the EUT to ascertain if any of the measured signals are ambient. When doing a broadband sweep over a large frequency range, one channel can be stored in memory and another channel used for the power-down condition. Using a display feature of the spectrum analyzer, one can subtract the two scans $(A - B)$. The resulting plot is assumed only from the EUT. This technique is not without its deficiency. The ambient signal could vary in amplitude or frequency or even be transitory in nature. It can be different from one scan to another and even between different sweeps.

Equipment-under-test signals that are masked by stronger ambient signals can be measured when ambient noise is canceled. Suppression of ambient signals deals with amplitude reduction. Even if the amplitude of a signal is decreased, the magnitude may still be greater than the desired signal. Suppression may only work at certain times and may not provide the benefit one desires when scanning the frequency spectrum.

One product that performs ambient cancellation is shown in Figure 9.43. The manner in which information is measured is illustrated in Figure 9.44.

Interference technology does work and has sound principles behind the theory. It is impossible to guarantee that *all* ambient signals can be canceled. Only a shielded chamber can guarantee this condition. The cancellation capability is strongly de-

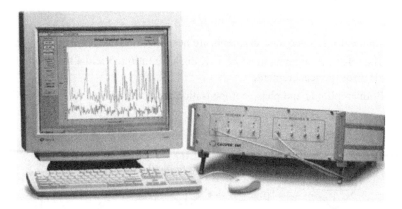

Figure 9.43 Commercially available noise cancellation system (CASSPER). CASSPER (Correlation Analyzer for Source Separation of Patterned EM Radiation) is a trade name of Scientific Applications & Research Associates (SARA) Inc. (Reprinted with permission of SARA Inc.)

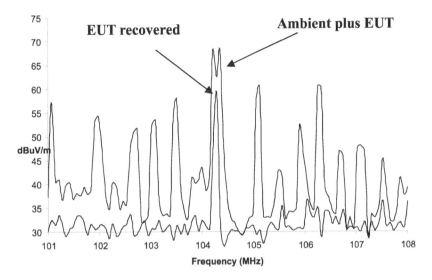

Figure 9.44 Diagram showing ambient cancellation.

pendent upon the quality of the RF inputs. For example, placing the reference antenna in the corner of a metal building will yield satisfactory results as if locating at an OATS. The two antennas should be of the same type and orientation and be pointed in the same direction. The system attempts to cancel or suppress signals common to both antennas. The overall amount of cancellation is also dependent upon signal strength. Cancellation increases with stronger ambient signals because the noise floor presents the lowest possible suppressed level.

Ambient cancellation technology uses two input ports to determine shielding effectiveness or transfer function measurements. Figure 9.45 shows a comparison between data taken with a true network analyzer and an ambient cancellation system. Another application for ambient cancellation is measuring conducted emission on both AC power input lines simultaneously. With some signal manipulation, one can break the conducted emission into separate common- and differential-mode components. This helps tremendously in power line filtering design and diagnostic work.

9.7 PRINTED CIRCUIT BOARD DIAGNOSTIC SCANNERS

Another diagnostic tool for PCB diagnostic work is a scanner containing a planar array of tiny near-field current probes arranged in a grid on a flatbed top interfaced to a personal computer and spectrum analyzer. The grid array of probes are embedded in a multilayer PCB and covered by a protective surface to prevent damage when a PCB under test is placed on the platform. The output of each current probe can be turned on or off by software, with resulting field strengths sent to a spectrum analyzer for a visual view of the field strength intensity of an entire PCB at the

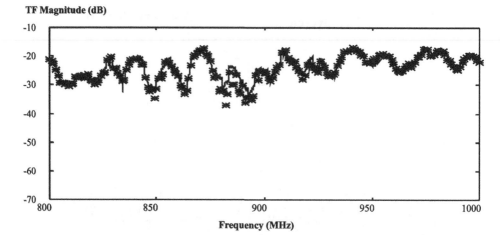

TF Magnitude (dB)

Frequency (MHz)

Figure 9.45 CASSPER in network analyzer mode (solid line) versus a commercially manufactured network analyzer (represented by an ×).

same time. The design and positioning of the individual loops enable them to read both the *x* and *y* currents. Each loop is addressed and measured. The computer controls the scanning and then stores all data in a digital format, performing data manipulation prior to displaying the data. The PCB under test is placed on top of this platform and powered up. Under software control each tiny near-field current loop can be selected and its output measured. A graphical display of the grid array is displayed on a monitor. One commercial system is shown in Figure 9.46.

The system shown in Figure 9.46 is able to provide a two-dimensional picture of electromagnetic energy radiated by circulating currents on a PCB. It provides a frequency-versus-amplitude plot at any specified location or an *x–y* coordinate versus near-field RF energy at a particular frequency. With this system, a designer can implement board fixes and immediately observe its impact, if any. This system can be an effective quality assurance tool by using it to evaluate production samples and comparing them to a known good board standard. Another useful aspect of this system is the ability to compare substitute components without the costly measure of going to an OATS or semianechoic facility for requalification purposes.

One shortcoming of this tool is that it only measures RF noise generated from the bottom of the PCB. This is reflective of the differential-mode current component. It cannot be used to measure noise being radiated due to an assembly of interconnecting PCBs or cables or through its relationship to the distributed impedances with respect to the chassis. Differential-mode current can mask a generation of common-mode currents by the higher level of differential-mode current in the near field. Common-mode radiation is primarily associated with current flow in attached cables to the PCB and must be measured by other techniques. This is not to say that measurement of a PCB is an essential step in diagnosing potential EMI problems. It is only *one* step in the process of diagnosis and troubleshooting EMI in a PCB.

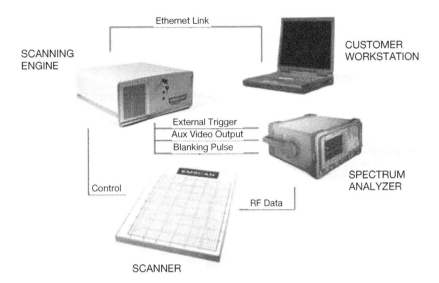

Figure 9.46   A PCB scanner—Emscan. (Photographs courtesy of Emscan Corporation.)

The output from the scanner is observed with a spectrum analyzer. Images of the electromagnetic field serve as a useful diagnostic tool. They enable the user to visually see where potential emissions are located. Measurements are performed in the reactive near-field region of the PCB. The energy is inductively coupled to the loop probes. High-current, low-voltage sources are associated mainly with magnetic fields. The low-impedance designs associated with most of today's logic families (i.e., 50 Ω or less) appear as high-current, low-voltage devices. Therefore, the small magnetic loop probes are ideal for measuring reactive magnetic near fields. The close proximity of the scanner to the PCB makes this measurement technique slightly invasive in nature.

Correlation of near-field scanner results to far-field distance can be accomplished by building a calibration fixture. Essentially, this fixture is comprised of an isolated trace with 50 Ω characteristic impedance between two connectors. With this fixture, one can obtain output data for a set level of current versus frequency. With knowledge of the current within the transmission line, one can then use Eq. (9.6) [12] to calculate the far-field radiation field strength. An example of both spectral and spatial plots is shown in Figure 9.47.

$$E = 263 \times 10^{-16}(f^2 A I_s)\left(\frac{1}{r}\right) \quad \text{(V/m)} \quad (9.6)$$

where A = loop area (m^2)
   f = frequency (Hz)
   I_s = source current (A)
   r = distance from radiating element to receiving antenna (m)

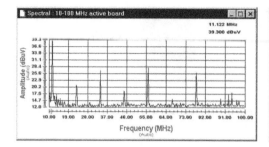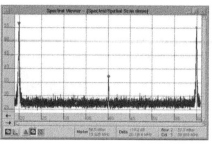

Spectral plots

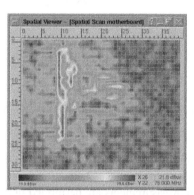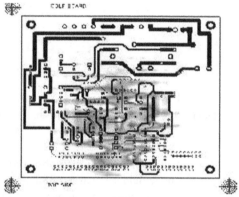

Spatial view

Figure 9.47 Typical plots from bottom side of PCB scanner. (Photograph and plots courtesy of Emscan Corporation.)

Figure 9.48 EMC scanner for top-sided electromagnetic field measurements. (Photograph courtesy of Detectus AB.)

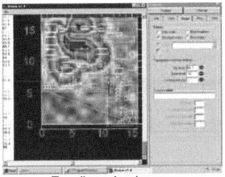
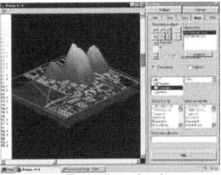

Two dimensional map   Three dimensional map

Figure 9.49   Typical plots from top side of PCB scanner. (Plots courtesy of Detectus AB.)

Another type of PCB scanner provides for the ability to measure emissions off the top a PCB (Figure 9.48). This is in contrast to the system mentioned above, which only measures emissions from the bottom side of the PCB. Similar to top-scanning systems, use of custom software allows one to view images in three dimensions (Figure 9.49). Use of a color spatial display allows observation on where a problem area might be located. It does not locate the actual source generator; however, the probability of determining the cause of radiated energy is simplified since most PCBs have components located on the top side. The field observed is electromagnetic. One is still unable to ascertain if the propagating field is electric or magnetic.

REFERENCES

1. Gerke, D. and W. Kimmel. 1995. *Electromagnetic Compatibility in Medical Equipment.* Piscataway, NJ: IEEE and Buffalo Grove, IL: Interpharm.

2. Montrose, M. I. 1999. *EMC and the Printed Circuit Board Design—Design, Theory and Layout Made Simple.* Piscataway, NJ: IEEE.

3. Montrose, M. I. 2000. *Printed Circuit Board Design Techniques for EMC Compliance— A Handbook for Designers.* Piscataway, NJ: IEEE.

4. Williams, T., and K. Armstrong. 2001. EMC Testing Part 1—Radiated Emissions. *EMC Compliance Journal,* February. Available: www.compliance-club.com.

5. Moriarty, F. 1988. EMI/RFI Diagnostic Tweezer Probe: A Construction Article. *EMC Technology* (Don White Publication), November/December.

6. Carsten, B. 1998. Sniffer Probe Locates Sources of EMI. *EDN Magazine,* June 4.

7. Smith, D. 1993. *High Frequency Measurements and Noise in Electronic Circuits.* New York: Van Nostrand Reinhold.

8. Agilent Technologies. *Making Radiated and Conducted Compliance Measurements with EMI Receivers.* Application Note 1302. Agilent Technologies (previously Hewlett Packard).

9. Mardiguian, M. 2000. *EMI Troubleshooting Techniques.* New York: McGraw-Hill.

10. Mardiguian, M. 1992. *Controlling Radiated Emissions by Design.* New York: Van Nostrand Reinhold

11. Mardiguian, M., and D. White. 1988. *Electromagnetic Shielding,* Vol. III. Gainesville, FL: emf-emi Control.

12. Ott, H. 1988. *Noise Reduction Techniques in Electronic Systems,* 2nd ed. New York: Wiley Interscience.

13. Paul, C. R. 1992. *Introduction to Electromagnetic Compatibility.* New York: Wiley.

14. Tsaliovich, A. 1999. *Electromagnetic Shielding Handbook for Wired and Wireless EMC Applications.* Boston: Kluwer Academic.

15. Williams, T. 1996. *EMC for Product Designers,* 2nd ed. Oxford: Newnes.

16. Williams, T., and K. Armstrong. 2000. *EMC for Systems and Installations.* Oxford: Newnes.

17. Williams, T., and K. Armstrong. 2001. EMC Testing Part 2—Conducted Emissions, *EMC Compliance Journal,* April. Available: www.compliance-club.com.

18. Williams, T., and K. Armstrong. 2001. EMC Testing Part 3—Fast Transient Burst, Surge, Electrostatic Discharge. *EMC Compliance Journal,* June. Available: www.compliance-club.com.

APPENDIX A

BUILDING PROBES

MAGNETIC FIELD PROBE: PAPER CLIP–WIRE LOOP*

Often, a simple probe can be used to successfully track down EMI problems in a circuit. A magnetic field probe is built from a paper clip or rigid wire. In addition to making voltage measurements in circuits, a paper clip probe can also be used to measure the relative phase or time delay between two signals. The application of this technique to impulsive signals has been made practical with the advent of relatively inexpensive, portable, and fast digitizing scopes. By measuring time delay, the location of a noise source can often be pinpointed in the circuit.

Figure A.1 shows the construction of a paper clip probe. Two minutes, some heat-shrink tubing, and a pair of needle-nose pliers are all that is needed. A BNC barrel adapter makes a convenient mounting for the loop. Connect one end of the wire or paper clip to the center pin of the connector and the other end of the wire or paper clip to the barrel with solder or other method of ensuring a solid bond connection. Being able to construct a probe quickly from available parts can be a real advantage when time is important during an investigation.

To measure radiated magnetic fields, a loop antenna is required.

A paper clip (or any stiff wire) is bent into a square shape and covered with insulation, forming an unshielded magnetic loop probe. Such a structure can be quite

*Details of this probe are derived from a *Technical Tidbit* article (August 1999, April 2000) provided at www.emcesd.com, courtesy of Doug Smith.

Testing for EMC Compliance. By Mark I. Montrose and Edward M. Nakauchi
ISBN 0-471-43308-X © 2004 Institute of Electrical and Electronics Engineers

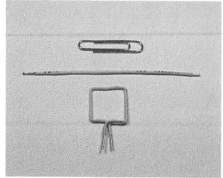

Raw material (paper clip and wire, bent into a loop)

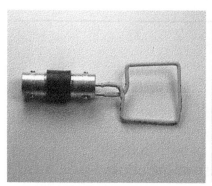

Bent wire installed on connector   Paper clip installed on connector

Figure A.1 Simple magnetic loop probe made from a paperclip or solid gauge wire.

useful in circuit troubleshooting and noise investigations. A conductor carrying current I will have a voltage drop across it given by

$$E_1 = L\frac{dI}{dt} + RI$$

where E_1 = voltage drop per unit length.
   L = inductance per unit length of conductor
   I – current within the transmission line being measured
   t = time period of observation
   R = resistance per unit length of conductor

The unit length of the conductor is physically small compared to typical wavelengths for most frequencies of interest. The difficult part about solving the above equation is knowing the real or actual value of inductance and resistance.

At frequencies above a few hundreds kilohertz, the resistive component becomes negligible and can be ignored. The equation simplifies to

$$E_1 = L\frac{dI}{dt}$$

A conductor that is physically nearby the current-carrying conductor, such as a side of the square loop above, will pick up an open circuit voltage of

$$E_2 = M\frac{dI}{dt}$$

where M is the mutual inductance between the current-carrying wire and the nearby wire per unit length.

Since L and M are constants, E_1 and E_2 will differ only by the value of the constant within the equation. The value of M must be smaller than L for two parallel conductors (due to magnetic flux that flows between the wires instead of enclosing both). Voltage E_1 is a lower bound estimate for the magnitude of E_2, which has the same wave shape.

Using this principle, a simple square loop antenna can estimate the voltage drop across the conductors. When the probe is held up to a conductor carrying high-frequency RF current, the probe's open-circuit output voltage is a lower bound for the voltage between the corners of the probe along the current-carrying conductor.

The loop probe must be connected to an oscilloscope or spectrum analyzer using a coaxial cable terminated in the characteristic impedance of the receiver. This resistive load on the loop in combination with the self-inductance of the loop forms a low-pass filter on the probe output. For a loop with sides of 1 cm, this corner frequency will be between 200 and 300 MHz.

SHIELDED MAGNETIC LOOP PROBE*

Magnetic loops are useful for many measurements, including the voltage drop across conductors and planes, current flowing in conductors, and magnetic fields within transmission lines. One such loop is shown in Figure A.2. Usually, shielded loops are made from semirigid coaxial cable (Figure A.3). The loop is formed by making a circle or square from semirigid coax with a gap placed symmetrically in the middle of the loop. The position of the gap is very important to the performance of the shield. If the gap is not in the middle of the loop, shielding effectiveness is compromised. Square loops are useful for making measurements of circuit voltage and current while either square or round loops are suitable for measuring magnetic fields in free space.

---

*Details of this probe are derived from a *Technical Tidbit* article (December 2000) provided at www. emcesd.com, courtesy of Doug Smith.

Figure A.2   Home-made shielded loop probe.

One problem in making a shielded probe in the laboratory is the availability of semirigid coaxial cable. Figures A.4–A.9 show construction steps for quickly making a shielded magnetic loop probe.

The starting material (Figure A.4) is a piece of stiff copper wire or rod. The best gauge is 16 AWG (American Wire Gauge) for fitting into a BNC connector. The wire is first covered with heat-shrink tubing with the tubing heat treated. Copper tape is then wound around the heat-shrink tubing with three layers of tape. The reason for using three layers is shown in Figure A.5. When the wire is bent to form a square loop, the outermost skin layer may crack slightly. Such a small crack is noticeable in Figure A.5. Since the crack is only on the outside of the bend and is small, it does not pose much of a problem. If three layers of copper tape are used, the underlying layers seal the gap. This is not as much of a concern for a round loop probe as it is for a square loop, owing to the relatively sharp corners of the square loop. The characteristic impedance of the coax formed by the foil and stiff wire is probably not 50 Ω, which is the typical value of the interconnecting coaxial cable. It does not matter much since the parasitic transmission lines formed by the two halves of the loop and their shields are unterminated at the gap.

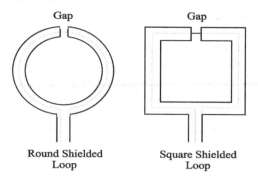

Figure A.3   Construction of shielded loop probes.

Figure A.4 Wire with heat-shrink tubing and copper foil tape added.

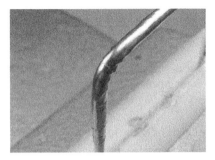

Figure A.5 Detail of bend in a corner. Three layers of tape are used to seal the crack on the outer portion of the wire.

Figure A.6 Second layer of heat-shrink tubing added over foil.

Figure A.7 Loop installed in BNC connector. Shield and wire are soldered to shell.

Figure A.8 Copper tape and solder connects two ends of loop foil together.

Figure A.9   Another layer of heat-shrink tubing finishes off the loop.

The next step is to put a second layer of heat-shrink tubing over the assembly (Figure A.6). The loop is then bent and one end is inserted into the center pin of the BNC connector (Figure A.7). *Both* the other end of the wire and copper tape are soldered to the side of the BNC connector per Figure A.7. Next, the end barrel of the BNC connector and two ends of the loop are covered in copper tape, which is soldered to the copper foil of the loop (Figure A.8). The added copper tape may also be soldered to the BNC connector as well. Finally, the assembly is covered with heat-shrink tubing to make the final product (Figure A.9).

The loop's shielding effectiveness may be tested by applying an electric field source to the loop. The loop should be least sensitive over the gap with the maximum sensitivity just off the gap in either direction. The sensitivity should gradually fall off as the electric field source is moved toward the side of the loop (at the BNC connector) opposite the gap.

The performance of a loop constructed in this fashion should compare favorably with one made from semirigid coaxial cable up to the first resonance of the loop. The first resonance occurs at the frequency where the circumference of the loop is one-half wavelength. At this frequency, the parasitic transmission lines formed by the two shields and the underlying stiff wire are one-quarter wavelength unterminated stubs.

Difference between Shielded and Unshielded Loop Probes.*   Different applications exist for shielded and unshielded loop probes. The gap in the shield prevents shield currents from flowing around the loop. Electric field shielding of the center conductor is still present. If electric fields were present in the shield, these fields would cancel out the incident magnetic field to a large extent. The voltage present would not be developed correctly in the center conductor in response to a magnetic field passing through the center of the loop.

The uniformity of the fields affects the quality of the electric field shield, with minimal effect for the magnetic component of the plane wave present. When this type of probe is held directly against a circuit to measure inductive drops, the fields measured may not be uniform. If the electric field couples onto one side of the loop probe with different amplitude than the other side of the loop relative to the gap, the

*From D. Smith, Signal and Noise Measurement Techniques Using Magnetic Field Probes, in *Proceedings of the IEEE International Symposium on Electromagnetic Compatibility*, IEEE, New York, 1999, pp. 559–563.

currents measured will be different on both sides of the gap. This will result in different inductive drops being generated in the shield on opposite sides of the gap. Because the mutual inductance between the shield and center conductor is equal to shield inductance, unequal voltages will be inducted into the center conductor and the electric field shield will not be effective.

For effective electric field shielding, it is necessary for the shielding to be symmetrical around the gap in the probe shield. For most circuit-level measurements, an unshielded loop probes works best. An unshielded loop probe has the advantage that the loop can be positioned closer to the circuit being measured with higher sensitivity without distorting both magnetic and electric fields present when measurements are made.

ELECTRIC FIELD PROBE: DC TO 1 GHz WITH 50-Ω TERMINATION*

The probe detailed herein is designed to operate from DC to 1 GHz and is as effective as probes commercially sold.

It is necessary to implement 50-Ω termination on the end of a female BNC connector for impedance matching to test instrumentation. Resistors are to be soldered to the rim of the BNC connector to create this termination. This can be a difficult and tedious process.

Step 1. (a) Incorporate 50-Ω termination using four 1206 surface-mount 200-Ω 1% resistors soldered into the BNC barrel adapter as shown:

Step 1a   50-ohm termination.

(b) To simplify incorporation of the resistors into the BNC connector, a fiber washer is used. This washer can be purchased at electronic or hardware supply stores. This washer fits neatly into the female BNC connector. The photograph be-

---

*Details of this probe are derived from an article (DC-1 GHz Oscilloscope Probe Plans) and a *Technical Tidbit* (March 2000) provided at www.emcesd.com, courtesy of Doug Smith.

low shows how to use the washer assembly. First, glue the resistors to the insulating washer or use double-stick tape trimmed to fit only under the resistor. Four 200-Ω 1206 1% surface-mount resistors are then placed on the tape (or glue) that holds them in place on the washer.

Step 1b   Resistors mounted on a fiber washer.

(c) The washer with attached resistors is placed into a female BNC connector, as shown below. Position the resistors on the washer so that they just contact or come very close to the walls of the connector.

Step 1c   Washer and resistors inserted into the BNC connector.

(d) To complete the assembly, a soldering iron is held against the outside of the BNC connector and a small amount of solder is applied to the inside wall. Just enough solder is used to ensure a good joint between each resistor and the wall of the BNC connectors (see below). The electrical characteristics of the washer and double-stick tape are not likely to affect the performance of the termination up to 1

GHz. It you want to be sure, the impedance of the termination can be measured after completion.

Step 1d   Resistor assembly soldered to the BNC connector (provides 50-Ohm termination).

Step 2. Add a central post 3.5 cm long using 16 AWG solid copper bus bar wire inserted into the hole located within the center of the resistor assembly, as illustrated below. Copper foil tape should be wound around the wire to enlarge the diameter of the post so that the foil just touches the resistors after insertion into the center hole of the connector. Solder is then used to connect the resistors to the foil tape and wire.

Step 2   Add central post.

Step 3. Add a tip resistor, as shown below. This resistor determines the probe factor and input impedance. A 976-Ω, 1%, 1206 surface-mount resistor works well, resulting in a 40 : 1 probe with 1000-Ω input resistance.

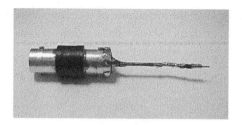

Step 3   Add tip resistor.

Step 4. Add heat-shrink tubing over the central post to insulate it from accidental contact to circuits and voltages not intended to be touched:

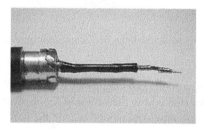

Step 4   Add heat shrink tubing.

Step 5. Cover the tubing with copper EMI tape, extending it over the BNC barrel, as shown below. This forms a distributed capacitor to compensate for the parasitic resistance of the tip resistor. The copper tape is trimmed until the probe has a flat frequency response (falling off a decibel or so at 1 GHz, which is actually desirable). The probe response can be measured on a spectrum analyzer containing a tracking generator. If instrumentation is not available, leave the copper tape at its full length to ensure the probe does not have increasing gain at 1 GHz.

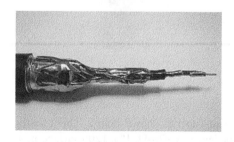

Step 5   Add frequency compensation.

Step 6. To complete the probe, solder a ground lead to the BNC connector and apply heat-shrink tubing over the entire assembly for insulation purposes:

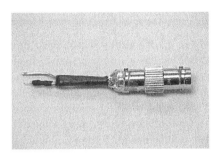

Step 6   The final probe.

Since this probe has resistive input impedance, the impulse response is enhanced (essentially no overshoot or ringing) over a probe with capacitive input impedance. This is especially true when the capacitive reactance of any probe is comparable to the inductive reactance of the measurement loop (including the probe tip, ground lead, and paths in the circuit being measured) at frequencies of interest. This condition is met in many common measurement situations.

Theory of Operation. The probe is a voltage divider comprised of a 976-Ω tip resistor and a 25-Ω load (the 50 Ω termination in parallel with the 50 Ω characteristic impedance of the coaxial cable). At 1 GHz, the parasitic capacitance of the resistors causes the divider to become an *RC* voltage divider, as shown in Figure A.10. Here, R_1C_1 is the 976-Ω tip resistor with its parasitic capacitance while R_2C_2 is comprised of the 25-Ω load and the parasitic capacitance of the 200-Ω resistors.

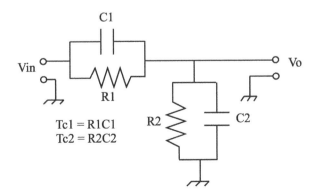

Figure A.10   *RC* voltage divider.

The transfer function for the RC voltage divider is given by

$$\frac{V_o}{V_{in}} = \frac{Z_2}{Z_2 + Z_1} = \frac{R_2}{R_2 + R_1 \dfrac{1 + SC_2R_2}{1 + SC_1R_1}} \qquad \text{where } S = j2\pi f$$

When R_1C_1 equals R_2C_2, all frequency dependence drops out of the transfer function and the divider behaves as a simple resistive divider. Since $R_1 = 976\ \Omega$, (approximately 39 times the value of R_2, 25 Ω), C_2 must be approximately 39 times the value of C_1. It is very unlikely that the four 200-Ω resistors will have that much capacitance across the 200-Ω resistors. If the parasitic capacitance of the tip resistor is 0.1 pF, then a total of 3.9 pF of capacitance across the 200-Ω resistors, including their parasitic capacitance, will be needed for the frequency response to be flat.

The copper tape provides additional required capacitance. By trimming its length, the right amount of capacitance can be added to give the probe a flat frequency response. Without the copper tape, typical surface-mount resistors will give the probe about 6 dB of unwanted gain at 1 GHz.

At 1 GHz, the input impedance of the probe drops due to the parasitic capacitance of the tip resistor. An unwanted gain of 6 dB at 1 GHz implies that the parasitic capacitance of the tip resistor is becoming comparable to its resistance of 976 Ω. Lowered input impedance at high frequencies becomes the main limiting factor for the useful frequency range of this probe. To build a probe usable to higher frequencies, resistors with lower parasitic capacitance must be used. Resistors used in microwave circuits are typically adjusted by grinding down the thickness of the film material rather than making a cut within the film material to lower parasitic capacitance. These resistors should allow for a higher frequency response for this probe. Because of reduced tip resistor capacitance, less copper tape foil is required.

RESISTIVE CURRENT PROBE*

The probe detailed in Figure A.11 does not look like a current probe. A resistor network built into a coaxial cable protected by heat-shrink tubing acts as the sensor. Before describing how it works and its uses, let's review a simple model of a coaxial cable.

A coaxial cable can be simply modeled, for this discussion, in two parts for frequencies above a few hundred kilohertz. First is the cable's input impedance. If a cable with 50-Ω characteristic impedance is connected to a 50-Ω load, in this case the input to a measurement instrument, then the input impedance into the opposite end of the cable is just 50 Ω. Under this condition, a 50-Ω resistor connected between the center conductor and shield connections can replace everything happen-

*Details of this probe are derived from a *Technical Tidbit* article (July 2000) provided at www.emcesd.com, courtesy of Doug Smith.

Figure A.11   Resistive current probe.

ing inside the cable assembly. Because skin effect makes the inside surface of the shield a different environment electrically than the outside surface of the shield, the outer surface of the shield must be included in the model as a thick wire. For the purpose of discussion, a 50-Ω coaxial cable is modeled as a 50-Ω resistor in series with a thick wire (shield) connected to the chassis of the measuring instrument.

Figures A.12 and A.13 show constructional details. First, the jacket of the cable is stripped to expose the shield and the center conductor (Figure A.12). A 47-Ω resistor connects the center conductor to the probe tip. Five 10-Ω resistors are connected from the tip back to the shield. The 10-Ω resistors should be equally spaced around the circumference of the cable (Figure A.13). Carbon composition resistors are ideal for this use, if available. These resistors form a 2-Ω circuit with reasonably low inductance.

Figure A.14 shows the equivalent circuit of the current probe. A 50-Ω resistor represents the input impedance of the coaxial cable. The 10-Ω resistor assembly is shown as a single 2-Ω resistor. The outer surface of the cable shield is illustrated as

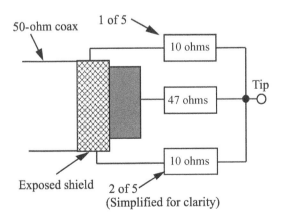

Figure A.12   Side view of current probe.

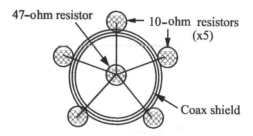

47–ohm resistor

10–ohm resistors
(x5)

Coax shield

Figure A.13 End view of current probe.

a wire. The junction of the 50-Ω and 47-Ω resistors is the node where the center conductor connects to the 47-Ω resistor.

The circuit operates as follows. Assume 1 A of current is flowing from the probe tip through the resistor network to the measuring instrument chassis on the cable shield. This current will generate approximately 2 V across the 2-Ω resistor (actually 1.96 V if the other two resistors are taken into account). Approximately one-half of the measured voltage level, about 1 V (actually 1.01 V), is developed across the 50 Ω input to the cable, which is delivered to the measuring instrument. Thus, the transfer impedance of this current probe is about 1 Ω. This gives an output of 1 V for a current of 1 A. The 1 Ω transfer impedance of the probe allows the scope vertical scale to be converted directly from volts to amperes.

One application for this probe is to add a wire to the probe tip (Figure A.15). Used in this way, the probe becomes an excellent ESD event detector, providing a reading of the current flowing on the wire as a result of radiated EMI from an ESD event.

Another use for this probe is finding noisy areas of ground on a circuit board or in a system. Only 10–20 μA of current above 30 MHz flowing on a cable at a particular frequency may cause radiated emissions to exceed Class A limits. If the probe is connected to a spectrum analyzer through a high-pass filter and protection network (to protect the input of the spectrum analyzer from destruction), it can be used to probe areas of ground on a PCB. Also recommended is a 0.01-μF DC blocking capacitor in series with the probe tip, in case the tip is touched to a power node by accident. When touched to the PCB ground, a signal is delivered to the

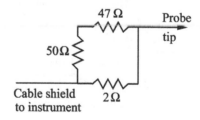

47 Ω

Probe
tip

50 Ω

Cable shield
to instrument

2 Ω

Figure A.14 Equivalent circuit of current probe.

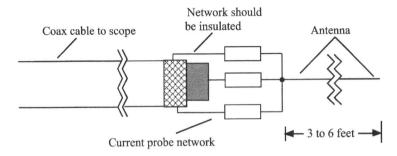

Figure A.15 Using a resistor current probe to detect and characterize radiation from an ESD event.

spectrum analyzer that is equivalent to the amount of high-frequency current flowing in the wire.

The probe can be applied to a PCBs 0-V reference plane in multiple locations, especially in the neighborhood of connectors. It can even be plugged into pins of connectors. A few tens of microamperes at any frequency becomes a dangerous location to install a cable connector. An external cable attached to this point on the PCB with large 0-V reference currents may carry undesired RF energy (current) into the outside environment, causing an emissions problem. Depending on the length and what is at the end of the cable, the current may be flowing at a different frequency (harmonic of a system clock). Regardless, there is a real possibility of an EMI event.

The probe and its resistor network could be built into a test probe similar to those used on multimeters for ease of use. Built with standard film resistors, the probe should work to about 300 MHz (the region where cable radiation starts to become a problem). For higher frequencies, resistors with lower inductance are needed. One approach would be to use ten 20-Ω resistors to lower the effective inductance, although resistor physical size now becomes an issue.

This probe is not a highly accurate way of predicting emissions problems or the best way to measure ESD events, but simple tools like this can make troubleshooting a circuit easier.

SPLIT-FERRITE CURRENT PROBE*

Split-ferrite cylinders (or clamp-on ferrite beads) can be used as a calibrated current probe when clamped over a cable of interest. Common-mode currents are useful for measuring both frequency and amplitude components that may cause radiated EMI to exist above acceptable levels. This probe produces a voltage output that is related

*Details of this probe are provided courtesy of Tim Williams and Keith Armstrong (www.compliance-club.com).

to the common-mode current within the cable and should be terminated in 50 Ω at the measuring instrument for accurate readings.

Figure A.16 illustrates this probe, constructed from ordinary coaxial cable. It uses a coaxial-based shield bonding screw secured clamp. This clamp allows access to the shield of the coax without damaging the dielectric or center conductor when a solder connection is made to the shield. The ferrite core is a typical split-cylinder configuration readily available from a number of manufacturers. The length or shape of the split-ferrite core is not critical. What becomes important is the ability to easily open the clamp without a tool and that it has an internal diameter large enough to pass both the cable to be tested and the coaxial cable at the same time. The calibration of such a probe will be specific for a particular ferrite part.

The formula $E = (12.6 \times 10^7)fLI$ is used to determine whether cable currents measured by this split-ferrite probe are likely to cause unacceptable radiated emissions. In the formula E is the radiated emission in volts per meter at 10 m distance. This value can be converted to dBμV and directly compared with the limits of emissions standards. Also in the equation, f is the measured frequency in megahertz, L is cable length in meters, and I is in microamperes (the current measured by the probe at the frequency f, having taken the probe's calibration factor into account).

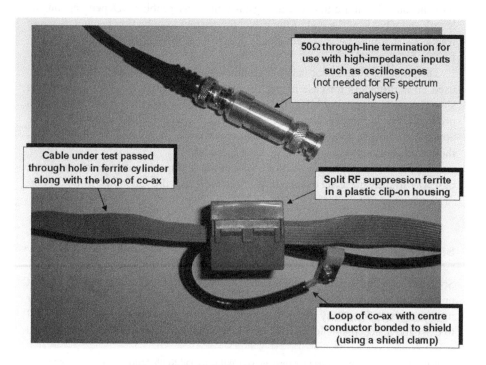

Figure A.16 Split-ferrite core current clamp.

Figure A.17   FET probe shorted by its ferrite equipped ground lead.

IMPROVING FET PROBE IMMUNITY TO UNWANTED NOISE PICKUP*

High-bandwidth, low-capacitance FET probes are often provided with an accessory kit, including various probing tips and ground connections. Accessory kits from several manufacturers may include a ground connection with a ferrite bead, similar to that shown in Figure A.17. The purpose of the bead is to dampen the resonance of the probe's input capacitance and the inductance of the measurement loop formed by the probe tip, ground connection, and circuit being measured.

However, the presence of the ferrite bead on the ground connection can cause the probe to become sensitive to outside interference that produces current flowing on the cable shield. Such currents can be the result of radiation from an ESD event nearby or from logic noise currents flowing from the logic ground of the circuit being measured over the probe cable shield to the scope chassis. To test the sensitivity of the probe to such currents, short the probe to its ground connection using a ferrite bead (Figure A.17) and inject a signal on the probe tip with a regular piece of wire. In this configuration, the probe is measuring the voltage drop across the ground connection and bead. This is a null experiment because it shows how much error there may be in the measurement because of ground currents. The measured result should be zero. In an actual measurement environment, the ground lead voltage drop is added to the intended signal, as far as the probed is concerned.

For the shorted probe tip of Figure A.17, Figure A.18 illustrates the result when 5-V square-wave output of a HC240 gate operating at 30 MHz is applied to the circuit. The circuit ground of the HC240 is connected with a test clip to the scope chassis to provide a ground return for the signal in addition to stray capacitance. Figure A.18 shows a signal of about 2-V peak–peak from the shorted probe. This is the voltage drop across the ground connection and ferrite bead (Figure A.18) caused by the signal current flowing into the scope chassis.

---

*Details of this probe are derived from a *Technical Tidbit* article (September 2001) provided at www.emcesd.com, courtesy of Doug Smith.

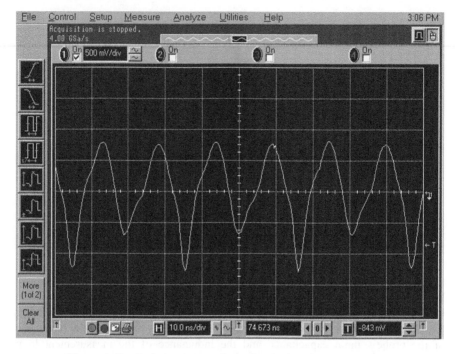

Figure A.18 Probe response to logic signal when shorted with bead.

Figure A.19 shows the ferrite bead ground lead replaced with a solid wire. The same signal is applied to the probe tip, with the result shown in Figure A.20.

The original signal of about 2-V peak–peak has been reduced to approximately 100-mV peak–peak! The reason is that solid wire has significantly less impedance than the ferrite bead, along with its short ground wire. The presence of the ferrite bead raises the ground lead impedance such that the ground lead looks much longer than it actually is.

Figure A.19 FET probe shorted with short wire.

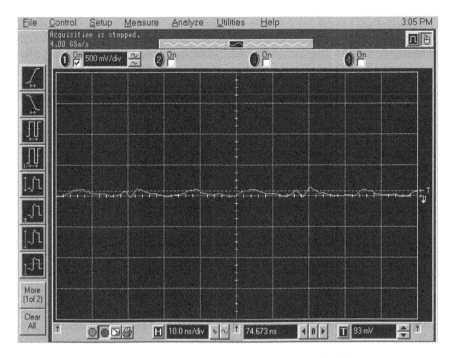

Figure A.20   Probe response to logic signal when shorted with short sire.

What does this mean to someone in the laboratory checking out a circuit? The presence of a ferrite bead in the ground lead, while improving probe response to the intended signal, increases the probe's response significantly to unwanted signals. For instance, a person touching an object across the room and generating an ESD event will cause the measurement to include a glitch due to the ESD event that is much larger than it would otherwise be. One might interpret the glitch as being a potential problem within a PCB layout when in fact it is not.

MEASURING CAPACITOR SELF-INDUCTANCE AND ESR*

The parasitic parameters of a capacitor—equivalent series resistance (ESR) and equivalent series inductance (ESL)—affect the way a capacitor behaves in circuits. Some applications are extremely sensitive to parasitics. For instance, a decoupling capacitor used between power and 0 V in a digital system must be able to supply current quickly to nearby active component. In addition, the transient response of a capacitor used to divert current pulses due to ESD may limit the ability of the capacitor to perform in an optimal manner.

*Details of this probe are derived from a *Technical Tidbit* article (February 2000) provided at www. emcesd.com, courtesy of Doug Smith.

If there is too much interconnect inductance, the capacitor will be unable to provide sufficient energy charge transfer in the time period that circuits require. In addition, the voltage drop across any inductance may lower the supply voltage below acceptable values. The summation of the internal interconnects (between capacitor plate and terminal) and trace inductance to power/0-V reference is referred to as ESL. The internal resistance of the dielectric material contributes to ESR.

It is almost impossible to know the value of ESL and ESR, as most capacitor manufacturers do not publish this information. With lack of information, how can the parasitic parameters of a capacitor be measured? One method is to connect the capacitor to a network analyzer and characterize the component. Network analyzers are extremely expensive. Other measurement equipment may not be available to physically measure ESL and ESR. Both instruments may not provide information in an easy-to-understand format. Using a pulse generator (preferably with 50 Ω output impedance) and an oscilloscope, one can easily measure the transient response of a capacitor. From these data, the capacitor's ESR and ESL can be determined.

First, construct the simple network shown in Figure A.21 at the end of a 50-Ω coaxial cable. This cable will then be fed from a 50-Ω pulse generator. The 50-Ω resistor in Figure A.21 terminates the coax during the rising edge transition and provides a total of 100 Ω of source impedance. The 51-Ω, 0.5-W carbon composition resistor has one lead trimmed so that the resistor just seats with the trimmed lead fully inserted into the BNC connector. It may be necessary to apply a little solder bump on the resistor lead so that it stays firmly in the BNC connector. The capacitor to be tested is then installed between the end of the resistor and the shell of the BNC connector. An oscilloscope is now connected directly across the capacitor using leads as short as possible. Probes with a resistive input impedance of 500–1000 Ω are recommended. Standard 10X Hi-Z probes often have rising edge effects that will distort the part of the waveform used for the calculations.

For a pulse length that is long with respect to the *RC* time constant, one will

Figure A.21   Test circuit for measurement of capacitor self-inductance and ESR.

see an exponential rise (in the measured data) to the open-circuit voltage of the pulse source. For purposes of this discussion, examine the first couple of hundred millivolts of a 5-V exponential rise time signal. An example of this appears in Figure A.22.

Figure A.22 shows the beginning of the exponential voltage rise across the capacitor when the generator pulse starts. The vertical scale is about 200 mV and the horizontal time is a small fraction of the *RC* time constant (100 Ω) for the capacitor being measured. Since the capacitor voltage is still very small, compared to the 5-V open-circuit output of the generator, the current through the capacitor may be presumed to be constant and equal to the generator open-circuit voltage divided by 100 Ω, which is 50 mA for this example.

The rise time of the current will be the same as the generator voltage. If the rise is a ramp with a constant slope and the capacitor had no inductance, the initial rise shown in Figure A.22 would follow the dotted line. The slope would change to the initial slope of the exponential rise determined by

$$\frac{dV}{dt} = \frac{I}{C} = 50 \text{ mA}/C \tag{A.1}$$

where *C* is the value of the capacitor at this low voltage and the rise time of the current is much less than *RC*.

The offset between the baseline and the beginning of the exponential rise is the voltage that the current develops across the ESR of the capacitor (50 mA). The ESR can be easily estimated by dividing the voltage offset (labeled ESR in Figure A.22) by 50 mA.

Parasitic inductance in the capacitor will cause the spike to occur in the waveform (Figure A.22), exceeding the value of the dotted line along its length. If the

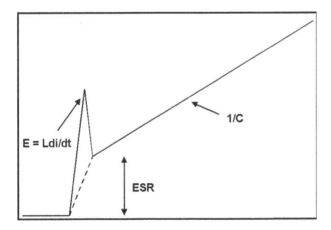

Figure A.22   Initial rise time of a capacitor.

current rise were in fact a ramp with constant slope and very sharp corners (high dI^2/dt), then the spike would be a square pulse with a value

$$E = L\frac{dI}{dt} \tag{A.2}$$

where L is the parasitic inductance of the capacitor.

The current rise from the generator used for this example was not a ramp with very sharp corners and constant slope. The characteristics of the generator combined with probe effects led to a peaked shape to the $L\ di/dt$ spike, as shown in Figure A.22. Using Eq. (A.2), the inductance of the capacitor can be calculated. Often, one does not need to calculate ESL or ESR but simply needs to choose a capacitor from those available that has the lowest inductance and/or ESR possible.

Soldering the components onto a BNC connector (Figure A.21) works up to 300 MHz. The inductive reactance of the loop formed by the capacitor and resistor will be approximately 20 Ω at 300 MHz (estimating inductance at 10 nH). This is small enough relative to the 100 Ω of resistance in the circuit to not significantly affect the initial current by any degree of magnitude. This test setup is optimal for signals with a 1–2-ns edge transition.

If you need to check the capacitor using faster rise times, it is best to build the test setup on a small circuit board with a ground plane and controlled impedances. At this point, the parasitic capacitance of the 50-Ω resistor would also be an issue to be taken into account. Fortunately, such accuracy is often not needed. This is especially true if one is just comparing the relative performance of several capacitors.

Data Analysis (Actual Test Results). Figure A.23 shows a sample rise from a test generator. The black square indicates vertical voltage and horizontal time scales. The open-circuit voltage is a little over 4 V with approximately 5-ns-rise-time transition. Figures A.24–A.26 show data obtained from several leaded capacitors (as opposed to surface mount). Two trace measurements were taken of each capacitor. The lower trace was measured at the capacitor body where the leads entered the capacitor. The upper trace includes the minimum amount of lead length to connect the capacitor to a PCB. The upper trace would not be needed for modern surface-mount

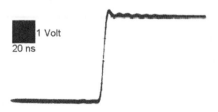

Figure A.23   Input f from pulse generator.

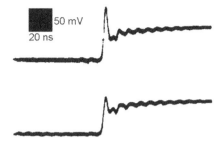

Figure A.24   Data from 4.0-μF electrolytic capacitor.

capacitors unless one wanted to model the connection inductance from the capacitor to the point of interest on a printed wiring board.

Figure A.24 shows data from a 4.0-μF electrolytic capacitor. The ESR offset is about 50 mV, yielding an estimated value of ESR of just over 1 Ω. Notice that there appears to be some oscillations on the $1/C$ part of the slope. This could be scope probe resonance or a resonance in the capacitor. The data was taken with a standard 10X Hi-Z probe. If planning to put a large capacitor in parallel with a smaller one, especially if they are constructed from different technologies, it is recommended to check the impulse response of the combination using this method. It is possible for the smaller capacitor to resonate with the inductance of the larger one, causing unexpected results

Figure A.25 shows results for a 1.0-μF capacitor of the same construction as the 4.0-μF capacitor. Note that inductance is similar to the 4.0-μF capacitor with slightly lower ESR. Since an analog scope was used, the waveform was repetitive and the slight slope on the left half of the waveform was the end of the exponential fall from the 5-V source signal. If a single pulse on a digital scope was used, the slope to the left of the $L(dI/dt)$ spike would be zero.

Figure A.25 shows the result for a 1.0-μF radial ceramic capacitor (square case package). Note the low inductance and undetectable ESR. Also, the slope of the $1/C$ exponential rise is flatter, indicating more capacitance than the 1-μF capacitor of Figure A.24. This slope may be due to the fact that the 1.0-μF electrolytic

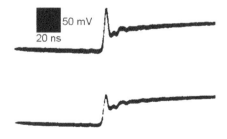

Figure A.25   Data from 1.0-μF electrolytic capacitor.

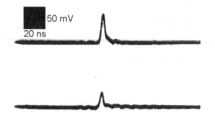

Figure A.26 Data from 1.0-μF ceramic capacitor.

capacitor (Figure A.24) may have lower capacitance near zero voltage than at its operating voltage, whereas the ceramic capacitor has a more constant capacitance value with voltage. The inductance corresponding to the lower trace is estimated to be 4.4 nH.

It is interesting to note that a 0.1-μF ceramic capacitor in the same size package as the 1.0-μF capacitor of Figure A.26 showed a slightly *higher* inductance in this test setup. This difference may be due to the fact that the smaller capacitor did not fill the package and internal lead inductance caused the effect. In this case, a 1.0 μF capacitor was a better choice than a 0.1-μF one!

One of the advantages of this test is that the output waveform is the transient response of the capacitor. The voltage developed across the capacitor in this test is directly related to what will happen in a real circuit if the current rise time from the generator is similar to what the capacitor will see in its intended application.

TIN CAN WAVEGUIDE ANTENNA (CANTENNA)

This unique antenna provides value when trying to determine if electrical or shielding changes on a wireless product, or other systems operating above 1 GHz, reduces unintentional radiated emission sidebands or other undesirable effects when a horn antenna is unavailable and log-periodic antennas cut off at 1 GHz. In addition, this antenna allows one to inject an RF signal into a system to determine if an immunity problem exists. This antenna can be built for practically no cost using recycled food containers made of tin, not aluminum.*

The Cantenna contains the following parts and requires only drilling or punching a hole in the can to mount the probe:

- N-series female chassis-mount connector
- Four small nuts and bolts
- A bit of thick solid wire (12 AWG)
- A tin can

*This antenna is in the public domain. Material is available on the Internet on numerous websites.

Parts List

Connector. An N-series female chassis-mount connector is required that permits insertion of a brass stub wire. This is the most expensive part of the antenna. A modified coaxial cable assembly is required for use with the Cantenna. One side of the coax cable will have the male end of the N connector while the other end of the coax contains an appropriate plug for interfacing to the wireless equipment. An illustration of N-series female chassis-mount connectors is shown in Figure A.27 with the location where the brass stub wire is to be soldered.

Nuts and Bolts. A #6 $\frac{1}{4}$-in. screw (bolt) with a #6 nut is required only during the drilling stage when one uses the flange mount connector shown on the left side of Figure A.27. If the N-connector is a screw-on type with integral nut, use of additional mounting hardware is not required.

Wire. This requires approximately 1.25 in. (3.18 cm) of 12 AWG solid copper wire. This wire will be inserted into the brass stub in the N-connector and soldered.

Tin Can. The most difficult part of the project is finding a tin can between 3 and 3-$\frac{2}{3}$ in. (7.6–9.3 cm) in diameter. The size does not have to be exact. Excellent antennas work with cans that are approximately 40 oz in size and 6 in. (15.2 cm) diameter. Try to use a tin can that is longer in length that diameter. Old fashioned fruit juice cans work well. Figure A.28 illustrates various cans that may be used. Be advised that the MJB can in the figure performs best, while the Pringle's potatoes chip can provides poor performance due to the nature of the metallic coating used inside the can.

Drill or Punch Holes in Can to Mount Probe. The N-connector assembly must be mounted in the side of the can. The placement of the hole and connector is critical. The location of the hole is derived from formulas commonly used in the design of microwave antennas:

Cutoff frequency in MHz for TE11 mode: 6917.26/(diameter of can) inches

Cutoff frequency in MHz for TE01 mode: 9034.85/(diameter of can) inches

Guide wavelength: $11,802.85/\sqrt{5,938,969 - (TE11 - TE01)}$ inches

    1–4-guide wavelength: wavelength 0.25

    3–4-guide wavelength: wavelength 0.75

Figure A.27   N-series coax connector for use with the Cantenna.

Figure A.28 Sample metal cans to be used for the Cantenna design.

The IEEE 802.11B networking equipment (Bluetooth) operates at a range of frequencies from 2.412 to 2.462 GHz. Ideally, the can size for TE11 cutoff frequency should be lower than 2.412 and the TM01 cutoff should be higher than 2.462. It is highly desirable to have the can longer than $\frac{3}{4}$-guide wavelength (calculation provided above). If the can is a little off in length or diameter, experiment with fine tuning the operating frequency.

Identify where a hole is to be drilled in the can with a ruler at a distance of the $\frac{1}{4}$-guide wavelength number, calculated in the above equation, starting from the rear of the can to the center-hole location (Figure A.29). This center-hole location is where the connector will be inserted. Make sure the can has been previously opened on only one end and the contents removed. A washed can is highly desirable for obvious reasons. Removal of a paper label, if provided, is optional.

With a drill, select a bit that matches the size of the center of your connector (the small brass insert). Start with a small bit and work the hole larger and larger. In lieu of a drill, a hammer and nail can be used with a file to get the hole to the required size. If using the bolt on an N-series connector with the four mounting screws, make four more holes for the bolts using the connector as a drilling guide. If using the connector without the flange, this step is not necessary.

Figure A.29 Drilling the hole in the can.

Figure A.30   Inserting the connector and probe into the can.

Figure A.31   Finished Cantenna on a tripod.

Assembling Probe and Mounting in Can. With the piece of wire and soldering iron, cut the wire and secure it to the center stud of the connector, as shown in the left side of Figure A.30. The total length of both the brass tube and wire sticking out past the connector is 1.21 in. (3 cm). Get as close to this length as possible.

Bolt or screw the connector assembly into the can. Put the heads of the bolts inside the can and the nuts on the outside to minimize the obstructions in your antenna. This elaborate antenna is now ready for use (Figure A.31).

APPENDIX B

TEST PROCEDURES

This appendix provides a summary of how to perform both emission and immunity tests on a step-by-step basis. The target audience is those already acquainted with a basic knowledge of instrumentation, test setups, transducers, auxiliary support equipment, and other areas related to EMC testing. Information herein is not intended to be inclusive as other techniques and procedures are available that are not documented or identified in *Generic, Basic,* or *Product Family* standards.

For *emissions,* information in this appendix targets products classified as ITE (Information Technology Equipment) and ISM (Industrial, Scientific, and Medical). The European standards that describe emission requirements are EN 50081-1 and EN 50081-2. The primary difference between these two standards is that EN 50081-1 targets light industrial (including commercial and residential environments) and EN 50081-2 is for the heavy industrial sector.

For *immunity* testing, the focus of the material is on EN 61000-6-1 and EN 61000-6-2, where again the primary difference between these standards is light industrial versus heavy industrial.

PART I: EMISSION TESTING

Pretest Preparations

With the aid and support of the system owner, for example, a person who knows the makeup of the system hardware, software, and how all subassemblies are interconnected:

Testing for EMC Compliance. By Mark I. Montrose and Edward M. Nakauchi
ISBN 0-471-43308-X © 2004 Institute of Electrical and Electronics Engineers

1. Define EUT system boundaries. Record the type, make, model, and serial numbers of devices or subsystems that define the unit to be tested. Include all support equipment and instrumentation used to perform the test.

2. Determine which devices and subsystems can be turned on and off during testing without requiring extensive preparation or boot-up time. Turning devices on and off helps to quickly locate emission sources should emissions be over the specification limit or require further investigation.

3. Make a list of RF-generating sources in the EUT. The list should include location and operating frequencies for the following

   • All printed circuit boards [central processing unit (CPU), video, network controllers, communication systems, intentional transmitters, and other digital circuits and control logic that use a clock signal, regardless of edge rate transitions or frequency; many 2-MHz oscillators have been known to cause EMI failures at 300 MHz]

   • All power supply internal switching frequencies (9 kHz and higher)

   • Ethernet or other networking/communications controllers, their interface specification, and frequency of data rate transfer

4. Diagram the EUT system, indicating location of all devices and support equipment relative to one another. Indicate placement, type, and routing of

   • AC power lines
   • I/O cables
   • Transducer cables
   • Alarm cables

The test engineer will use the information given above when trying to determine worst-case positioning of devices, cables, and support equipment for emissions testing or for troubleshooting problem areas.

Radiated Emissions (Enclosure Port)

Standard Used as Guide to Testing	CISPR 11 (EN 55011) and CISPR 22 (EN 55022)
Frequency Range	0.15 MHz to the highest frequency required per the standard, generally 1 GHz
Limits	Class A or B, as defined in the standard being referenced

Equipment Used to Perform Test

Antennas	Loop antenna: 0.15–30 MHz
	Biconical antenna: 30–200/300 MHz
	Log-periodic antenna: 200/300–1000 MHz
	Horn antenna: 1–40 GHz

EMC analyzer	9 kHz to highest frequency required with quasi-peak detector and optional tracking generator (*Note:* An EMC receiver can be used in lieu of a spectrum analyzer and in fact is preferred.)
Preamplifier	0.1 MHz to highest frequency required, 25 dB gain typical flat response
Cables	50-Ω coaxial cable, with a minimum length used between antenna and EMC analyzer
Probes	1- and 2-cm-diameter loop antenna, 5-cm rod antenna, or closed-field probe (for diagnostic or debug purposes)
Miscellaneous	Antenna tripod, measuring tape, coaxial cable connector adapters
Test location	Open-area test site (OATS), TEM/GTEM cell or anechoic chamber (full or semianechoic). Use of shield rooms is not recommended due to reflections off walls, floor, and ceiling.

Pretest Equipment Check and Calibration Procedures

Spectrum Analyzer. Note: The following information is generic in nature and does not reflect one manufacturer's analyzer over another. Certain key functions may be different between models and vendors.

1. Tune the analyzer to the unit's calibration frequency as indicated on the CAL-OUT connector, such as 100 MHz. Set amplitude units to dBm. This determines that the measured power level is within calibration limits.
2. Connect a short coax cable between the analyzer input connector and the calibrated signal's output connector (Figure B.1).

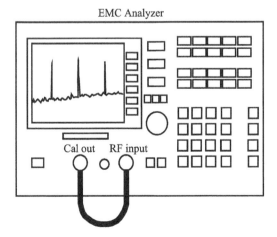

Figure B.1   Spectrum analyzer check.

3. Set the analyzer controls as follows:
   (a) Center frequency: Defined by the vendor's CALL-OUT frequency value.
   (b) SPAN: 10 MHz.
   (c) Amplitude Reference Level: 0 dBm (107 dBμV).
   (d) Hit peak search or stop the scan with the maximum value stored on the screen.
   (e) The marker on the display should be the value specified by the manufacturer. If the display shows a different value, refer to the spectrum analyzer manual for additional calibration procedures or have the unit repaired.

Measurement System Check

1. Arrange EMC analyzer, signal generator (or internal tracking generator of the analyzer), receive antenna, and transmit antenna as shown in Figure B.2. This test requires two identical antennas, usually the same ones used to perform site attenuation.
2. Take data at a minimum of six test frequencies. Take into consideration the receive antenna's loss factor, the loss within the coax, and amplifier gain when calculating actual received level of RF energy. This test validates if the antennas are working and if there are any problems in the test setup.

Antenna	Test Frequency (MHz)	Generator Output (dBm)	Received Level (dBμV)
Biconical	50	−10	As measured on the analyzer
	150	−10	As measured on the analyzer
	250	−10	As measured on the analyzer
Log periodic	400	−10	As measured on the analyzer
	600	−10	As measured on the analyzer
	800	−10	As measured on the analyzer

Performing Radiated Emission Tests. *Note:* Both EN 55011 and EN 55022 call for measurement equipment that complies with the requirements of CISPR-16. It is realized that most spectrum analyzers do not precisely meet the requirements in CISPR-16 with respect to overload protection, amplitude accuracy, and detector mode (i.e., quasi-peak readings are specified and spectrum analyzers generally use peak mode of detection). Even so, spectrum analyzers are preferred because of their portability and versatility and the large amounts of information that can be displayed quickly. The minor differences between spectrum analyzer amplitude accuracy and peak amplitude of EMC test receivers are not considered significant enough to preclude their use for most types of testing. If the measured peak signal is below quasi-peak levels, the EUT is deemed compliant.

1. Place the desired antenna at a test location specified by the regulatory standard or procedure being used to perform the test.

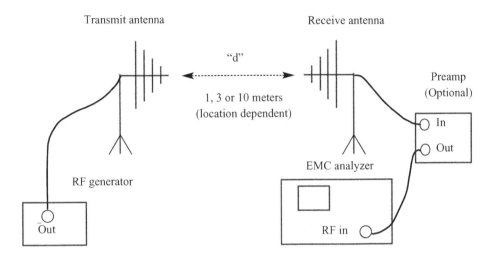

Transmit antenna   Receive antenna

"d"

1, 3 or 10 meters
(location dependent)

Preamp
(Optional)

In

Out

EMC analyzer

RF generator

Out

RF in

Figure B.2   Measurement system check.

2. Connect the preamplifier, spectrum analyzer, and antenna together. For both biconical and log-periodic antenna, select a particular polarity: horizontal or vertical. For a loop antenna, set the antenna parallel or perpendicular to the EUT. After the first round of testing, the polarity is to be changed. The distance d is the distance between antenna and EUT. The distance d is to be such that all other systems other than the EUT be more than 3 times d from the receive antenna. In practice, d typically ends up being on the order of 1 or 3 m, and sometimes 10 m (Figure B.3).

3. Enter the following spectrum analyzer settings:

Center frequency	40 MHz
SPAN	20 MHz
Resolution bandwidth (BW)	100 kHz
Video BW	100 kHz
Sweep	AUTO
Amplitude Reference Level	0 dBm (107 dBμV)
Amplitude Attenuation	AUTO
Scale	Log

4. Determine the source of any emissions detected by turning devices on and off where possible, and by referring to the list of known RF sources previously identified. To aid in determining if the signal is an ambient, move the antenna with respect to the EUT while monitoring emission levels. Usually,

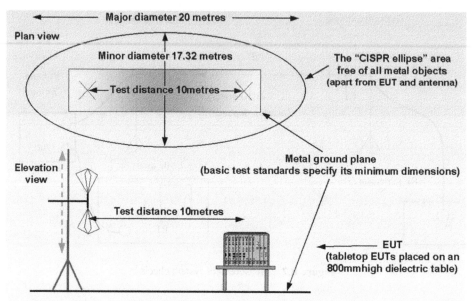

Figure B.3 RF radiated emissions test environment. (Illustration courtesy of Keith Armstrong.)

signal strength will increase when the antenna is moved closer to the source and will decrease when the antenna is moved further away from the EUT. Also, rotating the EUT will cause a variation in the received signal if it is emanating from the EUT.

5. A closed-field probe, or loop probe, can be useful in locating sources of emissions. Figure B.4 shows how the closed-field probe should be connected and used.

6. If the spectrum analyzer has a demodulator and audio output, use these features to determine which emissions are from TV and radio stations, two-way business radios, pagers, cell phones, and so on.

7. To determine the frequency and amplitude on the display, activate the marker function on the analyzer. Using the cursor dial knob, move the cursor to the emission signal in question to the peak level of the signal. Read the frequency and amplitude displayed on the analyzer's screen.

   The current display will show 30–50 MHz with 40 MHz as center frequency and a span of 20 MHz. To view emissions in the next 20 MHz of the spectrum, press CENTER FREQUENCY and then the STEP UP arrow or equivalent function key. The center frequency should now change to 60 MHz. Use the STEP arrows to investigate emissions up to 1000 MHz or desired end frequency.

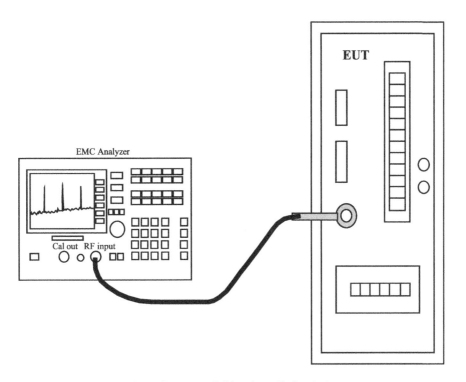

Figure B.4 Using a near-field probe to find emissions source.

8. Record the frequency and amplitude for each signal measured from the EUT. These are peak reading measurements. If the emission level at a given frequency is within 3 dB of the specification limit, repeat the measurements using the quasi-peak detector. The quasi-peak measurement procedure is as follows:

   (a) Place the marker over the peak of the emission.
   (b) Turn on the quasi-peak (QP) detector of the EMC analyzer.
   (c) The QP level will be displayed on the screen after a few seconds of set-up time.
   (d) Record the QP reading.
   (e) Return to the peak mode by turning off the QP detector.

9. Change antenna polarity (opposite from previous configuration) and determine whether the emission measured from the EUT is higher in amplitude. Technically, the FCC requires only the highest six emissions to be recorded in the test report. If a signal is present at both polarities and one polarity is higher than the other, only the highest amplitude and its respective polarity need to be recorded in the test report.

10. Repeat above steps for each antenna and both polarities throughout the frequency spectrum, based on the frequency range of operation for the antenna.
11. For in situ testing or when the EUT cannot be rotated on a turntable, physically move the antenna to alternate locations around the test sample with sufficient confidence level so that all angles of radiated emissions from the EUT can be investigated.
12. The maximum acceptable measured limit at distance d for the specific frequency range is determined using the equation

$$L_d = L_{30m} + D_{cf} - Af - CL + Amp$$

where L_d = maximum allowable spectrum analyzer reading at measurement distance d (set the display line for this dBμV level)

L_{30m} = maximum allowable field strength limit at frequency range as noted in standard

D_{cf} = distance correction factor, which is $20 \log(30m/d)$

Af = antenna factor for particular frequency range being tested (refer to antenna manufacturer's calibration chart)

CL = cable loss (refer to cable loss chart)

Amp = gain in decibels for preamplifier, if used

Check for Spectrum Analyzer or Preamplifier Overload. Strong signals may overload preamplifier or spectrum analyzer inputs. Strong signals *must* be checked for overload conditions by placing a known-value attenuator in series with the receive antenna cable. The measured reading on the analyzer should drop the same amount as the attenuator level. If the change in the reading is greater than the attenuator value by a significant amount (i.e., by a factor of 2 or more), then most likely the measurement system is overloaded at that frequency. Measurements should be taken with the preamplifier removed from the circuit whenever possible.

Note: A preamplifier may not be necessary if the spectrum analyzer noise floor is more than 20 dB below the emission limit for the frequencies being measured.

Conducted Emissions (Input AC Power Port)

Standard Used as Guide to Testing   CISPR 11 (EN 55011)
and CISPR 22 (EN 55022)

Frequency Range   0.15–30 MHz (or any other frequency range desired)

Limits   Class A or B, as defined in the standard being referenced

Equipment Used to Perform Tests

EMC analyzer   Spectrum analyzer or receiver, 9 kHz to >30 MHz. Internal tracking generator is desired. (*Note:* An EMC receiver can be used in lieu of a spectrum analyzer and in fact is preferred.)

Signal generator	Optional in lieu of tracking generator
Cables	Appropriate length of 50-Ω coaxial cable
Plotter (optional)	Accept screen print data from EMC analyzer
LISN	50 Ω/50 μH
Voltage probe	1500-Ω circuit
Current probe	50 Ω output, 0.150–30 MHz frequency response, or other frequency range of interest

Pretest Equipment Check and Calibration. Depending on the EUT, test environment, and other factors related to performing this test, one of three transducers may be used. When performing actual tests, the transducer most appropriate is the one that will be used.

The built-in tracking generator, if provided, is used to produce a test signal for pretesting both voltage and current probe (Figure B.5). If a LISN is used, which is the most common transducer for this test, calibration needs to be verified once per year with results compared to the vendor's calibration chart and the waveform specification detailed in the test standard. Only the pretest equipment check and calibration procedure for the voltage and current probe are provided. Use the test procedure provided by the LISN vendor.

(a) Adjust the amplitude of the tracking or signal generator by selecting AMPLITUDE UNITS. Set the reference level to 0 dBm (107 dBμV).

(b) Turn the generator's output ON. This is sometimes identified as "source power" for certain vendor systems.

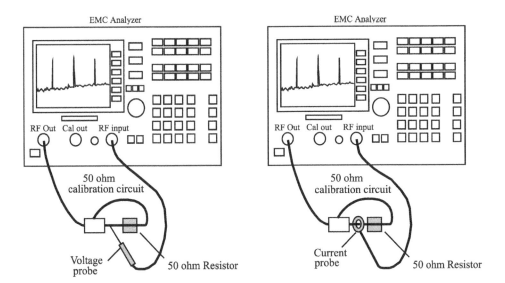

Figure B.5   Calibrating voltage and current probes.

(c) Adjust the tracking generator output level to 87 dBm.

(d) Select START FREQUENCY: 150 kHz.

(e) Select STOP FREQ: 30 MHz.

(f) Connect one end of the coaxial cable to the EMC analyzer's input and the other end of the coax to the tracking generator's output connector. A signal should appear at 87 dBμV, ±0.5 dB.

The tracking generator–EMC analyzer combination can now be used to perform pretest confidence verification for both the voltage and current probe in addition to validating calibration accuracy of the LISN.

Voltage Probe Method

1. Using a 50-Ω resistor (or terminator), make a bare-wire loop antenna (no jacket insulation; solid wire works best). Connect one side of the resistor to the center conductor of the coax connected to the tracking generator's output. Connect the other end of the wire loop to the return (shield) of the coax.

2. Place the voltage probe directly onto the wire loop antenna, ensuring the probe makes connection with the bare wire. Connect the voltage probe to the spectrum analyzer's input connector using a short length of 50-Ω coaxial cable.

3. Turn on the signal/tracking generator to produce -10 dBm at 7 MHz. The spectrum analyzer should read approximately -40 dBm (signal level plus probe loss, which is typically 30 dB across the operating frequency of the probe).

Current Probe Method

1. Create the same calibration fixture used for the voltage probe calibration procedure.

2. Place the current probe over the wire loop so that the 50-Ω resistor is in the center of the opening, with a short length of 50-Ω coax connected between probe and analyzer input.

3. Set the signal generator to produce 50 mV at 1 MHz. The spectrum analyzer reading should be within 2 dB of 60 dBμV + CF, where CF is the correction factor of the probe at 1 MHz.

4. Set the analyzer CENTER FREQUENCY to 1 MHz and the SPAN to 200 kHz. For a typical current probe with a –4 dB correction factor at 1 MHz, the level of the analyzer should read the value published by the manufacturer for the probe, dBμV ± 2 dB (dBμV = analyzer reading + correction factor of the probe).

Important Safety Warning. High levels of lethal earth leakage currents created by the 8-μF capacitor located inside the LISN exist. Each mains terminal must have two independent protective earth connections in place before the mains voltage is applied. All protective earth connection must be rated for the full value of the possible fault current (i.e., the maximum current if the live supply should short circuit directly to the chassis).

Training in the safe and correct use of LISNs/AMNs is required by authorized operators. It is also important to perform frequent safety inspections on the protective earth bonding to the facility's earth ground connection. Because of the high levels of continuous earth leakage currents associated with LISNs and AMNs, residual current circuit breakers (RCCBs) and other types of earth leakage protection devices cannot be used.

Warning labels on LISNs are also strongly recommended for

- Lethal high earth leakage current
- The need to maintain two independent protective earth connections at all times
- Use by only authorized and trained personnel

Use of Transient Limiters and High-Pass Filters. Transients limiters are recommended for use with LISNs/AMSs and are installed between the LISN and EMC analyzer to protect the analyzer from mains transient damage. Transient limiters require calibration. The LISN itself will attenuate most of the transients on the mains supply. The real problem when using LISNs is when connection and disconnection of the EUT to LISN occur. Surge currents can cause 400-V transients to appear at the RF output terminals.

Limiters usually consist of a −10 dB attenuator plus a pair of back-to-back small-signal diodes to clip the maximum signal to about ±1 V. Transient limiters should be calibrated and their correction factors taken into account along with the LISN's own transducer factors plus any correction factors for the cables whenever any measurements are made with transient limiters.

Some EMC receivers or spectrum analyzers have built-in transient limiters, in which case they are calibrated when the test instrument is calibrated.

High-pass filters are used to only permit signals of interest to reach the EMC analyzer. As with the transient limiter, any low-frequency or DC signal can cause harm. This filter prevents 50/60-Hz components from affecting the performance of the analyzer at the low end of the frequency spectrum without causing disruption in the higher frequency components that are being measured.

Performing Conducted Emission Test: Table-Top Equipment (Figure B.6). The EUT is placed on a nonmetallic table in a screened room or OATS. When using an OATS, a reference vertical and horizontal metal ground plane must be provided. The vertical plane must be at least 2 m × 2 m in size at a distance of 40 cm from the EUT. The horizontal reference plane (floor) must also be at least 2 m × 2 m in size. If the EUT is larger than these dimensions, the planes must be physically larger than the EUT by a distance of 0.5 m beyond the boundaries of the EUT. The EUT is positioned 0.8 m above the reference plane and at least 0.8 m

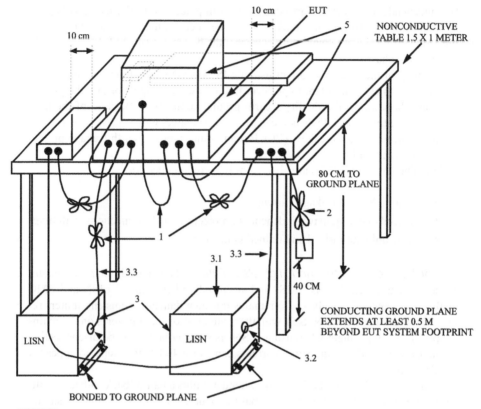

LEGEND:

1. Interconnecting cables that hang closer than 40 cm to the ground plane shall be folded back and forth forming a bundle 30 to 40 cm long.
2. I/O cables that are connected to a peripheral shall be bundled in the center. The end of the cable may be terminated, if required, using proper terminating impedance. The overall length shall not exceed 1 meter.
3. EUT connected to one LISN. Unused LISN measuring port connectors shall be terminated in 50 ohms. LISN can be placed on top of, or immediately beneath, reference ground plane.
   3.1 All other equipment powered from additionas LISN(s).
   3.2 Multiple outlet strip can be used for multiple power cords of non-EUT equipment.
   3.3 LISN at least 80 cm from the nearest part of EUT chassis.
4. Cables of hand-operated devices, such as keyboards, mouses, etc., shall be placed as for normal use.
5. Non-EUT component system being tested.
6. Rear of EUT, including peripherals, shall be aligned and flush with rear of tabletop.
7. Rear of the tabletop shall be 40 cm removed from a vertical conductive plane that is bonded to the ground plane.

Figure B.6 Test arrangement for conducted emissions (table-top configuration).

from any other metal surface. When preparing equipment for LCI tests, several configurations are possible.

If any cable extends to a distance closer than 40 cm to the reference ground plane, the excess cable shall be bundled in the center in a serpentine fashion using a 30–40 cm length in the center to maintain the 40 cm height above the reference plane. If the cables cannot be bundled because of physical limitations (bulk, length, or stiffness), they shall be draped over the back edge of the table unbundled in a manner such that all portions of the cable remain at least 40 cm from the reference plane. Interconnect cables between the EUT and peripherals shall be bundled in the center to maintain a distance of 40 cm height above the plane. The end of the cable may be terminated. The overall length of each bundled cable shall not exceed 1 m.

The power cord(s) of equipment other than the EUT do not require bundling. These cords should be placed in a typical configuration that one would find in actual use. Details on how to configure these accessory power cords are specified in the test procedure appropriate for the system being evaluated.

Performing Conducted Emission Test: Floor-Standing Equipment (Figure B.7). For floor-standing equipment, the setup is similar to table-top equipment except the EUT is on the floor. As a result, routing of cables is different. All excessive lengths of interconnecting cable shall be folded back and forth in the center to form a bundle between 30 and 40 cm in length. If the cables cannot be bundled because of physical limitations (bulk, length, or stiffness), the cables shall be arranged in a serpentine fashion.

Interconnect cables not provided to a peripheral may be terminated, if required, using correct terminating impedance or by connection into the auxiliary device that is itself powered on. Cables that are normally grounded shall be bonded to the reference plane for all tests. Cables normally insulated from ground shall be insulated from the ground plane by up to 12 mm of insulating material. For combined floor-standing and table-top equipment, the interconnecting cable to the floor-standing unit must drape to the reference plane with excess cable bundled. Cables not reaching the reference ground plane are draped to the height of the connector or 40 cm, whichever is lower. Chapter 6 contains setup configurations when performing conducted emission tests for AC line conducted emissions. The procedures presented next provide guidance on performing LCI tests.

Analyzer Control Settings

START frequency	150 kHz
STOP frequency	30 MHz
Resolution BW	10 kHz
Video BW	10 kHz
Sweep	AUTO
Amplitude Reference Level	0 dBm (107 dBμV)
Amplitude Attenuation	AUTO
Scale	Log

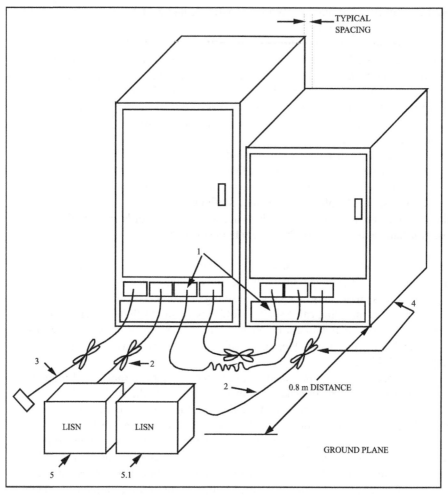

LEGEND:
1. Excess I/O cables shall be bundled in the center. If bundling is not possible, the cables shall be arranged in a serpentine fashion not exceeding 40 cm in length.
2. Excess power cords shall be bundled in the center or shortened to appropriate length. I/O cables that are not connected to a peripheral shall be bundled in the center. The end of the cable may be terminated, if required, using proper terminating impedance. If bundling is not possible, the cable shall be arranged in serpentine fashion.
3. The EUT and all cables shall be insulated from the ground plane by up to 12 mm of insulating material.
4. EUT connected to one LISN. LISN can be placed on top of, or immediately beneath the ground plane.
   4.1 All other equipment shall be powered from a second LISN or additional LISN(s)
   4.2 Multiple outlet strip can be used for multiple power cords of non-EUT equipment.

Figure B.7 Test arrangement for conducted emissions (floor-standing configuration).

LISN Method

1. Connect the EUT power cord to the LISN's output receptacle. Connect the LISN's input to the AC mains receptacle of the facility.

2. There are two or three coaxial connectors on the LISN, one for each line of the AC mains (L1 and L2, sometimes L3). Place a 50-Ω terminator on the unused BNC connector port(s) of the LISN.

3. Connect a coax between the LISN and EMC analyzer.

4. Auxiliary equipment and support equipment remote to the EUT should be provided with filtered AC mains power or be connected to a separate LISN (not the same LISN as the EUT).

5. Determine source of any emissions detected by turning devices on and off where possible and by referring to the list of known RF sources previously identified. Emissions in the 530–1705-kHz band can be checked against operating frequencies of known AM broadcast stations in the area or the audio output of the analyzer can be used to listen to the AM broadcast stations. If the EMC analyzer has the ability to detect AM and FM signals, use this detector to verify existence of audio within this frequency range.

   (a) Center the signal being investigated on the screen display.

   (b) Turn on the DEMODULATOR function.

   (c) Select AM or FM detector.

   (d) Turn the speaker on, if provided.

   (e) Change the SPAN to 0 Hz.

   (f) The AM station output should be heard from the speaker if the signal is indeed from a local radio station.

   (g) Be sure to turn off the DEMODULATOR when completed with this measurement.

6. Record the conducted emissions on each AC phase and neutral. Use of a digital camera or plotter/printer to record data is an acceptable means of displaying measured results.

7. Standards EN 55011 and EN 55022 specify both quasi-peak and average emissions limits. Any emissions measured in the peak mode that are below the quasi-peak and average limits are considered to meet the requirements, and no further testing is required at these frequencies. If, however, peak measurements indicate an emission is over the average limit and/or the quasi-peak limit, additional measurements will be required using the quasi-peak and/or the average detector functions of the analyzer:

   (h) Adjust the analyzer so that the emission to be measured is at the center of the analyzer display.

   (i) Set the SPAN to 50 kHz.

(j) Activate the quasi-peak detector function of the analyzer. Record results in the test data sheet.

(k) Disable the quasi-peak detector and activate the average detector function of the analyzer. Record result in the test data sheet.

If the recorded data in items 6 and 7 are below the quasi-peak and AVERAGE limits, respectively, the RF signal meets the limits requirement of EN 55011 and EN 55022.

Using a Voltage Probe (Figure B.8). If using a voltage probe, probe insertion losses and a 10-dB attenuator can be accounted for by using the spectrum analyzer's REF LEVEL OFFSET function. Refer to the calibration chart of the probe to

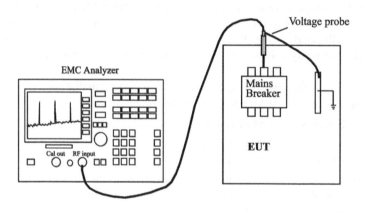

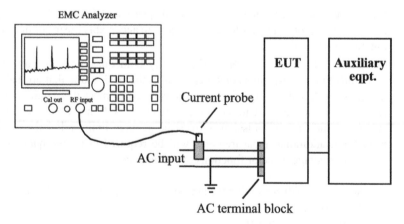

Figure B.8 Conducted emission test setup using voltage and current probes.

determine actual REF LEVEL OFFSET value. The offset will be 10 dB + probe correction factor (in decibels). *Note:* If the voltage probe correction is 30 dB, the total REF LEVEL OFFSET will be 40 dB when using an auxiliary 10-dB attenuator.

Using a Current Probe (Figure B.8). To determine the RF voltage on each AC mains line, from the current reading obtained when using a current probe, add the current probe correction to the reading obtained at that frequency, then add 34 dB [20 log(50 Ω)]:

$$\text{RF mains voltage (dB}\mu\text{V)} = \text{reading } [\text{dB}\mu\text{V} + \text{CF (dB)} + 34 \text{ dB}]$$

PART II: IMMUNITY TESTING

General Information

Unlike emissions, there is no generic "pass" or "fail" criteria when measuring system immunity to electromagnetic energy, either radiated or conducted fields. The generic immunity standards EN 61000-6-1 and EN 61000-6-2 make use of Performance Criteria which describe various levels of performance degradation. The manufacturer is permitted to define performance criteria based on the product specification for an acceptable level of EMC or degradation.

What the manufacturer classifies as a pass criterion may be defined by the test standard as a failure. An example of a modified criterion for a particular range of products is semiconductor manufacturing equipment. Performance Criterion A mandates a unit not shut down or require user intervention should an immunity event occurs. The manufacturer instead wants the system to shut down, automatically requiring user intervention. User intervention could take up to an hour to restart the machine. The reason why a system failure is acceptable is because the manufacturer wants to stop the machine from damaging very expensive silicon wafer(s) that are physically located inside the tool. These wafers must be manually removed if the system fails. Millions of dollars are at stake inside a machine, which is why total system failure is acceptable for this situation. When redefining a Performance Criteria, use of a Technical Construction File assessed by a European Competent Body is required if one wants to change acceptance definition. One cannot use the Standards Route for this type of equipment or assessment process.

Pretest Preparations

With the aid and support of the system owner, for example, a person who knows the makeup of the system hardware, software, and how all subassemblies are interconnected:

1. Define EUT system boundaries. Record the type, make, model, and serial numbers of devices or subsystems that define the unit to be tested. Include all support equipment and instrumentation used to perform the test.

2. Make note and distinguish between *user/operator* and *maintenance/installer* access areas. Electrostatic discharges are applied only to those points or surfaces of the EUT that are accessible to personnel during normal use, not maintenance.

3. Determine which devices and subsystems can be turned on and off during testing without requiring extensive preparation or boot-up time. Turning devices on and off helps to quickly locate the elements within the system that are susceptible to immunity events should abnormal operating conditions occur.

4. Diagram the EUT system indicating location of all devices and current equipment relative to one another. Indicate placement, type, and routing of

- AC power lines
- RF cables
- I/O cables
- Transducer cables
- Alarm cables
- Ethernet or other network cables

Note: The electromagnetic disturbances generated during immunity tests can trigger alarms and other personnel (life) safety systems if installed within a specific facility. Should a life safety sensor be activated, a costly and unnecessary building evacuation may occur. Coordination with the person in charge of the facilities prior to in situ immunity testing must occur to prevent activation of alarms or evacuation should the immunity event trigger an alarm.

A. Electrostatic Discharge (ESD) Immunity

Standard Used as Guide to Testing	IEC 61000-4-2 (EN 61000-4-2)
Test Severity	Per the product standard specification being referenced
Pass Criteria	Criterion B (definition in the Generic Standard)

Equipment Used to Perform Test

ESD simulator	Capable of producing contact and air discharges, wave shapes, and voltage levels described in IEC 61000-4-2 (EN 61000-4-2)
ESD table	Wood table, 0.8 m above a ground plane with a second ground plane on top. Both ground planes are connected together with two 470-kΩ resistors in series.
Coupling planes	Metal plate or screen, 2 m × 0.3 m × 0.65 mm (used for both horizontal and vertical coupling)

Preequipment Check and Calibration

1. Set ESD generator to produce 8 kV, air discharge. Connect the ground wire from the gun to the ground plane.

2. Discharge the ESD applicator to the ground plane. A spark approximately 0.5 cm long should be visible. If not, increase the voltage level until a spark is observed. Lack of a spark means the ESD gun is not operational.

3. For calibration, it is recommended that the ESD gun be sent to a qualified laboratory, as calibrating the ESD gun is a complicated and expensive process requiring a special Pelegrini calibration plate assembly.

Safety warning: The discharges defined in IEC/EN 61000-4-2 are not dangerous but may present harm to anyone fitted with a pacemaker or other implanted medical device. Custom-manufactured ESD simulators can be lethal. Standard EN 61010-1, *Safety Standard for Machinery,* provides guidance that voltages up to 15 kV peak or DC are considered unsafe if they can discharge >45 μC (microcoulombs) and voltages above 15 kV are unsafe if the energy associated with them exceeds 350 mJ (millijoules). For this reason, ensure that high-voltage protection measures are provided for test personnel, especially if these limits are to be exceeded. The value of 45 μC implies a maximum capacitance of 3 nF at a voltage of 15 kV (or 5.6 nF at 8 kV). For inductively generated ESD under 15 kV, it is recommended to apply the 350-mJ limit for personnel protection.

Test Setup. Before performing tests related to an ESD event, it is essential that the EUT be exercised in all modes of operation, and any test software should reflect the state of operating condition. Monitoring equipment should be adequately decoupled to prevent erroneous failures.

The following test points must be identified ahead of time:

- Points on metallic sections of a cabinet that are electrically isolated from ground
- Any point in the control or keyboard area and any other point of man–machine communication, such as switches, knobs, buttons, and other operator-accessible items
- Indicators, LEDs, slots, grills, connector hoods, and so on

Direct discharges should only be applied to those points and surfaces accessible to operators during normal usage, including maintenance personnel. Areas that are accessible to maintenance personnel internal to the system need not be tested, as the maintenance people should have already discharged themselves when approaching this section of system with their body. The ESD simulator should be gradually increased from a minimal voltage level to maximum to ascertain the threshold of failure. Fifty single discharges should be applied in the most sensitive polarity on the unit, allowing at least 1 second between discharges. The discharge points may be selected by exploration using a repetition rate of 20 discharges per second.

The EUT is located over a ground reference plane to which the ESD generator's ground wire is securely bonded. The ground plane shall be a nonmagnetic metal sheet (copper or aluminum) of 0.25 mm thickness. Other metals may be used with a minimal thickness of 0.65 mm. Overall size of the ground plane shall be 1 m × 1 m for table-top equipment. The final size will depend on the dimension of the EUT. In addition, the plane shall be connected to the safety earth system of the laboratory.

The ground plane must project at least 0.5 m beyond the EUT. The ground wire is calibrated with the generator and must not be changed. The ground wire must also be kept away from the EUT and other structures by 0.2 m minimum distance and the test engineer's body. The separation distance for the EUT above the ground plane is 10 cm for floor-standing systems and 80 cm for table-top equipment. Equipment cabinets shall be connected to safety earth ground directly to the ground reference plane through the earth terminal of the EUT.

Directly beneath the EUT for table-top equipment is a secondary plane, identified as the horizontal coupling plane (HCP). The HCP is connected to the primary ground plane by two 470-kΩ bleeder resistors. These resistors isolate the HCP during the actual discharge while allowing residual charge to bleed off after the event. The resistors are located at each end of the HCP. Specification on the bleeder resistors is found in the standard. The EUT is located on the HCP with its front face 10 cm from the edge of the plane. Cables are draped off the HCP (Figure B.9).

The ESD gun must be perpendicular to the surface being tested. If the gun is at any other angle with respect to the plane, stray capacitance between the front of the gun and the EUT can occur, affecting test results. For the air discharge test, the tip shall approach the system as fast as possible without causing mechanical damage to the EUT or coupling plane.

The power supply and input and output circuits shall be connected to appropriate interfaces. Cables supplied or recommended by the manufacturer shall be included in the test. In the absence of a recommended cable, unshielded cables are required.

Test Requirements (EN 55024, ITE Products). A static discharge shall be applied only to those points and surfaces of the EUT expected to be touched or exposed to a discharge event during normal operation. These areas include user access areas that a person may contact either intentionally or unintentionally. This includes ribbon and paper cartridge holders.

The test shall be performed in two ways: contact and air. *Contact* refers to direct injection of the charge into the system whereas *air* refers to a propagated electromagnetic field that enters the system from a remote source, causing harmful operation or disruption. There are two types of discharge probes used with the ESD gun—round and sharp point. The round tip is geared toward simulating a standard size finger that makes direct contact with the system. The sharp point is used for contact to the coupling planes to create a radiated electromagnetic field or to penetrate coatings that the rounded probe tip cannot penetrate.

(a) *Contact discharges are to be made only to conductive surfaces and to coupling planes:* The EUT shall be exposed to at least 200 discharges, 100 each

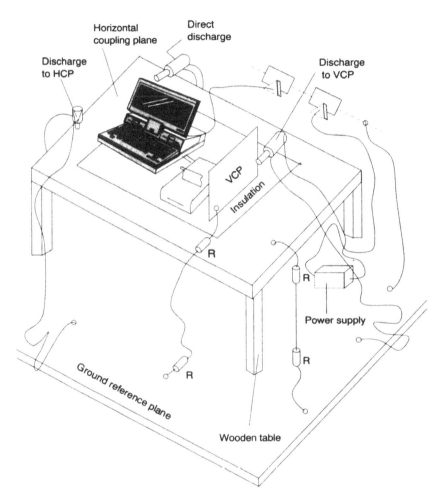

Figure B.9   ESD test setup for table-top equipment. (Illustration courtesy of IEEE Press, Kodali, 2001. *Engineering Electromagnetic Compatibility*. 2nd ed.)

in both the positive and negative polarities, at a minimum of four test points (a minimum of 50 discharges at each point). One of these test points shall be subjected to a minimum of 50 indirect discharges to the center of the front edge of the HCP. The remaining three test locations shall receive at least 50 direct contact discharges. If impossible to create a direct charge to the system (i.e., plastic front bezels prevent an ESD event to penetrate into the system from a direct charge), use of the vertical coupling plane (VCP) becomes mandatory. At least 200 direct discharges shall be applied to the VCP at different locations around the EUT. The maximum repetition rate shall be one discharge per second. If the metal enclosure has a paint finish or other coating that does not permit the ESD arc to occur, use of the probe tip that has a sharp point is required to penetrate the coating and to ensure that a discharge does occur.

(b) *Air discharge at slots, apertures, and insulating surfaces:* For products where direct discharge is not possible, the EUT should be investigated to identify user-accessible points where breakdown may occur. Examples include openings at edges of switches, covers of keyboards, and telephone handsets. These unique points are tested using the air discharge method and should be restricted to those locations normally handled by the user. A minimum of 10 single air discharges shall be applied to the selected test point for each area identified.

Test Procedure

Contact Method (IEC 61000-4-2/EN 61000-4-2)

1. Using the setup configuration detailed in the test standard, configure the EUT and ESD simulator as shown in Figure B.9 for table-top equipment and Figure B.10 for floor-standing equipment.
2. At each contact discharge test point previously identified in the test requirements matrix:

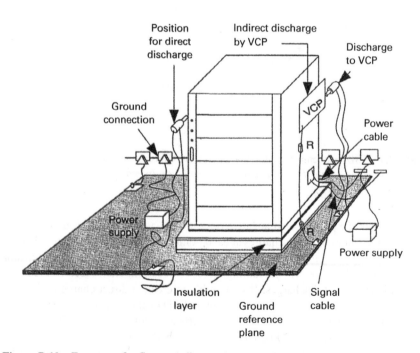

Figure B.10   Test setup for floor-standing equipment. (Illustration courtesy of IEEE Press, Kodali, 2001. *Engineering Electromagnetic Compatibility*. 2nd ed.)

(a) Set the ESD simulator to produce +0.5 kV discharge (or other voltage level deemed appropriate for the test being performed).

(b) Place the ESD simulator at a previously identified test point and hold the gun perpendicular to the test location before pressing the trigger. Press the simulator as hard as possible to the system without causing physical damage. If the pointed tip of the simulator does not permit a discharge to occur, substitute this tip with the pointed one to penetrate the paint or nonconductive surface.

(c) Discharge the simulator a minimum of 50 times, allowing at least 1 second between successive discharges. Do not move the gun from the test location. Most simulators can be programmed to provide a discharge automatically based on the number of discharges desired.

(d) Repeat the step above using higher voltage levels in increments of 0.5 kV, or other voltage levels deemed appropriate at each contact discharge test point identified in the test requirements matrix. The typical maximum voltage level is ±4 kV.

(e) Reverse the polarity of the discharge and repeat previous four steps.

(f) Record any EUT malfunctions, interrupts, resets, or failures.

Test Procedure—Air Discharge Method (IEC 61000-4-2/EN 61000-4-2). The following applies for both HCP and VCP. After completing one polarity of testing on a coupling plane, repeat using the other polarity.

1. Configure both HCP and VCP as shown in Figure B.9 for table-top equipment and Figure B.10 for floor-standing equipment (detailed in the test standard).

2. At each air discharge test point on the coupling plane, as detailed in the test requirements matrix:

   (a) Set the ESD simulator to produce +0.5 kV discharge (or other voltage level deemed appropriate for the test being performed).

   (b) Approach the center of the thin edge of the coupling plane with the ESD gun held perpendicular to the edge of the coupling plane. Upon approach, the gun will discharge into the plate. Discharge the simulator a minimum of 10 times allowing at least 1 second between successive discharges.

   (c) Record any EUT malfunctions, interrupts, resets, or failures.

   (d) Repeat above steps for higher voltages in 0.5-kV increments until the desired maximum voltage level is reached. The typical maximum voltage level is ±8 kV.

   (e) Reverse the polarity of the discharge and repeat above steps.

   (f) Record any EUT malfunctions, interrupts, resets, or failures.

When evaluating results of ESD testing, it is important to ascertain if the problem is permanent, latent, or temporary. In addition, manufacturers may define failure modes differently. What passes for one company may be classified as a failure for another. Typical severity levels V_{ESD} are defined by international standards and are as follows:

Severity Level	V_{ESD} Direct Discharge (kV)	Air Discharge (kV)
1	2	2
2	4	4
3	6	8
4	8	15

Criterion A. The apparatus shall continue to operate as intended. No degradation of performance or loss of function is allowed below a performance level specified by the manufacturer when the apparatus is used as intended. In some cases, the performance level may be replaced by a permissible loss of performance. If the minimum performance level or permissible performance loss is not specified by the manufacturer, then either of these may be derived from the product description and documentation and what the user may reasonably expect from the apparatus if used as intended.

Criterion B. The apparatus shall continue to operate as intended after the test. No degradation of performance or loss of function is allowed below a performance level specified by the manufacturer when the apparatus is used as intended. In some cases, the performance level may be replaced by a permissible loss of performance. During the test, degradation of performance is however allowed. If the minimum performance level or permissible performance loss is not specified by the manufacturer, then either of these may be derived from the product description and documentation and what the user may reasonably expect from the apparatus if used as intended.

Criterion C. The apparatus has degradation or loss of function or performance that is not recoverable due to permanent damage of equipment (components) or software or loss of data.

B. RF Electromagnetic Field Immunity

Standard Used as Guide to Testing IEC 61000-4-3 (EN 61000-4-3)

Test Severity   3 or 10 V/m (can be higher or lower as required)
                            80% AM depth at 1 kHz

Frequency Range   80–1000 MHz

Alternate Frequency Range 27/150/450/950 MHz (by arrangement with a European Competent Body)

Pass Criteria   Criterion A (definition in the Generic Standard)

Equipment Used to Perform Standard Test

Spectrum analyzer	80–1000 MHz (minimum); an internal tracking generator is highly desired
RF signal generator	80–1000 MHz (minimum); 1 V RMS output, automatic sweep/step capabilities
	Internal 1-kHz AM capability, adjustable 0–100%. Output level must match the input requirement of the power amplifier within a margin of a few decibels
RF amplifiers	15 W minimum linear RF amplifier(s); 80–1000 MHz
Isotropic field intensity meter	80–1000 MHz, 2 V minimum
Antennas	Biconical, log periodic or bilog

Preequipment Check and Calibration. The calibration procedure for radiated field immunity testing is detailed in the test standard and performed in an anechoic or semianechoic chamber. The procedure requires calibration at 16 points along the front plane where the EUT is to be positioned. The amount of radiated energy from the power amplifier is monitored by an isotropic field intensity meter and recorded. Use of automation software is valuable when performing this test. Constant monitoring of the field strength is important. Having software that will adjust the output level of the power amplifier or signal generator automatically, as measured by the field intensity meter, makes the job of testing more enjoyable. Manually adjusting the power level throughout the frequency spectrum is not recommended for obvious reasons.

A synthesized signal generator is used to cover the frequency range required in a series of discrete frequency steps under the control of the test system's software. The required frequency accuracy depends on whether the EUT exhibits any narrow-band responses to interference. Manual operation is necessary for examining discrete frequencies.

Field intensity meters are not particularly linear and require calibration at the same field strength at which they are expected to be used. Also, field intensity meters can provide erroneous readings on a modulated signal. To ensure accuracy, the level setting must only be attempted on an unmodulated field.

Bilog or biconical and log-periodic antennas must be protected against accidental damage during testing. The power-handling ability of antennas are limited by a balun transformer located at the antenna's feed point. This is usually a wideband ferrite core (1 : 1 transformer) that converts the *balanced* feed of the antenna to the *unbalanced* connection of the coax cable. It is supplied as part of the antenna. Antenna calibration includes a factor to allow for balun losses, which are usually minimal.

Test Setup (Anechoic Chamber Environment). The EUT is located in an anechoic chamber at a distance specified by the test standard. Use of anechoic material on the ground plane of the chamber is required to prevent reflected RF signals from being reinjected into the EUT. Reflections can cause an increase in field intensity levels at the EUT with the direct wave and reflected wave combining together. Testing a system outside an anechoic chamber is not permitted, even for preliminary, final, or troubleshooting purposes. If it is not possible to perform the test in an anechoic chamber, an alternate test method and procedure is required (next page).

Set up the equipment in accordance with the test standard as shown in Figure B.11. This figure provides only guidance. After preliminary tests have been performed at higher than desired RF levels to ascertain the ability of the EUT to withstand fields and sensitive configuration, the system setup must be documented and rigorously maintained throughout the compliance test. Changes in configuration during the test will invalidate test results along with lost test time. Equipment should be evaluated in conditions that are as close as possible to a typical installation, including wiring and cabling.

If the EUT is floor standing (such as a rack or cabinet), it must be placed on an insulator that is itself on the floor. If possible, use of a wooden table is recommended for this equipment, as the antenna will normally be placed at least 1 m from the EUT. A greater distance is possible consistent with generating an adequate field strength across the plane of the EUT. The preferred distance is 3 m. Too close a distance affects the uniformity of the generated field because of mutual coupling between the antenna and EUT.

For most types of electronic equipment, a total of eight sweeps are required—one in each polarization of the antenna. There are four sides of the EUT. If the equipment can be used in any orientation, such as portable headphones or entertainment devices, it is recommended to test in other positions, such as top and bottom.

The sweep rate is clearly defined in the standard, which permits one to evaluate the performance level of the system being disrupted. The signal generator should either be manually or automatically swept across the output range at 1.5×10^{-3} decades/second or slower, depending on the speed of response of the EUT. The

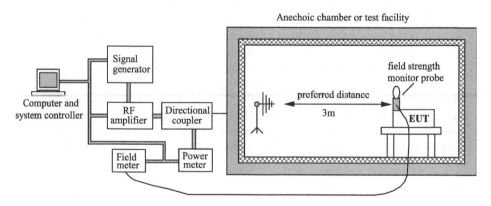

Figure B.11   RF radiated immunity sweep test.

generator can also be automatically stepped at this rate in steps of typically 1% of the current frequency. The dwell time for stepped application should provide sufficient time for the EUT to respond to a disruption. A need for a slower sweep rate translates directly to a longer test time.

Test Procedures Using RF Sweep Method

1. There should be a minimum of four test locations around the periphery of the EUT. If the EUT is not placed on a turntable, manual placement of antennas around the EUT is required or the EUT is manually moved to different positions relative to the transmit antenna. Refer to Figure B.11 for typical equipment setup.
2. Starting at 80 MHz with the biconical (bilog) antenna in the horizontal polarity, adjust the output level of the signal generator until a desired field strength is observed on the field strength meter located along the front plane of the EUT. Increment the frequency at a rate of no greater than 1% of the fundamental (800 kHz for the starting frequency of 80 MHz), with a dwell time of at least 1 second for each frequency tested.
3. Continue incrementing the fundamental frequency through the entire frequency range for the antenna being used.
4. Repeat steps 2 and 3 with the antenna in the opposite polarity.
5. Repeat steps 2–4 with the log-periodic antenna if a bilog antenna is not used.
6. Record changes in system behavior along with the frequencies and polarities at which they occur.

Alternate RF-Radiated Immunity Test Procedure (Hand-Held Radios). Two-way hand-held radios may be used as a substitute means of performing electromagnetic field immunity tests ONLY for *in situ* environments and with the approval of a European Competent Body. In the United States and Canada, it is illegal to transmit a radio signal into free space as disruption to communication services can occur to broadcast stations. For this reason, use of anechoic chambers is required. However, to perform in situ tests to Europe's EMC Directive in North America, an alternative means of performing this test must be available. This alternative test may not be accepted by a Competent Body. Consult with the Competent Body prior to use of this alternate procedure. The following hand-held radios (walkie-talkies) are used:

(a) 27-MHz band—citizen band frequencies, 3 W
(b) 156-MHz band—VHF business operations (police, fire, emergency services, etc.), 2 W
(c) 462-MHz band—GMRS and facility maintenance, 2 W
(d) 900-MHz band—cellular phone, 3 W

A photograph of typical radios and field intensity meter is shown in Figure B.12.

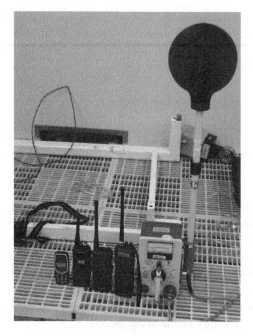

Figure B.12 Hand-held radios and field intensity meter.

Alternate Calibration Procedure (Hand-Held Radios)

Pretest Equipment and Calibration

1. Place the isotropic field intensity meter (FIM) and transmitter at a specified distance from each other (Figure B.13). If a FIM is not available, antennas may be substituted (biconical, log periodic, or bilog).
2. Key the transmitter at a known distance from the FIM and read the measured value.
3. Calculate the field strength of the radios using the following equation. If the reading is different, determine distance spacing between transmitter and sensor to achieve desired field strength.

$$V/m = 10^{(dB\mu V/m - 120)/20}$$

where $dB\mu V/m$ is the actual measured field strength value.

Test Procedures (Hand-Held Radios)

1. There should be a minimum of four locations around the periphery of the EUT where testing will take place. Refer to Figure B.14 for a photograph of an actual test being performed to illustrate how to perform this procedure with the radios.
2. Activate the push-to-talk (PTT) button on the hand-held transceiver five

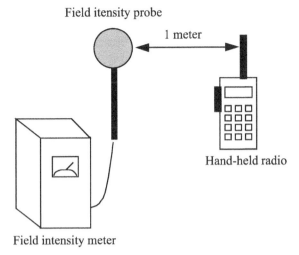

Figure B.13   Equipment check for radiated immunity.

Figure B.14   Radiated immunity testing with hand-held radio in a semiconductor manufacturing clean room.

times in quick succession followed by a sustained transmit time of 30 seconds during which the tester should speak into the microphone to modulate the signal. Keep the distance spacing between the transmitter and EUT at the calibrated distance between source and system.

3. Record changes in system behavior and the frequencies at which they occur.
4. Repeat step 2 with the antenna held horizontal, with respect to ground, as many times as required. Ensure the antenna is parallel to seams, door openings, and other discontinuities in the shielding structure of the EUT.
5. Repeat steps 2–4 for each location with each transceiver.

Evaluation of Test Results. The following table provides the standard test levels for ITE and the majority of other related products:

Severity Level (V/m)	Test Field Strength
1	1
2	3
3	10
X	Special

Note: X is an open test level. This level may be given in the product specification.

The following provides guidance on evaluating test results based on the severity level chosen:

Criterion A. The apparatus shall continue to operate as intended. No degradation of performance or loss of function is allowed below a performance level specified by the manufacturer when the apparatus is used as intended. In some cases, the performance level may be replaced by a permissible loss of performance. If the minimum performance level or permissible performance loss is not specified by the manufacturer, then either of these may be derived from the product description and documentation and what the user may reasonably expect from the apparatus if used as intended.

Criterion B. The apparatus shall continue to operate as intended after the test. No degradation of performance or loss of function is allowed below a performance level specified by the manufacturer when the apparatus is used as intended. In some cases, the performance level may be replaced by a permissible loss of performance. During the test, degradation of performance is however allowed. If the minimum performance level or permissible performance loss is not specified by the manufacturer, then either of these may be derived from the product description and documentation and what the user may reasonably expect from the apparatus if used as intended.

Criterion C. Temporary loss of function is allowed provided the loss of function is self-recoverable or can be restored by the operation of the controls.

C. Fast Transient/Burst Immunity

Standard Used as Guide to Testing	IEC 61000-4-4 (EN 61000-4-4)
Test Requirement	5 ns rise time/50 ns fall time; 5 kHz repetition frequency
Light industrial	AC mains: 1 kV
	DC power: 500 V
	Process and control: 500 V
Heavy industrial	AC mains: 2 kV
	DC power: 1 kV
	Process and control: 1 kV
Pass Criteria	Criterion B (definition in the Generic Standard)

Equipment Used to Perform Test

EFT generator	Capable of producing voltages and wave shapes specified in paragraph 6.1.1 of IEC 61000-4-4 (EN 61000-4-4)
Capacitive clamp	Described in paragraph 6.3 of IEC 61000-4-4 (EN 61000-4-4)
Reference ground plane	Copper or aluminum, 1 m × 1 m × 0.25 mm (for other metallic materials, minimum thickness 0.65 mm)
Miscellaneous	Tape or metallic foil for capacitive coupling to communications and I/O circuits as alternative to capacitive clamp or alligator clip for direct connection to AC control lines

Pretest Equipment Check and Calibration. Follow the manufacturer's instructions for pretest calibration check.

Test Setup Environment. This test is a wideband spectral event with frequency components up to hundreds of megahertz. For repeatability, the coupling of the burst energy is strongly dependent on the EUT's stray capacitance to its surroundings and test equipment setup. Test environment concerns include the following:

- *Coupling of the EUT, generator, and capacitive clamp to the ground plane:* The ground plane must be at least 1 m² and should extend beyond the EUT by at least 10 cm in all directions. Both burst generator and coupling clamp must be bonded to the ground plane by a short, wide-braid strap, not a regular piece of wire. Because of this grounding requirement, the EFT generator and clamp are normally located on the floor of a test facility and not on a table. The EUT is spaced away from the ground plane by 10 cm for floor-standing equipment or 80 cm for table-top equipment on a wooden table.

- *Clearance distance around the EUT:* The minimum distance from all conducting structure to the EUT must be greater than 0.5 m. The coupling clamp itself must also have a clear distance around it of at least 0.5 m.
- *Grounding:* If the EUT is normally provided with an AC mains ground, ensure the EUT is bonded to the ground plane in a representative manner; otherwise, do not connect the EUT to the ground plane.
- *Cable layout:* The distance between the EUT and coupling clamp must be 1 m or less. Long cables should be coiled into a 40 cm diameter loop 10 cm above the ground plane, which differs from the cable bundling requirements of radiated emissions. It is recommend to maintain cable separation at least 10 cm from the ground plane at all times.

The test requires at least 1 min of energy transference for each coupling mode or AC mains line. For a repetition period of 300 ms, this represents approximately 200 bursts. The basic or generic standard describing this event will identify which ports are to be tested and the voltage levels required. The basic test setup is shown in Figures B.15a–c, which illustrate EUT connection both with and without coupling clamp in different configurations.

Test Procedure

AC Mains Rated 230 V and Less Than 16 A Input Current (Figure B.15a)

1. Place the EUT POWER SWITCH to the ON position. Keep the burst generator in the OFF position.
2. Select OUTPUT receptacle option: "Power outlet receptacle or HV COAX OUTPUT." *Note:* Some generators may not be capable of accepting the AC mains plus configuration from the EUT; thus use of the coupling clamp may be required, connected to the generator with a special high-voltage coax.
3. For devices requiring less than 16 A input current, plug the EUT into the receptacle marked EUT POWER OUTLET on the test generator. Turn on the generator.
4. Adjust the voltage level to 0.5 or 1.0 kV, depending on minimum voltage test requirement.
5. Select the polarity to either positive or negative.
6. Test the product and observe the behavior of the EUT for 1 minute, noting any abnormalities or functional failures in the following order.
7. Selection COUPLING SELECTION: L.
8. Selection COUPLING SELECTION: N.
9. Selection COUPLING SELECTION: PE.
10. Selection COUPLING SELECTION: L + N.
11. Selection COUPLING SELECTION: L + PE.
12. Selection COUPLING SELECTION: N + PE.
13. Selection COUPLING SELECTION: L + N + PE.

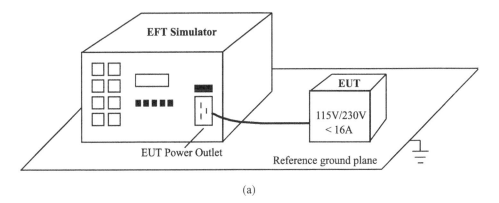

(a)

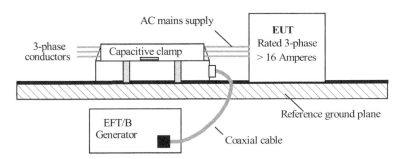

The coupling device shall be conductive tape or foil wrapped around as closely as possible to the cables to be tested. The coupling clamp described in the standard may be used if cable bundles will fit. If tape or foil coupling method is used, capacitance shall be equivalent to that of the coupling clamp (33 nF).

(b)

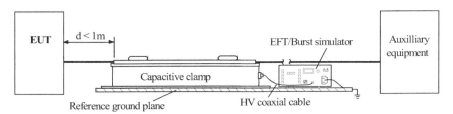

The coupling device shall be conductive tape or foil wrapped around as closely as possible to the cables to be tested. The coupling clamp described in the standard may be used if cable bundles will fit. If tape or foil coupling method is used, capacitance shall be equivalent to that of coupling clamp (33 nF).

(c)

Figure B.15 Fast transient /burst immunity: (*a*) AC mains for single phase < 16 A; (*b*) AC mains for three phases > 16 A; (*c*) communications and I/O lines.

14. Repeat steps 7–13 with the next higher voltage level required until the maximum voltage level is reached.
15. Change polarity of the test signal and repeat steps 7–14.
16. Note and record changes in EUT operation, fault conditions, or any other system behavior that is monitored for performance criteria.

AC Mains Test with Device Rated Three Phase or Greater Than 16 A (Figure B.15b)

1. Place the capacitive coupling clamp around one of the phases (preferred test method). If this is not possible, place the entire bundle in the capacitive coupling clamp.
2. Connect the capacitive coupling clamp to the HV COAX OUTPUT connector of the test generator using a short length of special high-voltage RF coaxial cable.
3. Select COAX option on the power output selector switch.
4. Adjust the voltage level to 0.5 or 1.0 kV, depending on minimum voltage test requirement.
5. Select the polarity to either positive or negative.
6. Test the product and observe the behavior of the EUT for 1 minute.
7. Change the polarity of the test signal and repeat test.
8. Increase the voltage to the next higher level required and repeat test until the maximum voltage level is reached.
9. Note and record changes in EUT operation, fault conditions, or any other system behavior that is monitored for performance criteria.

I/O and Communications Lines Greater Than 3 m in Length (Figure B.15c)

1. Place the capacitive coupling clamp around the I/O or communication lines.
2. Connect the capacitive coupling clamp to the HV COAX OUTPUT connector of the test generator using a short length of high-voltage RF coaxial cable.
3. Set the generator to the high-voltage coaxial mode output connector and select 0.5 kV or the minimum test level desired, positive polarity.
4. Run the test for at least 1 minute unless otherwise specified in the test requirements plan.
5. Repeat previous steps for the next voltage level described in the test plan.
6. Repeat previous steps for negative-polarity pulses.
7. Note and record changes in EUT operation, fault conditions, or any other system behavior that is monitored for performance criteria.

Evaluation of Test Results. The following provides guidance on evaluating test results based on the severity level chosen.

Criterion A. The apparatus shall continue to operate as intended. No degradation of performance or loss of function is allowed below a performance level specified by the manufacturer when the apparatus is used as intended. In some cases, the performance level may be replaced by a permissible loss of performance. If the minimum performance level or permissible performance loss is not specified by the manufacturer, then either of these may be derived from the product description and documentation and what the user may reasonably expect from the apparatus if used as intended.

Criterion B. The apparatus shall continue to operate as intended after the test. No degradation of performance or loss of function is allowed below a performance level specified by the manufacturer when the apparatus is used as intended. In some cases, the performance level may be replaced by a permissible loss of performance. During the test, degradation of performance is however allowed. If the minimum performance level or permissible performance loss is not specified by the manufacturer, then either of these may be derived from the product description and documentation and what the user may reasonably expect from the apparatus if used as intended.

Criterion C. Temporary loss of function is allowed provided the loss of function is self-recoverable or can be restored by operation of the controls.

D. Surge Immunity

Standard Used as Guide to Testing	IEC 61000-4-5 (EN 61000-4-5)
Test Waveform	1.2 μs rise time/50 μs fall time, combination wave
Test Severity	Refer to test standard for details based on environment
Light industrial	AC mains: 2 kV common mode, 1 kV differential mode
	DC power: 500 V common mode, 500 V differential mode
	Process and control: 1 kV common mode, 500 V differential mode
Heavy industrial	AC mains: 4 kV common mode, 2 kV differential mode
	DC power: 500 V common mode, 500 V differential mode
	Process and control: 2 kV common mode, 1 kV differential mode
Pass Criteria	Criterion B (definition in the generic standard)

Equipment Used to Perform Tests

Combination wave generator	Capable of producing voltages and wave shapes specified in IEC 61000-4-5 (EN 61000-4-5)
Backfilter protection	Rated for appropriate operation, single or three phase

Pretest Calibration and Equipment Check. Follow the surge generator and coupling network calibration procedure as detailed in the manufacturer's instruction manual. The surge waveforms in Chapter 6 must appear at the output of a compliant generator when calibrated with both short-circuit and open-circuit loads. The waveform through the mains coupling/decoupling network must also be calibrated and be unaffected by the network. This requires two calibration tests. There are no calibration requirements for coupling devices used for signal lines. The signal line coupling network includes a 40-Ω series resistor, which substantially reduces the energy in the applied surge. For mains coupling, the generator is connected directly through an 18-μF capacitor across each phase. For phase-to-earth applications, a 10-Ω resistor and a 9-μF capacitor are used. Under this configuration, the highest energy level available from the generator's effective source impedance of 2 Ω is applied only between phases.

Test Setup Environment. The test environment and setup for surge testing do not require special considerations since the frequency spectral content of the surge waveform is very low. For this reason, cable layout is not specified in the test standard, except that the length of the cable must be 2 m or less in length. Figure B.16 provides guidance on the test setup for surge testing.

When injecting the surge energy pulse, the signal line must be directly coupled to the test network. No clamp-on or noninvasive devices are available or specified for this test. For signal lines that are affected by a 0.5-μF capacitor, use of a gas dis-

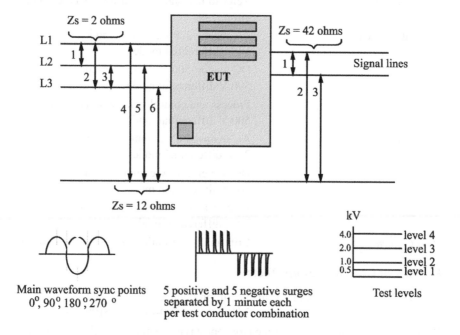

Figure B.16 Surge immunity test setup.

charge tube or surge arrester is permitted by the standard. This alternate coupling technique is applied longitudinally along the shield by coupling it directly to the EUT at one end of a noninductively bundled 20-m-long cable. The other end of the cable is connected to ground. This test is performed without the series resistor. This means that the surge current down the cable shield can be in the range of several hundred amperes.

If the last test consideration were implemented, surge testing would take 16 hours. It is the responsibility of the test engineer to ascertain the magnitude of this test and what shortcuts can be applied to minimize test time while ensuring conformity of the EUT.

Test Procedure. A test plan needs to be prepared which specifies the following:

- A list of all required equipment
- Input and output ports to be tested
- Representative operating condition of the EUT
- Sequence of application of the surge to the circuits
- Actual installation conditions, for example

    AC: neutral earthed

    DC: (+) or (−) earthed to simulate the actual earthing conditions

The test level for surging is based on installation conditions:

Class 0: Well-protected electrical environment, often within a special room.

Class 1: Partly protected electrical environment.

Class 2: Electrical environment where the cables are well separated, even at short runs.

Class 3: Electrical environment where cables run in parallel.

Class 4: Electrical environment where the interconnections are running as outdoor cables along with power cables and cables are used for both electronic and electric circuits.

Class 5: Electrical environment for electronic equipment connected to telecommunication cables and overhead power lines in a nondensely populated area.

Class X: Special conditions specified in the product specification.

The surge waveforms must appear at the output of a compliant generator when calibrated with short-circuit and open-circuit loads. The waveform through the mains coupling/decoupling network must also be calibrated and be unaffected by the network. For coupling networks used on signal lines, this requirement is waived. Three different source impedances are recommended, depending on the application of the test voltage and expected operating conditions of the EUT. The effective output impedance of the generator is defined as the ratio of peak open-circuit output voltage to peak short-circuit output current, which is 2 Ω.

The signal line coupling networks includes a 40-Ω series resistor, which reduces the energy in the applied surge substantially. For AC mains coupling, the generator is connected through an 18-µF capacitor across each phase with a 10-Ω resistor and 9-µF capacitor for phase-to-earth application. This means that the highest energy available from the generator's effective source impedance of 2 Ω is actually applied only between phases (Figure B.16). To perform surge testing, the following procedure is recommended:

- Apply a minimum of five positive and five negative surges at each coupling point.
- Wait at least 1 min between each surge to allow time for any protection device to recover.
- Apply the surges line to line (three combinations for three-phase, one for single phase) and line to ground (two combinations for single phase, four for three-phase).
- Synchronize the surges to the zero crossings and the positive and negative peaks of the mains supply (four possibilities—0°, 90°, 180° and 270° on the AC waveform).
- Increase the test voltage in steps up to the specified maximum level so that all lower test levels are satisfied. Table B.1 illustrates various surge test levels detailed in EN 61000-4-5/IEC 61000-4-5.

Table B.1 Surge Test Levels (Depending on Installation Conditions)[a]

	Test Levels							
	Power Supply, Coupling Mode		Unbalanced Operated, Circuits/Lines, LDB, Coupling Mode		Balanced Operated, Circuits/Lines, Coupling Mode		SDB, DB[b]	
Installation Class	Line to Line (kV)	Line to Earth (kV)	Line to Line (kV)	Line to Earth (kV)	Line to Line (kV)	Line to Earth (kV)	Line to Line (kV)	Line to Earth (kV)
0	NA	NA	NA	NA	NA	NA	NA	NA
1	NA	0.5	NA	0.5	NA	0.5	NA	NA
2	0.5	1.0	0.5	1.0	NA	1.0	NA	0.5
3	1.0	2.0	1.0	2.0	NA	2.0	NA	NA
4	2.0	4.0[c]	2.0	4.0[c]	NA	2.0[c]	NA	NA
5	d	d	2.0	4.0[c]	NA	4.0[c]	NA	NA
x								

[a]Abbreviations: DB = data bus (data line), SDB = short-distance bus, LDB = long-distance bus, NA = not applicable
[b]Limited distance, special configuration, special layout, 10 m to maximum 30 m; no test is advised at interconnection cables up to 10 m, only class 2 is applicable.
[c]Normally tested with primary protections.
[d]Depends on class of local power supply system.

For AC mains tests, use the direct-connect coupling network:

1. Using the common-mode coupling clamp, connect the combination wave generator, with appropriate backfilter to protect lines not to be surge tested, to the AC mains input.
2. Set the generator to produce 0.5 kV, positive-polarity surge, or the lowest test voltage level desired.
3. Apply the surge five times for each setting with a repetition rate of one surge per minute. Note and record changes in EUT operation, fault conditions, or any other system behavior changer per the performance criteria.
4. Increase the generator voltage level by 0.5 kV and repeat test up to 2 kV or desired end voltage level.
5. Repeat previous steps with negative polarity surges.
6. Repeat steps 3–5 for each AC phase and for the protective or function ground connection on the EUT cabinet. For single-phase AC mains, surge test the following lines:

   • Line to Protective Earth Ground
   • Neutral to Protective Earth Ground
   • Line–neutral to Protective Earth Ground

   For three-phase AC mains, surge test the following lines:

   • Line 1 to Protective Earth Ground
   • Line 2 to Protective Earth Ground
   • Line 3 to Protective Earth Ground
   • Lines 1–2 to Protective Earth Ground
   • Lines 1–3 to Protective Earth Ground
   • Lines 2–3 to Protective Earth Ground
   • Lines 1–2–3 to Protective Earth Ground

7. Replace the common-mode (CM) coupling network with the differential-mode (DM) coupling network.
8. Repeat previous steps until desired voltage level is reached.

The surges (and test generator) related to the different classes are as follows:

Classes 1–4: 1.2/50 μs (8/20 μs)
Class 5:   1.2/50 μs (8/20 μs) for ports of power lines and short-distance signal circuits/lines; 10/700 μs for ports of long-distance signal circuits/lines

All surges are to be applied synchronized with the voltage phase at the zero crossing and peak value of the AC voltage wave in both the positive and negative polarity. This means testing at 0°, 90°, 180°, and 270° positive and negative for all lines. All surges are to be applied line to line and line to ground, regardless of the number of phases. When testing line to ground, all phases must be tested successively. This means that the time it takes to test a system can be considerable, especially for a three-phase unit, which has seven successive test runs to perform in both positive and negative polarities and at all phase angles.

Evaluation of Test Results. The following provides guidance on evaluating test results based on the severity level chosen.

> *Criterion A.* The apparatus shall continue to operate as intended. No degradation of performance or loss of function is allowed below a performance level specified by the manufacturer when the apparatus is used as intended. In some cases, the performance level may be replaced by a permissible loss of performance. If the minimum performance level or permissible performance loss is not specified by the manufacturer, then either of these may be derived from the product description and documentation and what the user may reasonably expect from the apparatus if used as intended.
>
> *Criterion B.* The apparatus shall continue to operate as intended after the test. No degradation of performance or loss of function is allowed below a performance level specified by the manufacturer when the apparatus is used as intended. In some cases, the performance level may be replaced by a permissible loss of performance. During the test, degradation of performance is however allowed. If the minimum performance level or permissible performance loss is not specified by the manufacturer, then either of these may be derived from the product description and documentation and what the user may reasonably expect from the apparatus if used as intended.
>
> *Criterion C.* Temporary loss of function is allowed provided the loss of function is self-recoverable or can be restored by operation of the controls.

E. Conducted Disturbances Induced by RF Fields

Standard Used as Guide to Testing	IEC 61000-4-6 (EN 61000-4-6)
Test Severity	3 or 10 V RMS (can be higher or lower as required)
	80% AM depth at 1 kHz
Frequency Range	0.150–80 MHz (can be extended to 230 MHz if required)
Pass Criteria	Criterion A (definition in the Generic Standard)

Equipment Used to Perform Tests

RF power amplifier	15-W minimum linear RF amplifier(s), 0.150–80 MHz (can be higher in frequency if required)

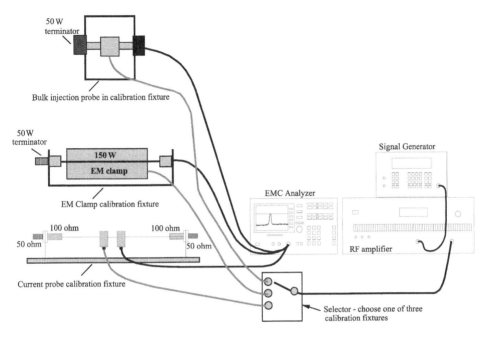

Figure B.17 Pretest calibration setup for conducted immunity (all configurations).

RF signal generator	0.150-230 MHz; 1 V RMS output, automatic sweep/step
	Internal 1-kHz AM capability, adjustable 0–100%
Coupling device	Per IEC 61000-4-6 (EN 1000-4-6), detailed in Chapter 6
Current probe	0.150–230 MHz, with ability to measure significant levels of RF current

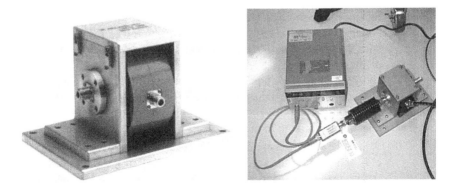

Figure B.18 Pretest calibration for conducted immunity test (bulk current injection config-uration).

Spectrum analyzer	0.150–230 MHz (internal tracking generator is highly desired) or;
Power meter	20 V RMS capability
Miscellaneous	Reference ground plane (metal screen), coaxial cables as required

Pretest Calibration and Equipment Check (Figures B.17 and B.18)

1. Install a 50-Ω coaxial terminator to one end of the calibration fixture.
2. Connect a coaxial cable to the other end of the calibration fixture and to the input connector of a spectrum analyzer or power meter.
3. Connect a coaxial cable from the output of the RF amplifier to the input connector of the injection device being calibrated (EM clamp, current injection probe, or bulk injection probe).
4. Connect a short coaxial cable between the RF signal generator output connector to the RF amplifier input port.
5. Place the clamp or probe in the appropriate calibration fixture using the setup from steps 1–3 above. Use the same cables, amplifier, and signal generator that will be used during actual product testing. For the EM clamp and current injection probe, locate the injection device in the center of the calibration wire between the two end terminals.
6. Turn the gain of the amplifier to minimum (all the way counterclockwise).
7. Turn on the signal generator to the following settings:

FREQUENCY	150 kHz
AMPLITUDE INCREMENT	0.3 dBm
AMPLITUDE	−40 dBm (suggested value)

   (a) International conducted immunity standards specify interference levels in terms of induced voltages. When interference is induced by current injection, one must be able to relate current or drive levels to induced voltage. The EM clamp fixture has built-in 150-Ω resistors in series with the through line. Instructions call for 50 Ω termination at one end of the fixture, and a 50-Ω-impedance spectrum analyzer at the other end. The total loop impedance is therefore 300 Ω. The analyzer sees 50/300 = 1/6 of the induced voltage. Using similar reasoning, the spectrum analyzer sees 50/100 = 1/2 the induced voltage in the bulk current injection calibration fixture.

   (b) Adjust the signal generator output level so that the required voltage level is achieved on the spectrum analyzer (for 3 V RMS, the EM clamp fixture reading is 0.5 V, and for bulk current injection, the value is 1.5 V).

8. Input the following into the spectrum analyzer.

CENTER FREQUENCY	150 kHz
SPAN	25 kHz
AMPLITUDE REFERENCE LEVEL	3000 mV (3 V)
DISPLAY LINE	500 mV (0.5 V) for 3 V RMS or 1000 mV (1.0 V) for 10 V RMS

9. Turn RF amplifier on.

10. When the RF amplifier completes self-diagnostics (optional feature for certain units), turn up the GAIN CONTROL to a reasonable value.

11. Press the AMPLITUDE UP and DOWN arrows on the RF signal generator to adjust the AMPLITUDE of the signal going into the power amplifier until the desired reading (i.e., 0.5/1.0 V) is obtained on the spectrum analyzer or power meter.

12. Record frequency and amplitude level (in dBm) as shown on the RF signal generator display panel.

13. Repeat the previous step for the frequencies 300 kHz, 500 kHz, 700 kHz, 1 MHz, 3 MHz, 5 MHz, 7 MHz, 10 MHz, 30 MHz, 50 MHz, 70 MHz, and 80 MHz and record data in a table with two columns: frequency and respective RF generator amplitude (in dBm). These will be the actual test levels from the RF signal generator and RF amplifier combination required for conformity testing. It is highly recommended to use more frequency points for greater accuracy throughout the frequency spectrum. Suggested step sizes include

- 100-kHz increments 300 kHz–1 MHz
- 1-MHz increments 1–30 MHz
- 2-MHz increments 30–80 MHz

During actual immunity testing, the RF drive levels are monitored to produce a desired interference level. Because the impedance of the EUT's cable assemblies are generally unknown, injection levels will most likely be different from that determined in the calibration fixtures. A second current probe is used to monitor the injected current level to prevent overtesting (current injection higher than theoretical or calibration limits). This current probe is shown as a second device in Figure B.17. Bulk current injection calibration is shown in Figure B.18.

Important note regarding calibration: The test voltage given in the specification for the EM clamp is the open-circuit (EMF) voltage to be applied. Paragraph 6.4.1 and Annex A, paragraphs A1 and A3 in EN61000-4-6, indicate how the test levels are set up in both 150/50-Ω and 50-Ω systems to compensate for the loading effect. Paragraph 7.3 of EN61000-4-6 indicates how the current injected into the line can be monitored to ensure that the equipment is not overtested.

If the current probe injection method is used, calibration data (input power to the probe versus test voltage) shall be included in the test report. The number of calibration points shall be selected to ensure that the voltage level between test frequencies measured at the port of the current probe calibration jig varies by no more than 1 dB with a constant input level to the power amplifier. A minimum of five points per decade shall be provided. For variations in excess of 1 dB, the input value to the amplifier shall be adjusted to restore the test voltage to its calibrated value. The frequency and the voltage measured on the forward power port of the coupler or power meter shall be noted in the calibration table.

When performing the actual conducted immunity test, for each frequency, the output voltage of the RF signal generator shall be adjusted to ensure that the forward power into the RF coupler matches the calibration value previously determined.

Test Preparation. To perform conducted immunity tests, the following guidelines should be followed:

1. In general, it is sufficient that only a limited number of current distributions through the EUT are excited, somewhere between two and five conductors at a time.

2. Testing is intended to be performed using the most sensitive cable configurations. Other cables connected to the EUT from the Auxiliary Equipment (AE) should either be disconnected (when functionally allows) or provided with a decoupling network to prevent responses from the AE to influence EUT performance.

3. If the EUT is comprised of a single unit (floor standing), the system shall be placed on an insulating support 0.1 m above the ground reference plane. For table-top equipment, the reference plane may be placed on a table.

4. For EUTs comprised of several units, the test method shall be either of the following:

    (a) *Preferred method:* Each subunit is treated and tested separately, considering all others as AE. Coupling and decoupling devices shall be placed on cables to be tested and on subunits considered as the EUT.

    (b) *Alternative method:* Subunits that are always connected together by a short cable (i.e., 1 m) and are part of the EUT to be tested can be considered as an integral part of the EUT. Testing is not required on associated interconnecting cables, as these cables are considered internal assemblies of the EUT.

5. The test shall be performed within an intended operating and climatic condition. Temperature and humidity need to be noted in the test report. Adherence to local regulations regarding interference must be complied with respect to the radiated fields from the test setup. If the radiated energy exceeds permitted levels, use of a shielded enclosure will be required to minimize disruption to communication systems.

6. The test shall be conducted with the test generator connected to each of the

coupling and decoupling networks in turn, while other nonexcited RF input ports of the coupling devices are terminated by 50-Ω load resistors.

7. Filters should be provided to prevent harmonics (higher order or subharmonic) from disrupting the EUT. Specifications for this filter are provided in the test standard.

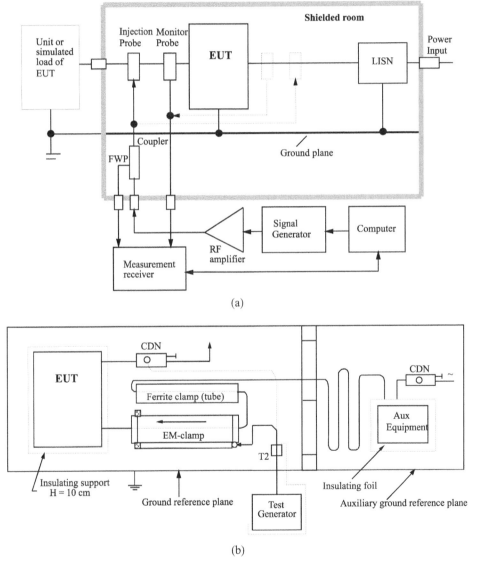

(a)

(b)

Figure B.19 Performing conducted immunity test: (*a*) BCI, current clamp or capacitive voltage probe; (*b*) EM injection clamp.

8. The frequency range is swept from the lower to the upper limit, generally 150 kHz–30 MHz (subject to the particular test plan) with an upper limit of 230 MHz (commonly used for specific systems and applications). The disturbance signal shall be modulated with an AM signal at 1 kHz with 80% deviation. The rate of the sweep shall not exceed 1.5×10^{-3} decades/second. When the frequency is swept incrementally, the step size shall not exceed 1% of the frequency range currently being tested.

9. The dwell time at each frequency shall be appropriate to allow the EUT to be exercised under full operating conditions, with the ability to respond to an abnormal condition by either error messages or system failure. Sensitive frequencies, such as clocks and their harmonics of dominant interest, shall be analyzed separately.

Test Procedure. Arrange the injection device (CDN, EM coupling clamp, current probe, or bulk current injection), RF signal generator and amplifier, reference ground plane, and measurement receiver/power meter as shown in Figures B.19*a*, *b*. The coaxial cables must be the same as those used in the calibration procedure.

1. Set the RF generator to 150 kHz, or a level sufficient to produce 3/10 V RMS (as determined via the calibration procedure).

2. Turn on the AM signal to produce 80% modulation depth with a 1-kHz sinusoidal signal. *Note:* The signal generator output must be adjusted so that current measured by the probe and spectrum analyzer does not exceed the predetermined voltage level (20 mA for 3 V RMS, 150 Ω configuration, which is 86 dBμA). Do *not* adjust the level set dial of the RF amplifier. For example,

   Reading on spectrum analyzer (dBuV) = 86 dBμA + CF of probe (dB)

3. Activate the RF signal generator to sweep the frequency range at a rate no faster than 0.0015 decades/seconds. While the generator is sweeping, adjust the RF signal generator output as required to maintain the desired signal injection levels per the calibration chart created prior to formal testing. Do *not* adjust the RF *amplifier's* control knob.

4. Note and record changes in EUT operation, fault conditions, or any other system behavior that is monitored per the performance criteria. Record the interference voltage level and frequency at which changes in operations, if any, occurred.

5. When the sweep is completed, return to those frequencies at which susceptibility problems were detected. Determine and record the threshold levels for each disturbance.

6. *Notes:*

   (a) The response times of some EUT processes may be such that dwell times of interference voltages may need to be increased to allow the system to react to the disturbance. If this occurs, it may be more appropriate to step through discrete-frequency increments manually rather conducting an automated sweep.

(b) Critical frequencies such as CPU clocks shall be analyzed separately at their frequency of operation.

Alternate Test Procedure When Using a BCI

1. Use the steps above, set the RF generator at a level sufficient to product a desired voltage level (V RMS), as determined by the calibration charts. Turn on the AM signal to produce 80% depth with a 1-kHz sinusoidal.
2. *Note:* The signal generator output should be adjusted so that current measured by the current probe and spectrum analyzer does not exceed 1.5 V RMS

F. RF Magnetic Field Immunity

Standard Used as Guide to Testing IEC 61000-4-8 (EN 61000-4-8)

Test Severity   3 A/m (light industrial)
                               30 A/m (heavy industrial)

Frequency   50 Hz

Pass Criteria   Criterion A (definition in the Generic Standard)

Equipment Used to Perform Tests

Test generator   1–100 A, divided by the coil factor, <8% total distortion
Induction coil   Unique per EUT
Magnetic field sensor   ±2% accuracy

Pretest Calibration and Equipment Check

1. Verify characteristics of test generator per paragraph 6.1.2 of IEC 61000-4-8 (EN 61000-4-8).
2. Calibrate induction coil and coil factor per paragraph 6.2.2 of IEC 61000-4-8 (EN 61000-4-8).
3. Locate field sensor at front plane of the EUT in a position visible to the EMC engineer.

Test Setup. The equipment to be tested can be evaluated in an open area without the need for a shielded enclosure. In addition, placement of a person next to the device under test will not affect results. The test setup contains the following components. Precaution shall be taken to ensure that the radiated magnetic field does not interfere with instrumentation or other sensitive equipment near the test setup.

- Ground reference plane (GRP)
- Equipment under test (EUT)

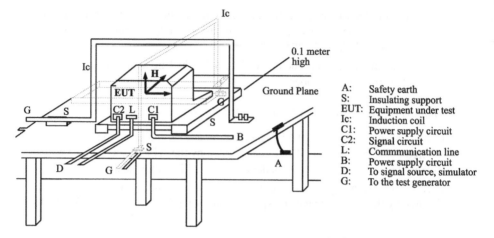

Figure B.20 Test setup for table-top equipment.

- Induction coil
- Test generator

There are two setup configurations: table top (Figure B.20) and floor standing (Figure B.21). The test procedure is identical; only the manner in which the test is

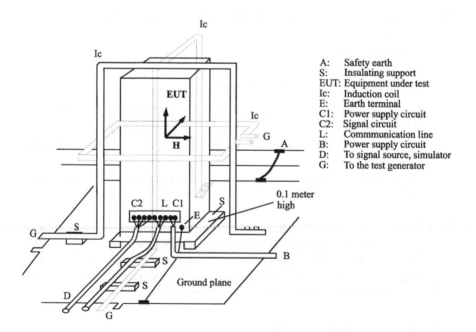

Figure B.21 Test setup for floor-standing equipment.

performed differs. According to IEC 61000-4-8, tests must be performed in compliance with the manufacturer's test plan and shall specify the following:

- How the test is carried out
- Verification of the laboratory reference conditions
- Preliminary verification of the correct operation of the EUT
- Evaluation of the test results

A ground reference place shall be placed with the EUT and auxiliary test equipment located on top, connected with secure bonding. The ground plane shall be a nonmagnetic metal sheet (copper or aluminum) 0.25 mm thick. Other metals may be used with a minimal thickness of 0.65 mm. Overall size of the ground plane shall be 1 m × 1 m for table-top equipment. The final size will depend on the dimension of the EUT. In addition, the plane shall be connected to the safety earth system of the laboratory.

The equipment is configured and connected in its functional operating condition, located on the ground reference plane with a 0.1-mm-thick insulating support (to ensure that the metal plane does not touch the EUT). Equipment cabinets shall be connected to safety earth ground directly to the ground reference plane through the earth terminal of the EUT.

The power supply and input and output circuits should be connected to appropriate interfaces. Cables supplied or recommended by the manufacturer shall be provided. In the absence of a recommended cable, unshielded cables shall be used of a type appropriate for the signals involved. All cable must be exposed to the magnetic field for a length of 1 m.

Backfilters, if any, shall be inserted in the circuit at a distance of 1 m from the EUT and connected to the ground reference plane.

An induction coil (Ic) having standard dimensions of 1 m per side for a rectangu-

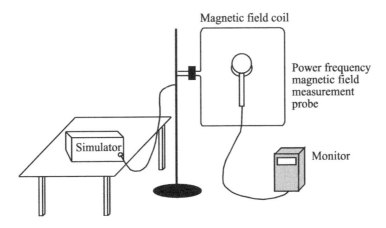

Figure B.22   Waveform verification test setup.

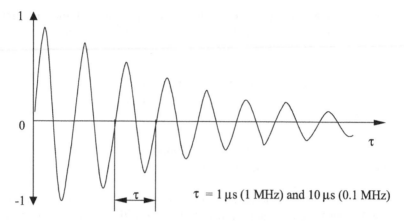

Figure B.23 Magnetic field waveform.

lar coil or 1 m diameter for a circular coil is to be used for testing small equipment. For large equipment, larger dimensions may be used; however, the coil should be able to envelop the EUT. The coil dimensions must give a minimum distance of coil conductors to EUT walls equal to one-third of the dimension of the EUT being tested.

Waveform Verification. IEC 61000-4-8/EN 61000-4-8 requires that the simulator output and magnetic field be verified periodically. Any oscilloscope is capable of verifying the power frequency alternating current to the coil. Verifying the actual field requires an AC field probe and appropriate monitor. Characteristics to be verified are the output current value and total distortion factor. Both items must have an accuracy of ± 2%. The waveform verification setup is shown in Figure B.22 while the required waveform is detailed in Figure B.23.

Test Procedure

1. Position test generator less than 3 m from the induction coil.
2. The induction coil shall enclose the EUT placed at its center. This coil, if provided in the vertical position, can be bonded directly to the ground plane and connected to the test generator identical to the calibration procedure.
3. Verify proper operation of the EUT prior to energizing the induction coil.
4. Energize the coil with the test generator, verifying 3 or 30 A/m with a field sensor in one plane.
5. Repeat the test by moving and shifting the induction coils in order to test the whole volume of the EUT for each orthogonal direction.
6. Repeat with the coil shifted to different positions along the side of the EUT in steps corresponding to 50% of the shortest side of the coil.
7. Rotate the induction coil 90° to expose the EUT to the test field with different orientations. Repeat steps 4–6.

Table B.2   Test Levels for Continuous Immersion

Level	Magnetic Field Strength (A/m)
1	1
2	3
3	10
4	30
5	100
x[a]	Special

[a]x is an open level. This level can be given in the product specification.

8. Record and document *all* unusual effects or anomalies noted during this test. The CRT display interference is allowed above 3 A/m.

Performing the Test. The test shall be performed per a test plan, including verification of performance criteria detailed in the product specification sheet. The power supply, signal, and other functional electrical portions of the system shall be operated at rated values. If actual operating signals are not available, they may be simulated.

Preliminary verification of EUT performance shall be carried out prior to applying the test voltage. The test magnetic field is then applied to the EUT, immersing the unit in a magnetic field operating at AC mains frequency.

For table-top equipment, the entire EUT is immersed by the induction coil. For floor-standing testing, the induction coil must be moved around the unit 360° in order to test the entire volume of the EUT for each orthogonal direction. This is a time-consuming test. Moving the induction coil in steps corresponding to 50% of the shortest side of the coil gives overlapping test fields. The induction coil shall then be rotated 90° in order to expose the EUT to the test field with different orientations using the same procedure as above.

Table B.3   Test Levels for Short Duration: 1–3 seconds

Level	Magnetic Field Strength (A/m)[a]
1	n.a.
2	n.a.
3	n.a.
4	300
5	1000
x[b]	Special

[a]n.a. = not applicable.
[b]x is an open level. This level can be given in the product specification.

The test specifications for various products based on installation and exposure are detailed in Tables B.2 and B.3.

G. Voltage Dips, Short Interruptions, and Voltage Variations

Standard Used as Guide to Testing   IEC 61000-4-11 (EN 61000-4-11)

Test Levels   ±10% rated voltage

30% for 10 ms; 60% for 100 ms; >95% for 5000 ms

Pass Criteria   Criteria B/B/C/C (definitions in the Generic Standard)

Equipment Used to Perform Tests

Power source   230 V AC, 50 Hz
Digital multimeter   Capable of measuring 600 V RMS
Control equipment   Capable of performing voltage sag, surge, and variations

Pretest Calibration and Equipment Check

1. Prepare power source for calibration measurements.
2. Using a digital mulitmeter, measure the output voltages between line to ground and line to line.
3. Adjust the power source output voltage until the output voltage is at the rated voltage of the system.
4. If possible, connect an oscilloscope or equivalent measurement device to determine accuracy of the control circuitry to provide proper voltage sag levels.

Test Setup and Procedures. The test setup is very simple (Figure B.24). The EUT is connected to the test generator with the shortest length power supply cable

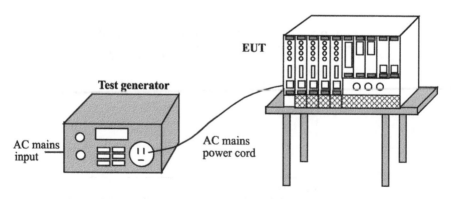

Figure B.24   Test setup for dips and short interruptions.

specified by the equipment manufacturer. If no cable length is specified, the short-est cable possible shall be used. To test three-phase equipment, three sets of test in-strumentation are required that are mutually synchronized.

There are two types of tests to be performed with this setup:

- Voltage dips and short interruptions
- Voltage variations with gradual transition between the rated voltage and the changed voltage (optional test)

There are no special requirements for the test environment except that the elec-tromagnetic environment should not disrupt operation of the EUT. The mains volt-age must be monitored and be within 2% of the desired value. The standard climat-ic conditions that must be met allow for full-compliance tests to be done on almost any test bench or floor in most places.

- Temperature   15°–35°C
- Humidity   25–75%
- Barometric pressure   860–106 kPA (860–1060 mbar)

Each dip or interruption test, defined by its dip level and duration as specified in the relevant test standard, shall be repeated three times with 10 seconds between each event. It is normal for the dip and interruption tests to start and finish at mains supply voltage zero crossings, in which case the zero-crossing accuracy of the test should be ± 10%. Requirements for starting and/or stopping at other phase angles are not specified at the time of writing.

The functions of the EUT must be monitored adequately to identify any degrada-tion both during and after each test. In addition, a full functional check shall be per-formed. The relevant generic or product-family harmonized standard will specify the performance degradation criteria to be applied in each case. In certain situa-tions, a momentary loss of displayed images or illumination is permitted.

Where the EUT's rated supply voltage range does not exceed 20% of its lowest voltage, any voltage within that range can be used for the test. When the EUT has a wider voltage range, two sets of tests are required: one at the lowest rated voltage and one at the highest. While it is acceptable for EMC compliance to test just for dips as specified by relevant harmonized standards, improved product reliability might be achieved by testing over the full range of voltage levels possible and for extended periods of time, that is, 60 seconds.

Typical mains supply voltage may be nonsinusoidal (clipped) at a maximum val-ue caused by power line harmonics. This clipping can affect the inrush current and the EUT's power supply charging voltage, affecting both the dips and short inter-ruption test results. Unfortunately, the test standard does not specify waveform pu-rity for the mains supply.

Test requirements are as follows:

- The test voltage must be within ±2% of the nominal voltage.
- The test frequency is 50 Hz ± 0.5 Hz.

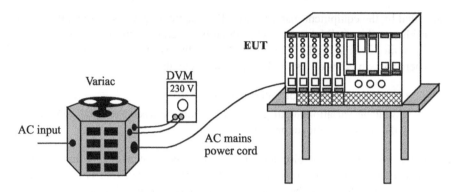

Figure B.25   Test setup for sag/voltage variation.

The percentage of total harmonic distortion (THD) of the supply voltage shall be less than 3%.

Test Procedure

Voltage Dips and Short Variations (Figure B.25)

1. Switch the main disconnect of the EUT to the off position.
2. Connect the EUT to the test unit generator.
3. Switch the main disconnect of the EUT to the on position and start up the system.
4. Program the power source to run the desired test for voltage dips and short interruptions.
5. Record and document *all* unusual effects or anomalies noted during this test.

Voltage Variation (Figure B.25)

1. Switch the main disconnect of the EUT to the off position.
2. Connect the EUT to the test unit generator.
3. Switch the main disconnect of the EUT to the on position and start up the system.
4. Program the system to run a representative process step or recipe.
5. While the system is running in its processing mode, reduce the output voltage of the variac to 90% of the rated voltage as quickly as possible.
6. Maintain the output voltage at 90% rated voltage for 30 seconds.
7. Repeat steps 6 and 7 two more times (total of three times).
8. Increase the output voltage from 90% rated voltage to 110% of the rated voltage as quickly as possible.
9. Maintain the output voltage at 110% rated voltage for 30 seconds.
10. Record and document *all* unusual effects or anomalies noted during this test.

AC Mains Supply Sags/Brownouts (Figure B.25)

1. Switch the main disconnect of the EUT to the off position.
2. Connect the EUT to a variac or programmable power source.
3. Switch the main disconnect of the EUT to the on position and start up the system.
4. Vary smoothly, or using a programmable power source with varied stepwise automated programming, the voltage level at a rate no greater than 10% of the nominal mains voltage. If using a programmable power source, ensure that the sag occurs at the zero crossing of the mains waveform.
5. Using a timer (a wrist watch is adequate), apply three separate tests with a 10-second delay between tests.
6. Determine if any degradation occurred both during and after each test.
7. It is recommended that two sag tests be performed, one at 40% and one at 0% of the supply voltage, each lasting for 1 second with the supply voltage ramped down and then up again over a period of 2 seconds in each test.

Three-Phase Equipment—Compliance Testing. If a three-phase test generator or three-phase variac is unavailable, three-phase equipment can be tested for dips, interruptions, and voltage variations using three sets of single-phase test equipment. Phase-by-phase testing is preferred, although some equipment may require simultaneous testing. This requires a common or synchronized control of the three single-phase test systems. The test standard (generic or product family) will specify whether phase-by-phase or simultaneous control is required.

When using simultaneous control for dip and interruption testing, the $\pm 10\%$ specification for the zero-crossing performance can only be met for one phase. The test generators must be of the type that can switch at any phase angle.

PERFORMING TESTS IN CELLS (TEM AND GTEM)

Measurements Using a TEM Cell

Numerous procedures and methods exist worldwide. Each varies with the approach and interpretation of results for obtaining accurate data. The most common test procedures are detailed below. It is recommended that the test engineer consult the manufacturer of the cell on how it would test products with accurate results based on the type and construction of the EUT. There are two functions for a TEM cell, immunity and emission testing. Both test procedures are provided below. Figure B.26 illustrates simple test setups.

For Radiated Immunity

Step 1. *Positioning the equipment centrally in the upper half of the TEM cell.* The EUT is placed on the floor of the cell directly if the EUT casing is to be connect-

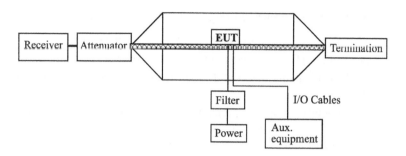

Typical Test Set Up - Emissions

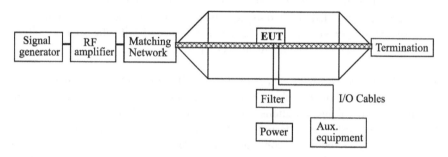

Figure B.26 Typical setup configurations for a TEM cell.

ed to the chassis or frame ground. If the EUT casing (cabinet) must be isolated electrically, a sheet of insulating material with a dielectric constant as close to unity as possible is inserted between the EUT and bottom plate of the TEM cell. In order to not expose I/O interconnect leads to the test field, a thin dielectric sheet can be placed between the EUT and bottom plate, especially if close location to the bottom plate is required. If it is desired to locate the EUT halfway between the bottom plate and septum, dielectric closed-cell foam (with a dielectric constant close to unity) can be inserted under the EUT. Note precisely the EUT orientation relative to field polarization. The radiated field might change with orientation. For accuracy of measurement, rotate the EUT several times and repeat the test.

Step 2. *Installation of AC mains power and I/O cables.* All interconnects that exit the chamber must be filtered, including AC mains. Filtered cables prevent RF leakage into the cell in addition to preventing RF fields internal to the chamber from corrupting measured results. A shielded filter compartment is generally provided for housing filters. It is recommended that all cables be the same length as in actual usage. Special circumstances may require use of fiber-optic cables to prevent perturbation of the test environment. It is important to pay attention to the manner in which the cables are routed, especially inside the TEM cell. Avoid or minimize cross coupling of fields. Certain cables located on the bottom of the cell can be covered with conductive tape if exposure of these cables to high levels of RF fields is to be avoided. All cables will be exposed simultaneously to the RF energy when located internal to the cell.

Step 3. *Connection of the TEM cell to the EUT.* Depending on the test plan, connection to an appropriate RF power source (including the amplifier) must occur to establish required field levels inside the TEM cell. Some TEM cells have provision for field intensity monitors to be secured inside the chamber. Other cells require location of a sensor adjacent to the EUT. Regardless of the method chosen, it is important to know the field intensity level. This power level can also be measured at the input terminal of the TEM cell and calculated per the manufacturer's instructions.

Step 4. *Conducting the test per the test plan or product specification.* Activate the EUT in its worst-case mode of operation. Verify that monitoring sensors and equipment are functional. Turn on power and set to the desired power level. The output level of the amplifier is adjusted to ensure a constant level at all frequencies. Sufficient dwell time must occur at each frequency and power level to guarantee that any response from the device under test has time to respond to the high-intensity field levels. This is the most difficult part of the test, especially if there is no window to observe the unit visually. Conduct testing in different orientations inside the TEM cell as required by the test standard and procedure. After any engineering modifications have been implemented, the entire sequence of tests must be repeated.

For Radiated Emissions. The purpose of a TEM cell for radiated emission testing is to ascertain the magnitude of energy that is propagated in the TEM mode from the EUT. The exact location from which the energy is coming cannot be determined. By measuring the energy level, quantitative analysis on the spectral profile becomes possible. Limitations regarding the size of the EUT and useful upper frequency of the TEM cell are applicable for both immunity and emission testing. A simple procedure for conducting radiated emission testing is provided below, based on the configuration of Figure B.26:

Step 1. Position equipment centrally in the upper half of the TEM cell in a typical configuration with worst-case cable placement (same as for immunity testing).

Step 2. Install AC mains power and I/O cables (same as for immunity testing).

Step 3. Connect the TEM cell to the monitoring equipment. The type of information desired from testing determines the type of instrumentation required. To ascertain the equivalent free-space radiated electric field from the EUT, the setup in Figure 4.21 can be used. Typical receivers include a spectrum analyzer, precision RF voltmeter, or power meter. For time-domain information, the receiver is replaced by a simple oscilloscope or receiver/recorder.

Step 4. Record the radiated emission. For accuracy in determining the location of the leak point of the EUT, EMI tests must be performed on all six axes of the EUT. This requires manual manipulation of the system after each test.

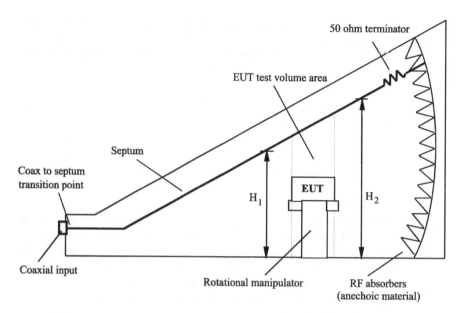

Figure B.27   GTEM with location of EUT positioned inside the chamber.

Measurements Using a GTEM Cell

The procedures for using a GTEM cell are almost identical to that of a standard TEM cell. Refer to the prior section for details. Important areas of concern are detailed below.

For Radiated Immunity. The EUT is placed inside the GTEM cell in a volume between the bottom of the cell and the septum. The useful test volume is bound by height $H_1/3$ and $H_2/3$ from the bottom of the cell. This is detailed in Figure B.27. The field strength uniformity can be maintained within ±1 dB with this setup.

An appropriate signal input from an amplifier for higher power levels required during immunity tests connects to a coaxial connector at the tip of the cell. A septum (flat metal bar) takes the RF energy from the coaxial connector and propagates a radiated RF field up to the 50-Ω terminator. The input source is set to a desired frequency and power level. It is important to monitor the field strength either by a power meter between the amplifier and chamber or by a field probe located inside the chamber adjacent to the EUT. Additional instrumentation may be used to determine the precise field strength at various locations of the EUT.

As with the TEM cell, the EUT must be tested on all six axes. Many chamber manufacturers provide a rotational system that is nonconductive called a rotational manipulator. Rotation is controlled either manually or by computer. This provides a significant advantage to the test engineer if the EUT is heavy.

To ensure accuracy of performance, the sweep time of the injected signal must be slow enough to allow the EUT time to respond. The length of time required for the sweep is specified in most test standards, which is usually not long enough to accurately evaluate the EUT for abnormal operation.

For Radiated Emissions. Measurement of radiated emissions is nearly identical to that of the TEM cell. Here, again, the size of the EUT must fit within the test volume (Figure B.27). Emissions from the EUT are coupled onto the septum and propagated to the receiver in the TEM mode. Suitable voltage or power and frequency measuring instrumentation is connected to the coaxial connector. The orientation of the EUT must be tested in all six axes. Similar to dwell time for immunity testing, the receiver must have a response or sweep time set to permit all RF energy to be recorded. Not all RF energy is periodic in nature and may be quite random, which accounts for the longer sweep time.

Due to the unique nature and characteristics of the GTEM cell, recorded emissions will not correlate to what would be measured at an OATS. Special software is required, generally provided by the manufacturer or by third parties. A computer must be connected to the receiver and other instrumentation using a bus configuration. With proper software, the recorded emissions can be translated to a distance of 3 m or 10 m from the EUT with excellent accuracy.

GLOSSARY

Absorbing clamp. A device used for the measurement of radiated RF power on cable assemblies in the frequency range of 30–1000 MHz that is nondestructive to the specimen. An absorbing clamp contains current transformers using ferrite split rings. These rings are split to allow for cable insertion.

AC Impedance. The combination of resistance, capacitive reactance, and inductive reactance seen by AC, or time-varying, voltage.

Alternating current (AC). A current level that varies with time. This label is commonly applied to a power source that switches polarity many times per second, such as the power supplied by utility companies. It may take a sinusoidal shape but could be square or triangular.

Amplitude. The height or magnitude of a signal measured with respect to a reference, such as signal ground, usually measured in decibels (dB).

Anechoic chamber. A test facility shielded from the electromagnetic environment using RF-absorbing material. Radio-frequency energy does not enter or leave this room.

Antenna. A device used for transmitting or receiving electromagnetic signals or power. Designed to maximize coupling to an electromagnetic field.

Antiresonance. The opposite of resonance, or the parallel combination of a capacitor and an inductor. The frequency at which, neglecting dissipation, the impedance of the object under consideration is infinite.

Artificial mains network (AMN). A network inserted in the supply mains load of apparatus to be tested that provides, for a given frequency range, a specified load impedance for the measurement of disturbance voltages and that may isolate the apparatus from the supply mains in that frequency range. Also identified as a *line impedance stabilization network (LISN)*.

Testing for EMC Compliance. By Mark I. Montrose and Edward M. Nakauchi
ISBN 0-471-43308-X © 2004 Institute of Electrical and Electronics Engineers

Attenuation. A general term used to denote a decrease in signal magnitude in transmission from one point to another due to losses in the media through which it is transmitted, measured in units of decibels (dB).

Average detector. A detector whose output voltage is the average value of the magnitude of the envelope of an applied signal.

Backward crosstalk. Noise injected into a quiet line that is placed next to an active line, as seen at the end of the quiet line at the signal source.

Bandwidth. The range of frequencies within which performance, with respect to some characteristic, falls within specific limits.

Biconical antenna. An antenna consisting of two conical conductors that have a common axis and vertex and are excited or connected to a receiver at the vertex. When the vertex angle of one of the cones is 180°, The antenna is called a discone.

Bilog antenna. A single antenna that combines the features and electromagnetic characteristics of both biconical and log-periodic antennas into one assembly.

Bonding. The permanent joining of metallic parts to form an electrically conductive path that will ensure electrical continuity and the capacity to conduct safely any current likely to be imposed.

Brownout. A slow varying decrease in the voltage level on the AC mains input to a system, sometimes over an extended period of several cycles to hours.

Bulk current injection. A method used to evaluate the electromagnetic susceptibility (immunity) of a wide range of electronic devices by magnetically coupling an RF field using a clamp or probe placed around cables.

Capacitance. A measure of the ability of two adjacent conductors separated by an insulator (dielectric material) to hold a charge when a voltage is impressed between them. Measured in units of farads.

Characteristic impedance. The impedance of a parallel conductive structure to the flow of alternating current. Usually applied to transmission lines in printed circuit boards and cables carrying high-speed signals. Normally, characteristic impedance is a constant value over a wide range of frequencies.

Circuit. Multiple devices with source impedance, load impedance, and interconnect. For digital circuits, multiple sources and loads may be part of one circuit where all devices are referenced to the same point or may use a common signal return conductor. Circuits usually originate in one location and terminate in another.

Circuit referencing. The process of providing a common 0-V reference for multiple circuits that allows communication between two devices. Circuit referencing is the most important reason for providing a 0-V reference. This reference point is not intended to carry functional current.

Coax. A term used to describe conductors that are concentric about a central axis. Takes the form of a central wire surrounded by a conductor tube that serves as a shield and ground. May have a dielectric other than air between the conductors.

Coherence factor. Coherence is a measure of the relationship between two signals. It requires two signals to be simultaneous in time. The coherence function is a measure of how or if the signals are related or if they are from the same source. If the signals are perfectly correlated, the coherence factor will be unity. If the signals are uncorrelated, the coherence factor will be zero.

Common-impedance coupling. Occurs when both source and victim share a trans-mission line path through a fixed impedance. The most obvious location for common impedance is found within a shared conductor resulting in an undesired voltage.

Common mode. The instantaneous algebraic average of two signals applied to a balanced circuit, with both signals referred to a common reference.

Common-mode current. The component of a signal current that creates electric and magnetic fields which do not cancel each other. For example, a circuit with one signal conductor and one 0-V reference conductor will have common-mode current as the summation of the total signal current flowing in the same direction on both conductors (in phase). Common-mode currents are the primary source of EMI (electromagnetic interference).

Common-mode interference. Interference that appears between both signal leads and a common reference, causing the potential of both sides of the transmission path to be changed simultaneously and by the same amount relative to the com-mon reference point.

Conducted emissions. The component of RF energy that is transmitted through a medium as a propagating electromagnetic wave, generally through a wire or in-terconnect cable. LCI (line-conducted interference) generally refers to RF ener-gy in a power cord or AC mains input cable. Conducted signals do not propagate as fields but propagate as conducted currents.

Conducted immunity. The relative ability of a product to withstand electromag-netic energy that penetrates through external cables, power cords, and I/O inter-connects.

Conducted susceptibility. Electromagnetic interference that couples from outside of the equipment to the inside through I/O interconnect cables, power lines, or signal cables.

Containment. A process whereby RF energy is prevented from exiting an enclo-sure, generally by shielding a product within a metal enclosure (Faraday cage) or by using a plastic housing with RF conductive paint. Reciprocally, we can also speak of containment as preventing RF energy from entering the enclosure.

Correlation analyzer. An instrument similar to a spectrum analyze but with a syn-chronized second input. This second input allows for determining coherence be-tween the two inputs and can be used to perform signal or ambient cancellation in real time.

Coupling. The association of two or more circuits or systems in such a way that power or signal information is transferred from one to another.

Coupling/decoupling network (CDN). A *coupling* network is used to measure or inject RF power to a transmission line without degrading signal integrity. A *de-coupling* network is used to ensure that the disturbance signal does not influence auxiliary equipment. Commonly used for immunity testing.

Crosstalk. Unintended electromagnetic coupling between traces, wires, trace-to-wire, cable assemblies, components, and any other electrical component subject to electromagnetic field disturbance.

Current. The flow of electrons within a wire or a circuit; measured in amperes.

Current probe. A transducer that measures current level within a transmission line. This probe consists of a magnetic core material that detects the magnitude of flux present and presents this field measurement to a receiver.

Decibel. A standard unit for expressing the ratio between two parameters using logarithms to the base 10. Decibels provide a convenient format to express voltages or powers that range several orders of magnitude for a given system.

Decoupling. Preventing noise pulses inserted in the power distribution network by digital components switching logic states. Switching noise injected into the power distribution network can disturb other digital and analog logic components using the distribution supply. Decoupling is generally performed with capacitors.

Detector. An electrical circuit that performs detection (extraction of signal or noise from a modulated input) and a weighting function (extraction of a particular characteristic of the signal or noise being measured).

Dielectric constant. The property that determines the electrostatic energy stored per unit volume for unit potential gradient (generally given relative to a vacuum; *see* Permittivity).

Differential mode. The element of data communication where a signal is propagated from a source to a load with a return path containing an equal amount of energy.

Differential-mode current. The components of RF energy present on both signal and return paths that are equal and opposite to each other.

Differential pair. Parallel routed signals exhibiting mutual inductance between both lines, typically 50–150 Ω. The signal on each pair is generally equal and opposite in amplitude with the same time reference.

Direct current (DC). Current produced by a voltage source that does not vary with time.

Dynamic range. The maximum ratio of two signals simultaneously present at the input of a receiver or analyzer that can be measured to a specific accuracy or range of performance.

Edge rate transition. The rate of change in voltage with respect to time of a digital logic signal transition, usually referenced to 0 V. Expressed in volts per nanosecond.

Effective radiated power (ERP). The relative gain of a transmitting antenna with respect to the maximum directivity of a half-wave dipole multiplied by the net power accepted by the antenna from the connected transmitter.

Effective relative permittivity (ε'_r). The relative permittivity that is experienced by an electrical signal transmitted through a conducted path or an actual dielectric.

Electric field. A vector field of electric field strength or of electric flux density that has a significant magnitude. The electric force acts on a unit electric charge independent of the velocity of the charge. Spatial gradients in the field between conductors at different potential that have capacitance between them allow the electric field to propagate.

Electromagnetic compatibility (EMC). The capability of electrical and electronic systems, equipment, and devices to operate in their intended electromagnetic environment within a defined margin of safety, and at design levels or performance without suffering or causing unacceptable degradation as a result of electromagnetic interference.

Electromagnetic interference (EMI). Electromagnetic energy from sources external or internal to electrical or electronic equipment that adversely affects equipment by creating undesirable responses. EMI can be divided into two classes: continuous wave (CW) and transient.

Electrostatic discharge (ESD). A transfer of electric charge between bodies of different electrostatic potentials in proximity to each other or through direct contact. This event is observed as a high-voltage pulse that may cause damage or loss of functionality to susceptible circuits. Although lightning qualifies as a high-voltage pulse, the term ESD is generally applied to events of lesser amperage and more specifically to events that are triggered by human beings.

Electrostatic field. The element of static electric charge that is not time variant, containing a high energy level that will eventually discharge from one electrode to another at a lower potential level.

EMI filter. A circuit or device containing components that block the flow of certain high-frequency RF and allow the flow of desired RF energy within a specific frequency range This filter may also be used to protect a particular circuit from electromagnetic field disturbance.

Equipotential ground plane. A reference used as a common connection point for power and signal referencing. This plane may not be at equipotential levels for RF due to its electrically large size.

Faraday shield. A term referring to conductive shielding used to contain or control electromagnetic fields. This shield may be located between the primary and secondary windings of a transformer or may completely surround a circuit (or system) to provide both electromagnetic and electrostatic shielding. No functional earth ground is necessary for a Faraday shield to work.

Far field. A region in space where the electromagnetic field from a radiator appears as a propagating field plane wave (Poynting vector). This field contains both electric and magnetic field components. These field components are at right angles to each other, and is sometimes referred to as a transverse electromagnetic field, (TEM). The ratio, or impedance, of a propagating wave in the far field is $377\ \Omega$.

Ferrite components. Powered magnetic (permeable) material in various shapes used to absorb conducted interference on wires, cables, and harnesses. Acting as a lossy resistance and increased self-inductance, ferrites convert an EMI magnetic flux density field into heat (an exothermic process). One benefit of this process of operation, in contrast to filters that perform by reflecting EMI in their stopbands, is that ferrites dissipate rather than reflect EMI, which otherwise could enhance radiation and disturb other victim components or circuits.

Ferrite material. A combination of metal oxides sintered into a particular ceramic shape with iron as the main ingredient. Ferrites provide two key features: (1) high magnetic permeability that concentrates and reinforces a magnetic field and (2) high electrical resistivity that limits the amount of electric current flow. Ferrites exhibit low energy losses, are efficient, and function at higher frequencies (1 MHz–1 GHz).

FET probe. A high-impedance transducer used to measure signal characteristic, usually by an oscilloscope, without adding capacitive loading or affecting performance of the signal.

Filter. A device that blocks the flow of RF current while passing a desired frequency of interest. For communication or higher frequency circuits, a filter suppresses unwanted frequencies and noise or separates channels from each other.

Flicker. A perceptible change in electric light source intensity due to a fluctuation of input voltage.

Forward crosstalk. Noise induced into a quiet line placed next to an active line, as seen at the load end of the quiet line farthest from the signal source.

Frequency-domain analysis. The study of electrical signals to determine the characteristics of signal propagation with regard to frequency.

Frequency span. The difference between two frequency points (low frequency to high frequency) on an EMC analyzer.

Gain (antenna). The ratio of power increase delivered to an isotropic (omnidirectional) antenna that is required to develop a given field at a given distance to the power delivered to an antenna acting as a receiver in the direction of maximum radiation or boresight efficiency.

Ground. A term used to describe the terminal of a voltage source that serves as a 0-V measurement reference for all voltages in the system. Often the negative terminal of the power source but sometimes may be the positive terminal. The word *ground* must be prefixed by a descriptor that describes the type of ground system being referenced (e.g., analog, digital, chassis, frame, earth, signal, common, RF).

Grounding methodology. A chosen method for directing return currents in an optimal manner appropriate for an intended application.

Ground loop. A potential interfering condition formed between circuits when interconnected by a conducting element (plane, trace, wire) assumed to be at ground potential through which return currents pass. At least one ground loop will exist in a circuit. Although a ground loop is acceptable, the severity of the problem of currents flowing through the loop depends on the unwanted signals that may be present and those that can cause system malfunction.

GTEM (gigahertz transverse electromagnetic) cell. A tapered enclosure used in laboratory environments for both emissions analysis and radiated immunity testing. This chamber has a very large operating frequency range.

Harmonics. The component of a signal that is observed in the frequency spectrum at both even and odd whole-number multiples of the primary frequency.

Horn antenna. A radiating or receiving element having the shape of a horn. Generally used in the frequency range above 1 GHz.

Hybrid ground. A grounding methodology that combines single-point and multipoint grounding topologies simultaneously, depending on the functionality of the circuit and the frequencies present.

Immunity. The ability of a device, equipment, or system to perform without degradation in the presence of an electromagnetic disturbance.

Impedance. The resistance to the flow of energy (e.g., voltage, current, power) within a transmission line. May be resistive, reactive, or both.

Inductance. The property of a conductor that allows for storage of electrical energy in a magnetic field induced by current flowing through coils. Measured in henrys.

Insertion loss. The ratio between power received at a load after the insertion of a filter for a given frequency or how much power loss to the propagating signal a filter provides for its intended function.

Line coupling. The electromagnetic coupling between two transmission lines caused by mutual inductance and capacitance between the lines.

Line impedance stabilization network (LISN). A network inserted in the supply mains load of an apparatus to be tested that provides, for a given frequency range, a specified load impedance for the measurement of disturbance voltages and which may isolate the apparatus from the supply mains in that frequency range. Also identified as an artificial mains network (AMN).

Load capacitance. The capacitance seen by the output of a logic circuit or other signal source within a transmission line. Usually the sum of distributed line capacitance and input capacitance of all load circuits.

Log-periodic antenna. A class of antennas having a structural geometry such that its impedance and radiation characteristic repeat periodically as the logarithm of frequency. Generally used in the frequency range of 200–1000 MHz.

Loop antenna. A loop antenna is sensitive to magnetic fields and is shielded against electric fields. This antenna is in the shape of a coil. A magnetic field component perpendicular to the plane of the loop induces a voltage across the coil that is proportional to frequency according to Faraday's law.

Magnetic field. A condition in a medium produced by a magnetomotive force such that, when altered in magnitude, a voltage is induced in an electric circuit linked with the flux. The field surrounding any current-carrying conductor.

Modulation. The process in which the characteristics of a carrier are varied in accordance with a modulating wave in both amplitude and time.

Multipoint ground. A method of referencing different circuits together to a common equipotential or reference point. Connection may be made by any means possible in as many locations as required. In reality, a more accurate definition of multipoint grounding is "multiple connection points or locations to a single-point reference, such as a metal chassis."

Near field. A region in space, close to a radiator, where far-field conditions do not exist.

Network analyzer. A measurement instrument commonly used during the engineering design cycle to characterize physical components and their performance characteristics.

Open-area test site (OATS). A test facility located outdoors used for radiated emission testing and sometimes conducted emission. A wooden table is placed on a conductive turntable, which in turn is electrically connected to a metal ground plane. An antenna mast is located a specific distance away which records propagated fields from the item under test.

Oscilloscope. An instrument primarily used for viewing the instantaneous value of one or more rapidly varying electrical quantities as a function of time or of another electrical quantity.

Parasitic capacitance. The capacitive leakage across a discrete component (e.g., resistor, inductor, filter, isolation transformer, optical isolator) that adversely affects high-frequency performance. Parasitic capacitance is also observed between active components (or printed circuit boards) and sheet metal mounting plates or chassis enclosures.

Peak detector. A detector enabling the capture and recording of the peak amplitude of a time-varying signal.

Permeability. The extent to which a material can be magnetized, often expressed as the parameter relating magnetic flux density induced by an applied magnetic field. Pemeability is not a constant value but varies with electrical frequency at which the measurement is made and the temperature of the environment.

Permittivity (dielectric constant ε_r). The ratio of the incremental change in electric displacement per unit electric field of a material to that of free space. This term is preferred to the words *dielectric constant.* Permittivity is not a constant value but varies with several parameters, including electrical frequency at which the measurement is made, temperature of the environment, and extent of water absorption in the material carrying a propagating electromagnetic field.

Power factor correction (PFC). The process of altering the amount of lost electrical power input, in watts, from a low value to a maximum value for AC mains input. Power factor refers to the ratio of total power input in watts to the total volt-ampere input.

Power/return bounce. Digital components, during a logic transition, will consume direct current plus line charge current required to supply capacitively loaded currents. During this period of time, inductance in the interconnect between component and power/return may require more direct current than available and/or cause a voltage drop according to $L(di/dt)$. When this occurs, signal integrity problems may result as well as development of EMI. Power and return planes will modulate at RF frequencies whenever current demand exceeds that of the distribution network. This modulated wave can cause harmful disruption to other digital components.

Propagation delay. The time required for a signal to travel through a transmission line or the time required for a logic device to perform its desired function from input to output.

Quasi-peak detector. A detector having a specified electrical time constants that, when regularly repeated pulses of constant amplitude are applied to it, delivers an output voltage that is a fraction of the peak value of the pulses. The fraction increases toward unity as the pulse repetition rate is increased with an integration time constant. Used in receivers and spectrum analyzers for EMI testing.

Radiated emissions. The component of RF energy that is transmitted through a medium as an electromagnetic field. The RF energy is usually transmitted through free space; however, other modes of field transmission may occur.

Radiated immunity. The relative ability of a product to withstand electromagnetic energy that arrives via free-space propagation.

Radiated susceptibility. Undesired propagating electromagnetic fields radiating through free space into equipment from external electromagnetic sources.

Radio frequency (RF). The frequency range containing coherent electromagnetic radiation of energy; roughly 10 kHz–300 GHz used for communication purposes. This energy may be transmitted as a byproduct of an electronic device's undesired mode of operation. The RF energy is transmitted through two basic modes: radiated and conductive.

Reflections. Radio-frequency energy that is sent back toward the source as a result of encountering a change in impedance in the transmission line on which it is traveling or bouncing off a metallic wall back to its source generator.

Relative permittivity. The amount of energy stored in a dielectric insulator per unit electric field, and hence a measure of the capacitance between a pair of conductors in the vicinity of the dielectric insulator, as compared to the capacitance of the same conductor pair in a vacuum.

Resonance. A condition within a transmission line or on metal structures where inductive reactance and capacitive reactance are equal. This condition occurs at only one frequency in a circuit with fixed constants.

Reverberation chamber. A facility consisting of a rectangular chamber with bare walls and a mode-stir paddle that disrupts RF fields internal to the enclosure. The mode-stir paddle causes a change in boundary conditions and intentionally generates a high-Q cavity effect. This chamber can be used as an alternate facility for conducting radiated immunity tests.

RF ground. A ground reference using a specific methodology in order to allow a product to comply with both emissions and immunity requirements; radiated or conducted.

Safety ground. The process of providing a return path to earth ground to prevent the hazard of electric shock through proper connection and routing of a permanent, continuous, low-impedance, adequate fault capacity conductor that runs from a power source to a load.

Shield ground. Providing a 0-V reference or electromagnetic shield for both interconnect cables or main chassis housings.

Shield room. A room containing metal walls. Used for EMI analysis.

Signal analyzer. A system that is a combination of an oscilloscope and a spectrum analyzer. It displays signals in both the time and frequency domain.

Signal integrity. The engineering discipline that investigates the propagating quality of a signal (electromagnetic field) to ensure that minimal energy loss occurs and that the desired waveform arrives at its destination within a desired period of time without degradation, ringing, reflection, and crosstalk.

Single-point ground. A method of referencing many circuits together to a single location to allow communication to occur.

Skin depth. The distance to the point inside a conductor at which the electromagnetic field, and hence current, is reduced to 37% of the surface value.

Skin effect. The effects of AC propagation within a transmission line in the areas of lowest impedance. With DC levels, current flow travels in the center of the conductor. At higher frequencies, AC flow will migrate to the outer skin of the transmission line. Above a certain frequency, signal loss occurs due to excessive amount of current traveling in a very thin portion of the transmission line.

Sniffer probe. Any small transducer used to isolate or locate radiating RF energy. Calibration of measurement is not a concern.

Spectrum analyzer. An instrument primarily used to display the power distribution of an incoming signal as a function of frequency. Useful in analyzing the characteristics of electrical waveforms by repetitively sweeping through a frequency range of interest and displaying all components of the signal being investigated.

SPICE (Simulation Program with Integrated Circuit Emphasis). A computational program for analysis of transmission line behavior in an effort to ensure optimal signal integrity in the time domain.

Suppression. The process of reducing or eliminating RF energy that exists without relying on a secondary method, such as a metal housing or chassis. Suppression may include shielding and filtering as well.

Susceptibility. The inability of a device, equipment, or system to resist an electromagnetic disturbance. Susceptibility is the lack of immunity.

Swell. A momentary increase in the power frequency voltage delivered by the mains, outside normal tolerances, with a duration of more than one cycle and less than a few seconds.

TEM (transverse electromagnetic) cell. A small enclosure used in normal laboratory environments for both emissions analysis (non–product qualification) and radiated immunity (product qualification).

Time-domain analysis. The study of electrical signals in a transmission line with regard to time, or how long it takes for a signal to propagate from one point to another.

Time-domain reflectometry (TDR). An instrument primarily used to verify the proper functioning of physical components of a network or transmission line with a sequence of time-delayed reflected electrical pulses.

Transducer. A device providing a means where energy can flow from one or more transmission systems or media to another transmission system or medium. The energy transmitted may be of any form (electric, mechanical, or acoustic) and may be of the same or different form in various input and output systems or media. *Antenna* or *probe* is another name for a transducer.

Transmission line. Any form of conductor used to carry a signal from a source to a load. The transmission time is usually long compared to the speed or rise time of the signal such that coupling, impedance, and terminators are important in preserving signal integrity.

Velocity of propagation. The speed at which data is transmitted within a conductive medium or dielectric. In free space, velocity of propagation is the speed of light. In a dielectric medium, the velocity of the transmitted electromagnetic wave is slower (approximately 60% the speed of light, depending on the dielectric constant of the material).

Voltage fluctuation. A variation in the voltage level that is above or below the desired range of operation.

Voltage probe. A transducer that measures the voltage level within a transmission line. This probe consists of a series resistor, an AC-blocking capacitor, and an inductor to provide a low-impedance input to a receiver. Used for direct connection and is unaffected by the current level in the transmission line.

Voltage standing-wave ratio (VSWR). The ratio of the magnitude of the transverse electric field in a plane of maximum strength to the magnitude at the equivalent point in an adjacent plane of minimum field strength. It is also commonly referred to as the ratio of forward power from a transmitter to reflected power back to the transmitter.

BIBLIOGRAPHY

Adam, S. F. 1969. *Microwave Theory and Applications.* Englewood Cliffs, NJ: Prentice-Hall.

Agilent Technologies. Application Note 1302. *Making Radiated and Conducted Compliance Measurements with EMI Receivers.* Agilent Technologies (previously Hewlett Packard).

ANSI C63.16. Multiple dates. *American National Standard, Guide for Electromagnetic Discharge Test Methodologies and Criteria for Electronic Equipment.*

ANSI C63.4. Multiple dates. *American National Standard for Methods of Measurement of Radio-Noise Emissions from Low-Voltage Electrical and Electronic Equipment in the Range of 9 kHz to 40 GHz.*

Armstrong, K. 2002. "Adding Up Emissions." *Conformity Magazine,* Vol. 7, No. 6 (July).

Badawi, A., and G. Ramsay. 2000. "Current Mapping on PCB Structures." Emscan Corp. Available: www.emscan.com.

Bakoglu, H. B. 1990. *Circuits, Interconnections and Packaging for VLSI.* Reading, MA: Addison-Wesley.

Booton, R. C. 1992. *Computational Methods for Electromagnetics and Microwaves.* New York: Wiley.

Boxleitner, W. 1988. *Electrostatic Discharge and Electronic Equipment.* New York: IEEE.

Brown, R., R. Sharpe, W. Hughes, and R. Post. 1973. *Lines, Waves and Antennas.* New York: Ronald Press.

Carr, J. J. 1999. *Practical Radio Frequency Test & Measurement: A Technician's Handbook.* Oxford: Newnes.

Carsten, B. 1998. "Sniffer Probe Locates Sources of EMI." *EDN Magazine,* June 4.

Christopoulos, C. 1995. *Principles and Techniques of Electromagnetic Compatibility.* Boca Raton, FL: CRC Press.

CISPR 11 (multiple dates). *Industrial, Scientific and Medical (ISM) Radio Frequency Equipment—Radio Disturbance Characteristics—Limits and Methods of Measurement.*

CISPR 16-1 (multiple dates). *Specification for Radio Disturbance and Immunity Measuring Apparatus and Methods—Part 1: Radio Disturbance and Immunity Measuring Apparatus.*

CISPR-16-2 (multiple dates). *Specification for Radio Disturbance and Immunity Measuring Apparatus and Methods—Part 2: Methods of Measurement of Disturbances and Immunity.*

CISPR-22 (multiple dates). *Limits and Methods of Measurements of Radio Interference Characteristics of Information Technology Equipment.*

Coombs, C. F. 1996. *Printed Circuits Handbook,* 4th ed. New York: McGraw-Hill.

DiBene, J. T., and J. L Knighten. 1997. "Effects of Device Variations on the EMI Potential of High Speed Digital Integrated Circuits." In *Proceedings of the IEEE International Symposium on Electromagnetic Compatibility.* New York: IEEE, pp. 208–212.

Drewniak, J. L., T. H. Hubing., T. P. Van Doren, and D. M. Hockanson. Power Bus Decoupling on Multilayer Printed Circuit Boards. *IEEE Transactions on Electromagnetic Compatibility,* Vol. 37, No. 2, pp. 155–166.

Ediss, R. "Investigating and Visualizing Emission from PCB Structure." Paper written for EMSCAN Corp. by Philips Semiconductors. Available: www.emscan.com.

EN 55011: 1998. *Industrial, Scientific and Medical (ISM) Radio Frequency Equipment-Radio Disturbance Characteristics—Limits and Methods of Measurement.*

EN 55022: 2002. *Limits and Methods of Measurements of Radio Interference Characteristics of Information Technology Equipment.*

EN 61000-3-2: 1995. *Electromagnetic Compatibility (EMC), Part 3-2. Limits for Harmonic Current Emissions (Equipment Input Current Up to and Including 16A Per Phase).* Basic EMC publication.

EN 61000-3-3: 2000. *Electromagnetic Compatibility (EMC), Part 3-3. Limitation of Voltage Fluctuations and Flicker in Low-Voltage Supply Systems for Equipment with Rated Current $\leq 16A$.* Basic EMC publication.

EN 61000-4-2. 1999. *Electromagnetic Compatibility. Part 4. Testing and Measurement Techniques. Section 2. Electrostatic Discharge Immunity Test.* Basic EMC Publication.

EN 61000-4-3. 1997. *Electromagnetic Compatibility. Part 4. Testing and Measurement Techniques. Section 3. Radiated Radio Frequency Electromagnetic Field Immunity Test.*

EN 61000-4-4. 1995. *Electromagnetic Compatibility (EMC), Part 4. Testing and Measurement Techniques. Section 4. Electrical Fast Transient/Burst Immunity.* Basic EMC publication.

EN 61000-4-5. 1995. *Electromagnetic Compatibility (EMC), Part 4. Testing and Measurement Techniques. Section 5. Surge Immunity Test.* Basic EMC publication.

EN 61000-4-6. 1996. *Electromagnetic Compatibility (EMC), Part 4. Testing and Measurement Techniques. Section 6. Immunity to Conducted Disturbances, Induced by Radio Frequency Fields.* Basic EMC publication.

EN 61000-4-8. 1994. *Electromagnetic Compatibility. Part 4. Testing and Measurement Techniques. Section 8. Power Frequency Magnetic Field Immunity Test.*

EN 61000-4-11. 1994. *Electromagnetic Compatibility (EMC), Part 4. Testing and Measurement Techniques. Section 11. Voltage Dips, Short Interruptions and Voltage Variations Immunity Tests.* Basic EMC publication.

Evans, A. J., and G. McWhorter. 1994. *Basic Electronics.* Chicago, IL: Master Publishing.

Gerke, D., and W. Kimmel. 1994. "The Designers Guide to Electromagnetic Compatibility." *EDN* January 10.

Gerke, D., and W. Kimmel. 1995. *Electromagnetic Compatibility in Medical Equipment.* Piscataway, NJ: IEEE and Buffalo Grove, IL: Interpharm.

German, R. F., H. Ott, and C. R. Paul. 1990. "Effect of an Image Plane on Printed Circuit Board Radiation." In *Proceedings of the IEEE International Symposium on Electromagnetic Compatibility.* New York: IEEE, pp. 284–291.

Ghose, R. N. 1996. *Interference Mitigation—Theory and Application.* New York: IEEE.

Goel, A. 1994. *High-Speed VLSI Interconnections, Modeling, Analysis and Simulation.* New York: Wiley.

Hall, S., H. Hall, W. Garrett, and J. McCall. 2000. *High-Speed Digital System Design: A Handbook of Interconnect Theory and Design Practices.* New York: Wiley.

Hartal, O. 1994. *Electromagnetic Compatibility by Design.* West Conshohocken, PA: R&B Enterprises.

Haykin, S. 1996. *Adaptive Filter Theory,* 3rd ed. Upper Saddle River, NJ: Prentice-Hall.

Hsu, T. 1991. "The Validity of Using Image Plane Theory to Predict Printed Circuit Board Radiation." In *Proceedings of the IEEE International Symposium on Electromagnetic Compatibility*, pp. 58–60.

Hubing, T., et. al. 1995. "Power Bus Decoupling on Multilayer Printed Circuit Boards." *IEEE Transactions on EMC,* Vol. 37, No. 2, pp. 155–166.

Hubing, T., M. Xu, and J. Chen. 2000. *NCMS Embedded Capacitance Project—Electrical Model and Test Results for Embedded Capacitance Boards.* University of Missouri-Rolla.

Hubing, T., T. P. Van Doren, and J. L. Drewniak. 1994. "Identifying and Quantifying Printed Circuit Board Inductance." In *Proceedings of the IEEE International Symposium on Electromagnetic Compatibility*, pp. 205–208.

IEC 61000-3-2. 2001. *Electromagnetic Compatibility (EMC), Part 3-2. Limits for Harmonic Current Emissions (Equipment Input Current Up to and Including 16A Per Phase).* Basic EMC publication.

IEC 61000-3-3. 2002. *Electromagnetic Compatibility (EMC), Part 3-3. Limitation of Voltage Fluctuations and Flicker in Low-Voltage Supply Systems for Equipment with Rated Current* $\leq 16A$. Basic EMC publication.

IEC 61000-4-2. 1998. *Electromagnetic Compatibility, Part 4. Testing and Measurement Techniques. Section 2. Electrostatic Discharge Immunity Test.* Basic EMC publication.

IEC 61000-4-3. 2002. *Electromagnetic Compatibility, Part 4. Testing and Measurement Techniques. Section 3. Radiated Radio Frequency Electromagnetic Field Immunity Test.*

IEC 61000-4-4. 2001. *Electromagnetic Compatibility (EMC), Part 4. Testing and Measurement Techniques. Section 4. Electrical Fast Transient/Burst Immunity.* Basic EMC publication.

IEC 61000-4-5. 2001. *Electromagnetic Compatibility (EMC), Part 4. Testing and Measurement Techniques. Section 5. Surge Immunity Test.* Basic EMC publication.

IEC 61000-4-6. 2001. *Electromagnetic Compatibility (EMC), Part 4. Testing and Measurement Techniques. Section 6. Immunity to Conducted Disturbances, Induced by Radio Frequency Fields.* Basic EMC publication.

IEC 61000-4-8. 2001. *Electromagnetic Compatibility (EMC), Part 4. Testing and Measurement Techniques. Section 8. Power Frequency Magnetic Field Immunity Test. Section 8.* Basic EMC publication.

IEC 61000-4-11. 2000. *Electromagnetic Compatibility (EMC), Part 4. Testing and Measure-*

ment Techniques. Section 11. Voltage Dips, Short Interruptions and Voltage Variations Immunity Tests. Basic EMC publication.

IEEE 100-1996. *IEEE Standard Dictionary of Electrical and Electronics Terms,* 6th ed. New York: IEEE.

Johnson, H. W., and M. Graham. 1993. *High Speed Digital Design.* Englewood Cliffs, NJ: Prentice-Hall.

Kodali, P. 2001. *Engineering Electromagnetic Compatibility,* 2nd ed. New York: IEEE.

Kraus, J. 1984. *Electromagnetics.* New York: McGraw-Hill.

Laverghetta, T. S. 1996. *Practical Microwaves.* Englewood Cliffs, NJ: Prentice-Hall.

Magnusson, P. C., G. C. Alexander, and V. K. Tripathi. 1992. *Transmission Lines and Wave Propagation,* 3rd ed. Boca Raton, FL: CRC Press.

Mardiguian, M. 1992a. *Controlling Radiated Emissions by Design,* New York: Van Nostrand Reinhold.

Mardiguian, M. 1992b. *Electrostatic Discharge, Understand, Simulate and Fix ESD Problems.* Gainesville, VA: Interference Control Technologies.

Mardiguian, M. 2000. *EMI Troubleshooting Techniques,* New York: McGraw-Hill.

Mardiguian, M., and D. White. 1988. *Electromagnetic Shielding,* Vol. III. Gainesville, FL: emf-emi Control.

Marino, M. 2000. Ambient Cancellation: "A Technology Whose Time Has Come." *Conformity,* Vol. 5, No. 6.

Montrose, M. I. 1999. *EMC and the Printed Circuit Board Design—Design, Theory and Layout Made Simple.* Piscataway, NJ: IEEE.

Montrose, M. I. 2000. *Printed Circuit Board Design Techniques for EMC Compliance—A Handbook for Designers,* 2nd ed. Piscataway, NJ: IEEE.

Moriarty, F. 1988. "EMI/RFI Diagnostic Tweezer Probes: A Construction Article." *EMC Technology* (Don White Publication), November/December.

Motorola. 1988. *MECL System Design Handbook (#HB205).* Chapters 3 and 7.

Motorola Semiconductor Products. 1990. *Transmission Line Effects in PCB Applications.* #AN1051/D.

Ott, H. 1988. *Noise Reduction Techniques in Electronic Systems,* 2nd ed. New York: Wiley Interscience.

Paul, C. R. 1989. "A Comparison of the Contributions of Common-Mode and Differential-Mode Currents in Radiated Emissions." *IEEE Transactions on Electromagnetic compatibility,* Vol. 31, May, No. 2, pp. 189–193.

Paul, C. R. 1992a. *Introduction to Electromagnetic Compatibility.* New York: Wiley.

Paul, C. R., 1992b. "Effectiveness of Multiple Decoupling Capacitors." *IEEE Transactions on Electromagnetic Compatibility,* Vol. 34, May, pp. 130–133.

Prentiss, S. 1992. *The Complete Book of Oscilloscopes,* 2nd ed. Blue Ridge Summit, PA: Tab Books.

Product Safety Standard, *Information Technology Equipment, Including Electrical Business Equipment.* Published under the following international numbers: Underwriters Laboratories (UL 1950), Canadian Standards Association (CSA C22.2 No. 950), International Electrotechnical Commission (IEC 950/IEC 60950), and European Normalized Standard (EN 60950).

Sadiku, M. 1992. *Numerical Techniques in Electromagnetics.* Boca Raton, FL: CRC Press.

Smith, A. 1998. *Radio Frequency Principles and Applications.* New York: IEEE/Chapman & Hall.

Smith, D. 1993. *High Frequency Measurements and Noise in Electronic Circuits.* New York: Van Nostrand Reinhold.

Smith, P. H. 1969. *Electronic Applications of the Smith Chart.* New York: McGraw-Hill.

Swainson, D. 1988. "Radiated Emission and Susceptibility Prediction on Ground Planes in Printed Circuit Boards." In *Proceedings of the Institute of Electrical Radio Engineers EMC Symposium.* York, England: pp. 295–301.

Tihanyi, L. 1995. *Electromagnetic Compatibility in Power Electronics.* Sarasota, FL: J.K. Eckert & Company.

Tsaliovich, A. 1999. *Electromagnetic Shielding Handbook for Wired and Wireless EMC Applications.* Boston: Kluwer Academic.

Waller, M. 1994. *Harmonics.* Indianapolis, IN: PROMPT Publications.

Whitaker, J. 1996. *The Electronics Handbook.* Boca Raton, FL: CRC Press.

Williams, T. 2001. *EMC for Product Designers,* 3rd ed. Oxford: Butterworth-Heinemann.

Williams, T., and K. Armstrong. 2000. *EMC for Systems and Installations.* Oxford: Newnes.

Williams, T., and K. Armstrong. 2001a. EMC Testing Part 1—Radiated Emissions. *EMC Compliance Journal,* February. Available: www.compliance-club.com.

Williams, T., and K. Armstrong. 2001b. EMC Testing Part 2—Conducted Emissions. *EMC Compliance Journal,* April. Available: www.compliance-club.com.

Williams, T., and K. Armstrong. 2001c. EMC Testing Part 3—Fast Transient Burst, Surge, Electrostatic Discharge. *EMC Compliance Journal,* June. Available: www.compliance-club.com.

Williams, T., and K. Armstrong. 2001d. EMC Testing Part 4—Radiated Immunity, *EMC Compliance Journal,* August/October. Available: www.compliance-club.com.

Williams, T., and K. Armstrong. 2001e. EMC Testing Part 5—Conducted Immunity. *EMC Compliance Journal,* December. Available: www.compliance-club.com.

Williams, T., and K. Armstrong. 2002a. *EMC Compliance Information.* Nutwood, UK.

Williams, T., and K. Armstrong. 2002b. EMC Testing Part 6—Low-Frequency Magnetic Fields (Emissions and Immunity); Mains Dips, Dropouts, Interruption, Sags, Brownouts and Swells. *EMC Compliance Journal,* February. Available: www.compliance-club.com.

Williams, T., and K. Armstrong. 2002c. EMC Testing Part 7—Emissions of Mains Harmonic Currents, Voltage Fluctuations, Flicker in Inrush Current; and Miscellaneous Other Tests. *EMC Compliance Journal,* April. Available: www.compliance-club.com.

Witte, R. 1991. *Spectrum and Network Measurements.* Englewood Cliffs, NJ: Prentice-Hall.

INDEX

Absorbing clamp, 130
AC mains testing, 192
AC power source, 207
Active transducers, 114
Air discharge, 237
Aliasing, 50
Alternate test sites–OATS, 95
Aluminum foil, 288
AM/FM radio, 147
Amateur radio transmitters, 321
Ambient assessment, 11
Ambient cancellation/suppression, 339
AMN. *See* artificial mains network
Amplifier, 99
Amplitude, 5
Anechoic chamber, 97
Antennas
  biconical, 154
  biconical (definition), 5
  bilog, 155
  bilog (definition), 5
  dipole, 153
  dipole (definition), 5
  factors, 152
  horn, 155
  horn (definition), 5
  log periodic, 154
  log periodic (definition), 5
  loop, 155
  loop (definition), 5

polarization, 153
positioner, 80
Artificial Mains Network (definition), 124
Attenuators, 186
Average detector, 62, 67

Bandwidth, 50, 66
Bare hand troubleshooting, 293
Baron Jean Baptiste Joseph Fourier, 13
BCI. *See* bulk current injection probe
Biconical antenna, 154
  definition, 5
Biolog antenna, 155
  definition, 5
Biological harm, 104
Broadband field sensor, 148
Broadband signal, 66
Brownouts, 195
Bulk current injection, 134
Bulk current injection probe, 135, 187

Cable configuration, 85
Cables and interconnects
  differential-mode, 305
Calibration jig, 188
Capacitive coupling clamp, 171
CDN. *See* coupling/decoupling network
Chambers, 96
Clamp injection, 187
Closed-field probes, 137

Testing for EMC Compliance. By Mark I. Montrose and Edward M. Nakauchi
ISBN 0-471-43308-X © 2004 Institute of Electrical and Electronics Engineers

ABOUT THE AUTHORS

Mark I. Montrose has more than 25 years of experience in the field of regulatory compliance, electromagnetic compatibility, and product safety. His experience includes extensive design, test, and certification of information technology equipment (ITE) and industrial, scientific, and medical (ISM) equipment. He specializes in the international arena of compliance, including the European EMC, Machinery and Low-Voltage Directives for light industrial, residential, and heavy industrial equipment. In addition to his consulting work, Mr. Montrose is an assessed EMC test laboratory by a European Competent Body and is accredited by NARTE.

Mr. Montrose graduated from California Polytechnic State University (1979), San Luis Obispo, California, with a B.S. degree in electrical engineering and a B.S. degree in computer science. He holds a Master degree in engineering management from the University of Santa Clara in California.

Mr. Montrose is a member of both the Board of Directors of the IEEE EMC Society and IEEE Press, a senior member of the IEEE, and life member of the American Radio Relay League (ARRL) with the Amateur Extra Class License, K6WJ. In addition, he is a past distinguished lecturer for the IEEE EMC Society (1999–2000), a director of TC-8 (Product Safety Technical Committee), founder of TC-10 (Signal Integrity) and TC-11 (Nanotechnology) both within the IEEE EMC Society, Regional Interest Group Chairman of the IEEE Nanotechnology Council, and cofounder and first president of the IEEE Product Safety Engineering Society. He is also a member of the dB Society.

He has authored and presented numerous papers and seminars in the field of electromagnetic compatibility (EMC) and signal integrity for high-technology products, printed circuit board (PCB) design, and EMC theory at international EMC symposiums and colloquiums in North America, Europe, and Asia. Mr. Montrose is

Testing for EMC Compliance. By Mark I. Montrose and Edward M. Nakauchi
ISBN 0-471-43308-X © 2004 Institute of Electrical and Electronics Engineers

an adjunct professor at several universities in the United States and Asia, providing PCB design and layout seminars to corporate clients worldwide.

Mr. Montrose has authored several textbooks published by Wiley and IEEE Press: *EMC and the Printed Circuit Board—Design, Theory and Layout Made Simple,* 1999 (translated into Chinese and Japanese), and *Printed Circuit Board Design Techniques for EMC Compliance—A Handbook for Designers.* 2000. In addition, he is a contributing author for Chapter 6 in *Electronics Packaging Handbook,* 2000, copublished by CRC and IEEE Press.

Edward M. Nakauchi has a BSEE and MSEE from Northrop University and Columbia Pacific University. He has more than 30 years experience in analog, power, and digital design. For the last 20 years, he has spent a majority of his time in the EMI/EMC/EMP and ESD areas for both military/aerospace companies and commercial audio/computer/medical companies. He has published numerous technical papers and magazine articles as well as seminars on various EMI/EMC/EMP/ESD topics. He teaches EMI courses through the University of California Irvine Extension program. Mr. Nakauchi was the primary author of a shielding design guideline for the U.S. Army and was an EMI consultant to the Air Force's Space and Missile Command on its customer off-the-shelf program. Some of the systems that he has worked on are the Space Shuttle, Global Positioning Satellite, Splash Mountain/Rocket Rods for Disneyland, and the B-2 bomber. He is a NARTE Certified EMC/ESD engineer with senior membership in the IEEE. Mr. Nakauchi is also active in supporting education and serves as a mentor teaching robotics and electronics to high school students.

Printed and bound by CPI Group (UK) Ltd, Croydon, CR0 4YY

16/04/2025

14658598-0005